Advances in Mathematics Education

For further volumes:
www.springer.com/series/8392

Michael N. Fried • Tommy Dreyfus
Editors

Mathematics & Mathematics Education: Searching for Common Ground

Springer

Editors
Michael N. Fried
Ben Gurion University of the Negev
Beer Sheva, Israel

Tommy Dreyfus
Tel Aviv University
Ramat Aviv, Tel Aviv, Israel

ISSN 1869-4918 ISSN 1869-4926 (electronic)
Advances in Mathematics Education
ISBN 978-94-007-7472-8 ISBN 978-94-007-7473-5 (eBook)
DOI 10.1007/978-94-007-7473-5
Springer Dordrecht Heidelberg New York London

Library of Congress Control Number: 2013956009

Printed on acid-free paper

Springer is part of Springer Science+Business Media (www.springer.com)

Dedicated to Polly Eisenberg
A devoted teacher and wonderful human being
She would have been proud to see this book
in honor of Ted

Preface and Acknowledgements

This book is dedicated to Ted Eisenberg. It grew out of a symposium with the same title held in May, 2012 on the occasion of Ted's retirement. The venue was Ben Gurion University of the Negev where Ted spent over thirty years as a professor of mathematics education in the mathematics department. Ted received plenty of praise during the event, praise, needless to say, that was well-deserved. However, for a scholar, teacher, and human being like Ted, it is a far greater tribute to discuss ideas that matter to him. Indeed, nothing could be more disheartening for a scholar like Ted than to feel his concerns are ignored or belittled. Ted has confessed to us he has often felt just that in recent years.

Ted's sense of neglect has not been entirely unfounded regarding one concern that has preoccupied him with particular acuteness over the years, namely, the growing distance between mathematicians and mathematics educators. For Ted, to be a mathematics educator one must know and care about mathematics itself. It is a position that, for him, is axiomatic and uncompromising. On the other hand, it is also a fact Ted painfully admits that the development of mathematics education as an academic field has allowed some of its practitioners at times to put mathematical knowledge aside or even pronounce it as irrelevant.

But "at times" does not mean "always" and "some" does not mean "all." The truth of the matter, as we see it, is that mathematics education has developed to a point where the place of mathematics within the field simply cannot be taken for granted: the question of the relationship between mathematics and mathematics education needs to be explored deeply and better understood. And as counterweight to Ted's own sense that there is a lack of urgency about this in the field, the very stature of the mathematics educators and mathematicians who participated in the symposium and who contributed to this book underlines the tremendous interest there actually is on all sides about this question which Ted holds dear.

We are thus grateful to the participants of the symposium who subsequently contributed to this volume, showing how far Ted's concern is their concern. And it must be added that these participants were not only those who presented papers and took part in planned panel discussions, but also those in the audience who asked astute questions and brought up issues enriching the general conversation then and the ideas in this book now. Many of those audience members were especially invited because we thought that they would indeed enrich the event. So, our first thanks are to all those who came and made that symposium such a great success. Of course, though, there were others whose contribution it is important to us to acknowledge:

- Daniel Berend, Miriam Cohen and Michael Lin from the Department of Mathematics at Ben Gurion University, for their excellent organization of the symposium and their success in finding sources for funding it;
- Abraham Arcavi (Weizmann Institute of Science), Hannah Perl (Israel Ministry of Education), and Norma Presmeg (Illinois State University) who have, together with the editors of this volume served as program committee for the symposium;
- Ina Aviv who has been indefatigable and effective in making sure that all participants were cared for as best as one could imagine before, during and after the symposium;
- The Israel Science Foundation, the Trump Foundation, the Chief Scientist's Office at the Israeli Ministry of Education, the Center for Advanced Studies in Mathematics at Ben Gurion University, the Faculty of the Natural Sciences, and the Faculty of Humanities and Social Sciences at Ben Gurion University, all of whom have provided generous financial support.

This volume would never have become possible without the symposium. It is an amplification and expansion of the proceedings of the symposium, on which the authors have worked during the year following the symposium. This work was done happily not only because of the importance of its central question but also because it was an opportunity finally to present ideas to the public that we have discussed so often with Ted. More than the symposium itself, in this way, we owe this work to Ted's passionate concerns. So, as Francis Lowenthal, one of Ted's old friends, wrote at the top of his own contribution and throughout it, we also say with pleasure, Thank you, Ted!

Beer Sheva, Israel Michael N. Fried
Tel Aviv, Israel Tommy Dreyfus

Contents

Introduction

Chapter 1
Mathematics & Mathematics Education: Searching for Common Ground

Michael N. Fried

> *Between these two groups… there is little communication and, instead of fellow-feeling, something like hostility. (C.P. Snow, The Two Cultures, p. 59)*

Prologue

If being mathematically educated could be summed up simply as a familiarity with certain key mathematical ideas—integer, algebraic equation, function, proof—their applications, and a facility in working with them, one could state unequivocally what the interests, foundations, and goals of mathematics education as a field should be. Not too long ago, only the conditional form of this statement would strike one as curious and odd. For what else could one mean by being mathematically educated, and what else could one place higher on the agenda of mathematics education research than the teaching and learning of these key mathematical ideas? And, with that, one could hardly imagine challenging the close and natural alignment between mathematics education and mathematics as academic disciplines.

However, over the last quarter century or so, and for better or for worse, this simple notion of where the core of mathematics education lies has been offset by goals and interests allying it, as an academic field, more closely with psychology of learning, cultural differences, and social justice, among others, than with mathematics itself. Thus, while the first two-thirds of the twentieth century could boast of great mathematicians such as Felix Klein, Jacques Hadamard, George Pólya, and Hans Freudenthal making contributions to mathematics education, today, not only are such figures rare in the field, they have also been to an extent alienated by it.

In the spring of 2012 a symposium concerning the relationship between mathematics and mathematics education was held at Ben Gurion University of the Negev. The symposium was in honor of Ted Eisenberg, who over the years has lamented profoundly the growing divide between the mathematics community and the mathematics education community. It has always been his opinion, shared by the editors

M.N. Fried (✉)
Science and Technology Education, Ben-Gurion University of the Negev, Beer Sheva, Israel
e-mail: mfried@bgu.ac.il

M.N. Fried, T. Dreyfus (eds.), *Mathematics & Mathematics Education: Searching for Common Ground*, Advances in Mathematics Education,
DOI 10.1007/978-94-007-7473-5_1,

of this volume, that the divide between the two communities is wasteful and unhealthy for both. The work at hand, which grew out of that symposium, confronts this disturbing gap. By examining areas of commonality as well as disagreement we hope to define more clearly the role mathematics as a discipline plays in mathematics education and mathematics education research and will try to establish a basis for fruitful collaboration between these disciplines. We can only hope that in the end readers will be left with a clearer sense of the mutual benefit both communities stand to lose by failing to strengthen the natural bonds between them.

With the exception of the first part, where we have pieces by Ted, Michael N. Fried, and Norma Presmeg set together in a kind of general dialogue, the various parts of the book take up particular subjects, such as proof, history of mathematics, and educational policy, among others, in which mathematicians and mathematics education researchers either both have a stake or a common interest. It is important to remark that in reading the contributions by the mathematicians and mathematics education researchers one should consider not only what is said but also the ways in which the different communities approach their respective tasks. While we have tried to maintain a certain uniformity in format, we have allowed considerable freedom in other regards. This comes out of the recognition that although we stress common interests and shared concerns, there are nevertheless differences between the communities of mathematicians and mathematics education researchers. One must confront these differences and try to understand them. Thus, to introduce the work and frame its theme, we expand a bit more about the distinctions, divisions, and possibility of cooperation between these two communities. Following that, we shall describe the main parts of the book in brief.

Distinctions and Connections

The moment one broaches the possibility of conflict or tension or misunderstanding between the mathematics and mathematics education communities the difficulty immediately arises, how are these to be distinguished, if at all? Not only this, but also a whole set of distinctions that, previously, one could write off as merely academic, become relevant—not only "Mathematics vs Mathematics education," but also "Mathematician vs Mathematics educator" and "Mathematics educator vs Mathematics education researcher" and "Mathematics education vs Mathematics education research." These distinctions are at the heart of the entire problem we are considering in this work. Granted, the distinctions may not be new, but their problematic character is. In the past, the problem of mathematics vs mathematics education, the main distinction we are considering, could only be viewed as a non-problem, a false dichotomy. One could then easily say that mathematics and mathematics education simply belonged to different categories: a whole, "mathematics," and a part, "mathematics education." Asking about the distinction between mathematics and mathematics education would have been like asking about the distinction between mathematics and geometry.

Nor does that view necessarily vanish with mathematics education's becoming a separate academic field (see Kilpatrick 1992 for a very good exposé of how that happened). However, with that change in place, the relationship between mathematics and mathematics education became no longer obvious and necessary: it has now become a question. One must ask, at very least, what justifies the formal, academic distinction between mathematics and mathematics education in the first place? While the separation may be merely bureaucratic and not essential, members of the new field do need to consider their own identity as mathematics educators. At some level, this itself is a bureaucratic necessity, albeit one also requiring genuine introspection—as a separate discipline, a basis has to be established for hiring and promoting mathematics educators: what is it a mathematics educator has to do well, what is that makes a mathematics educator an expert? This runs together with the next distinction, namely, between a mathematics educator and a mathematician. The question of the identity of the discipline thus becomes one of the identity of the practitioner: Is one a mathematician first before one is a mathematics educator? Is a mathematics educator a kind of mathematician?

Of course this begs the further question of what makes one a mathematics educator—in particular, how one should distinguish a mathematics educator from a researcher in mathematics education. Here, since one has the term "mathematics education researcher," one can treat a mathematics educator simply as a mathematics teacher. At the university level, naturally, the difference between a mathematician and mathematics teacher is not nearly so pronounced as it may be at school level since, besides the obvious fact that it is typically mathematics researches teaching mathematics students, university level mathematics already begins to have the feel of mathematics as the mathematician knows it. One could go further and argue that the difference between doing and teaching mathematics is actually never very great in that mathematicians must always communicate their thinking. Consider, in this connection, Andrew Wiles' "graduate seminar" taught principally to fellow mathematician Nick Katz when Wiles was working on Fermat's last theorem. As Simon Singh (1997, p. 242) relates:

> Virtually everything Wiles had done was revolutionary, and Katz gave a great deal of thought as to the best way to examine it thoroughly: "What Andrew had to explain was so big and long that it wouldn't have worked to try and just explain it in his office in informal conversations. For something this big we really needed to have the formal structure of weekly scheduled lectures, otherwise the thing would just degenerate. That's why we decided to set up a lecture course."

Although the seminar was also a ploy to hide Wiles' secret work on Fermat's last theorem, nevertheless, when all the graduate students had dropped out leaving Wiles and Katz alone, the teacher-student structure remained, as Katz emphasized.

In more ways than one, then, being a mathematician is being a mathematics teacher and communicator, which is a kind of teacher. The converse, however, is far from clear. It is even not entirely clear that a teacher should have a mathematician's training. Surely, mathematics teachers should know what they teach, but saying that begs the question at least and really is a mere platitude. In fact, the question of requisite knowledge for teachers and others is a true question, and an old one. Plato asked a similar question regarding the sophists and teachers of rhetoric. He also asks it most delightfully in a little dialogue centered on a rhapsodist, a reciter of Homer, Ion, after whom the dialogue is named. Socrates claims—and Ion agrees—that an expert rhapsodist must understand what he recites if he is to produce worthy interpretations of the epics, similarly, if he is to distinguish a good rhapsodist from a bad one. To use one of Socrates' examples, if the subject were numbers, one would expect that only an expert in the "arithmetical art," the arithmetical *techne* (*Ion*, 537e), would be able to judge whether the subject was being discussed well. In reciting Homer, Ion speaks about soldiers, generals, and even doctors: is Ion such an expert in these that he can speak so well about them? Ion is no general or doctor. Socrates teases him, saying it must be divine inspiration that he can do so. Yet, as in all Platonic dialogues the issue remains open in the end, for Socrates well knows that rhapsodists *are* successful at what they do, even they are not skilled generals and doctors.

This is true too about mathematics teachers. For this reason, in informal settings and casual conversation, one often hears their success explained by saying that teaching is an art—ironically meaning something closer to Plato's divine inspiration than what the Greeks mean by art, *techne*, a skill informed by knowledge! There may, nevertheless, be some truth to that, though research as to what makes a good mathematics teacher is much more circumspect and far from definitive. As the National Science Foundation report on science and engineering indicators remarks, "No research has conclusively identified the most effective teachers or the factors that contribute to their success, but efforts to improve measures of teaching quality have proliferated in recent years" (National Science Foundation 2012, Teachers of Mathematics and Science, side bar 6).

Be that as it may, the specific relationship between mathematical knowledge and mathematics teaching is equivocal. On the one hand, there is something as powerful as it is inexplicable about simply being in the presence of teachers who have thought deeply about their subjects. The philosopher and literary critic, George Steiner, describes this beautifully reflecting on his own experiences at the University of Chicago:

> Once a young man or woman has been exposed to the virus of the absolute, once she or he has seen, heard, 'smelt' the fever in those who hunt after disinterested truth, something of the afterglow will persist. For the remainder of their, perhaps, quite normal, albeit undistinguished careers and private lives, such men and women will be equipped with some safeguard against emptiness. (Steiner 1997, p. 44)

On the other hand, an early finding of modern mathematics education research showed that a *direct* connection between the depth of teachers' mathematical knowledge and their students' level of achievement cannot be fully maintained. In the course of his work with the School Mathematics Study Group (SMSG), Edward Begle (1972) had shown that empirically there was no significant relationship between teachers' knowledge of advanced algebra and their students' achievement in algebra. This was a result that Ted Eisenberg himself strengthened with a follow up paper in 1977 that controlled for potentially biased factors in Begle's original report.[1]

Although the issue is still open to certain extent, it is hard to doubt that success in mathematics teaching demands *some* combination of subject and non-subject dependent knowledge. The utterly integral character of that combination was driving force of Lee Shulman's (1986) now-standard concept of pedagogical-content-knowledge. But even without the concept, the necessary and simultaneous attention to content and pedagogy is evident in accounts of great mathematics teachers. Thus, Jeremy Kilpatrick says this of Pólya as a teacher:

> One of the things I learned from Pólya was if someone in class had trouble following the presentation, then you slow the class down. Pólya was always willing to slow the class down, but he could still make it interesting. That's one of the remarkable talents he had. He could move at a slower pace so that students could follow his presentation; even the slowest member of the class could get something out of it. Yet, at the same time, what he was presenting was interesting enough and rich enough that the people who understood what was happening could also learn something. He was not interested in getting someplace in the discussion where he felt he should be; he was interested in making what he was doing as illuminating as possible. (Kilpatrick, quoted in Taylor and Taylor 1993, p. 107)

In his writings about mathematics education, the Berkeley mathematician Hung-Hsi Wu agrees that school mathematics teachers need pedagogical-content knowledge (and he uses the term explicitly) and not just content knowledge (Wu 2011). Wu objects to what he calls the "Intellectual Trickle-Down Theory," which holds that extensive mathematical knowledge will effortlessly trickle down into teaching competence; he believes that while elementary school teachers simply need more mathematical knowledge, secondary school teachers need better developed means in order to make solid mathematical knowledge more presentable and understandable for their young, not-yet-mathematically-mature students.

Divisions

If mathematicians like Wu recognize these limits of mathematical knowledge and the concomitant need for insight into teaching and learning, do they also recognize the need for mathematics education research, which is supposed to study the teaching and learning of mathematics? More pointedly, do they recognize the need for mathematics education *researchers*? The answer is yes and no. One sign on the positive side is that there are mathematicians who themselves engage in mathematics education research in the company of other professional mathematics education researchers, Hyman Bass, for example; another is the active field of tertiary mathematics education research which has developed in recent years and is pursued at

---

[1] Recent work by Ruhama Even (2011), however, has shown that *in their own view*, teachers see advanced mathematical work helpful on three fronts: that it is a knowledge resource; that it improves their understanding of mathematics and what it is; that it provides a model for what learning mathematics feels like.

least with the cooperation of mathematics departments (see, for example, Holton 2002).

For the negative side, we can return to Wu. One could find other examples where a dismissive attitude towards mathematics education researchers is more clearly evident, examples bordering on rancor (the continuing attacks on Jo Boaler by James Milgram and Wayne Bishop, as documented in Boaler 2012, come to mind). But it is more informative to look at Wu (in the context of Wu 2011) in part because he is a mathematician genuinely concerned about mathematics education and somewhat informed about research touching on mathematics education; Wu's case shows the subtle ways one can recognize the need for mathematics education research but not for mathematics education researchers.

To be fair, Wu does refer to non-mathematician mathematics education researchers, such as Deborah Ball, and not unfavorably, especially when she recognizes the poor mathematical backgrounds of teachers. However, what he sees as the important task of mathematics education is "the *customization* of abstract mathematics for use in schools" (Wu 2011, p. 378, emphasis in the original), and this is a task for mathematicians. Wu describes the paper as "a call for action," namely, a call for mathematicians to recognize the "urgent need of active participation in the education enterprise" (p. 372).

On the face of it, there is nothing wrong with this. In fact, is it not what we ourselves are asking for in this book? The problem is that while Wu bemoans the "communication gap between mathematicians and educators" (p. 382), it is not hard to see that, for him, the gap consists in educators' not taking account of mathematicians rather than mathematician's missing the views and knowledge of educators. It is telling that he chooses to describe the dangers of the communication gap by recalling how Watson and Crick in their work on the DNA molecule benefitted crucially from the visit of a professional crystallographer, Jerry Donohue. And he summarizes the story and its moral as follows:

> ... but for the fortuitous presence of someone truly knowledgeable about physical chemistry, Crick and Watson might not have been able to guess the double helix model, or at least the discovery would have been much delayed.
> The moral one can draw from this story is that, if such misinformation could exist in high-level science, one should expect the same in mathematics education, which is much more freewheeling. This suggests that real progress in teacher education will require both the education and the mathematics communities to collaborate very closely and to be vigilant in separating the wheat from the chaff. In particular, given the long years during which incorrect information about mathematics has been accumulating in the education literature and school textbooks, there should be strong incentive for educators to seek information about the K-12 mathematics curriculum anew and to begin some critical rethinking. (pp. 382–383)

Although Wu speaks about collaboration explicitly, *knowledge* is placed squarely in the mathematicians' camp. He may object to the "intellectual trickle-down theory," but, when it comes down to it, whatever is wrong with the "theory," it is still the mathematicians who must correct it. It is hard to see where mathematics education researchers have a role, other than to sit quietly and listen. Indeed, the paper is addressed to mathematicians, and it appears in the mathematics journal, the *Notices of the AMS* (American Mathematical Society). It is not a call for collaboration: it is, as Wu says, a call for mathematicians to take action, not necessarily for *them* to listen.

But it is not only mathematicians who are to blame for dividing communities that should collaborate. Mathematics education researchers can also be dismissive of what mathematicians might bring to the floor. It can be asked, equally, whether mathematics education researchers recognize a need for mathematicians in mathematics education. Again, one can cite examples of open opposition to mathematicians having a central role in mathematics education research (see the account of the ICMI centenary in the first chapter of *Dialogue on a Dialogue* below). More often, however, what one finds is an agenda that leaves little room for mathematicians, a tendency to give precedence to areas hardly any mathematician would call mathematics and which certainly no mathematics department would include in its program. This is especially manifest when social justice issues are brought into mathematics education. Accordingly, Sriraman, Roscoe and English note that,

> Numerous scholars like Ubiratan D'Ambrosio, Ole Skovsmose, Bill Atweh, Alan Schoenfeld, Rico Gutstein, Brian Greer, Swapna Mukhopadhyay among others have argued that mathematics education has everything to do with today's socio-cultural political and economic scenario. In particular mathematics education has much more to do with politics, in its broad sense, than with mathematics, in its inner sense. (Sriraman et al. 2010, p. 627)

And in her commentary on Sriraman et al. (2010), Keiko Yasukawa confirms this by concluding:

> If we believe that mathematics learning can be a resource to increase democratic participation in society, to increase equity and social justice, then mathematics learning cannot be divorced from learning the politics of the world in which we live. Has the study of politics in mathematics education gone far enough? Evidently not. Can it go further? Yes, through critical mathematics education that will awaken learners to the ways in which mathematics is concealed but active in the dominant discourses that are influencing the ways we think about the fundamental principles of equity and fairness. (p. 643)

If understanding the nature and role of mathematics, not only in science and engineering but also in students' everyday lives, should be considered part of mathematics education, then these kinds of political investigations are not out of place. However, even these authors would have to admit that there is something merely accidental about mathematics' place in the political superstructure. There are other elements of the superstructure, and there *could* be other areas attaining the same prominence as mathematics *if they happened* to be valued in the same way. In other words, this key place of mathematics is not related to mathematics "in its inner sense," to use Sriraman et al. (2010) phrase. In fact, once one puts on the glasses of critical mathematics education, *every* mathematical notion becomes suspect and must be examined for its socio-political function: every mathematical idea has an ulterior meaning. This is almost axiomatic in "critical theory" (which dominates the thought of the authors mentioned by Sriraman, et al.' above), and it may reflect a true state of things. But if so, the mathematical meaning of any given mathematical concept, as the mathematician understands it, becomes not only secondary but also the very thing one must learn to move beyond. Mathematicians, in this way, can be of no help, and critical mathematics education may well see them, if they are not "liberated," as part of the problem.[2]

Distinctions Once Again and the Possibility of Cooperation

The kind of divisive positions we have just described—and there are others—not only distance the possibility of cooperation between mathematicians and mathematics education researchers, they also lead to a lack of coherence in mathematics education, regardless whether it is the mathematicians or the mathematics education researchers who take charge. Mathematicians dismissive of mathematics education researchers and mathematics education researchers dismissive of mathematicians must both find themselves edging towards inconsistency: the first wants to customize advanced mathematics for use in the schools but gives little credence to those who research the conditions and nature of learning; the second wants to teach mathematics and wants it taught while showing that it is only part of a superstructure concealing the non-mathematical political forces.

The truth is these two communities cannot be completely divorced. Even if they feel pushed to declare their loyalty to one camp, they will inevitably have one foot in the other. As argued above, university mathematicians typically and often necessarily take on the role of a teacher, that is, a mathematics educator, and as such must take an interest in how students learn and how best to teach. Mathematics education researchers, on the other hand, still insist that they are interested in *mathematics* education. And there are areas that interest both communities in ways that are quite similar, for example, visualization and problem-solving. Remember Pólya's interests in problem-solving and his work as a mathematician were joined almost seamlessly: consider for example his book with Gábor Szegö on problems and theorems in analysis, a serious book in which the problems were organized according to their solution strategies (see Taylor and Taylor 1993, pp. 24–25).

With these common areas of interest in mind, one would expect far more cooperation and collaboration than one typically finds between the communities of mathematicians and mathematics education researchers. True, as remarked above, there are instances of open enmity between these communities that would poison

[2] I am referring to mathematicians and mathematics teachers who, lacking the "critical" outlook, devote themselves to teaching mathematics as if it were a neutral subject. For proponents of critical theory, they, unwittingly, support the power structure rather than reveal it. Thus in his well-known article "Ideology and ideological state apparatuses" (Althusser 1971), Louis Althusser writes: "I ask the pardon of those teachers who, in dreadful conditions, attempt to turn the few weapons they can find in the history and learning they 'teach' against the ideology, the system and the practices in which they are trapped. They are a kind of hero. But they are rare and how many (the majority) do not even begin to suspect the 'work' the system (which is bigger than they are and crushes them) forces them to do, or worse, put all their heart and ingenuity into performing it with the most advanced awareness (the famous new methods!). So little do they suspect it that their own devotion contributes to the maintenance and nourishment of this ideological representation of the School..." (p. 157).

any attempt to work together. However, such *open* enmity is not the rule: most of the time, it is rather only a vague dismissal of one or the other or simply lack of acknowledgement. Moreover, as we also remarked, there is not a total lack of collaboration.[3] So why is there not more?

That question is one of the preoccupations behind this book. Paradoxically, though, having finally arrived at the question of cooperation and collaboration, we must return to the question of how these communities are distinct. For collaboration is a relation between groups that complement one another, and being complementary presupposes difference—difference in focus, in method, in worldview. Without such difference, the communities are thrown into a relation not of collaboration but of competition, as, unfortunately, the relation is all too often perceived.

For our case, a key source of difference between research in mathematics education and in mathematics is the alignment of mathematics education, as part of general education, with the social sciences or even the humanities, and mathematics, with the exact sciences. The emphasis on research is important. For when one considers mathematics education research, one must consider not only its methodology, but also, at a deeper level, what kind of knowledge it generates. Recall how Wu's treatment of the problems of mathematics teaching rested on what kind of knowledge teachers possessed and needed to possess. The social sciences and humanities and the exact sciences have their own sense of knowledge, what it means to know something and what one needs to do to know something. The possibility of cooperation and collaboration, therefore, comes with an appreciation of the more fundamental difference between these two streams of thought: cooperation and collaboration must be premised on coexistence of such different kinds of knowledge and modes of pursuing knowledge.

Of course one can deny this and embrace the tempting assumption that these different kinds of knowledge and modes of pursuing knowledge are, *mutatis mutandis*, the same for the humanities and social sciences on the one side and the exact science on the other. It is the assumption that on both sides there are facts and universal immutable laws which can be verified by methods each side can accept and understand. To be sure, it is not assumed that a law of "learning science" would be a law of physics, but that there *would be* laws; nor is it assumed that, say, a particular experimental technique would be the same in both cases, but that there *would be* experimental techniques whose warrants for accepting or rejecting a claim could be explained each in the other's terms.

[3]One good example of collaboration that does exist is the Klein Project developed and implemented by ICMI. The project was commissioned in 2008 by the International Mathematical Union (IMU) and the International Commission for Mathematical Instruction (ICMI). Its guiding idea was to revisit Felix Klein's book "Elementary Mathematics from an Advanced Standpoint" and produce a book for secondary teachers communicating the breadth and vitality of mathematics as research discipline while connecting it to the secondary school curriculum. An international design team for the project was appointed led by two ICMI presidents: Michèle Artigue and Bill Barton and a book is under preparation. In the meantime, Klein Project has produced a set of "vignettes" for teachers and students. The rationale for this phase of the project and examples of the vignettes already produced can be found at the website: http://blog.kleinproject.org/.

As tempting as this assumption may be, it leads to a whole variety of mutual misunderstandings and false expectations. And more than anything else it is what allows the relationship between the communities to slip into one of competition. It can also obscure self-understanding—particularly in the social sciences (and, there, particularly in educational research)—as can be seen in the common tendency towards "physics envy,"[4] "desiring this [other] man's art... [Ourselves] almost despising," as Shakespeare would say (see *Sonnet xxix*). Yet, the existence of "physics envy" as well as the unreflective use of such terms as "hard science" and "soft science" are only signs that the assumption we are speaking of is adopted widely, even if it be so unconsciously or unacknowledged.

Still, this way of thinking in which the methods and rigor of an intellectual pursuit, indeed, the value of its knowledge, are judged according to its closeness or distance from sciences like physics and chemistry has deep roots. Its greatest expression is in the work of Auguste Comte. Comte's *Cours de Philosophie Positive*, composed between 1830 and 1842, is little read today; yet, despite enormous revisions in how philosophers and historians have come to think about the sciences, including the social sciences, the spirit of this work of Comte haunts the world of research.

Comte invented the word "sociology," and what he meant by that is best seen in the other term he employed, "social physics." He really meant that, as he goes on to describe "social statics" and "social dynamics"! Comte believed that the evolution of society and, therefore, its improvement could be charted by laws comparable to those of physics. In fact, he thought that laws of social phenomenon were incorporated into a greater system of laws including physics. Thus he writes:

> It is the exclusive property of the positive principle to recognize the fundamental law of continuous human development, representing the existing evolution as the necessary result of the gradual series of former transformations, by simply extending to social phenomena the spirit that governs the treatment of all other natural phenomena. This coherence and homogeneousness of the positive principle is further shown by its operation in not only comprehending all the various social ideas in one whole, but in connecting the system with the whole of natural philosophy, and constituting thus the aggregate of human knowledge as a complete scientific hierarchy. (Comte 1975a, p. 211)[5]

---

[4]This is the lament of a recent opinion piece in the New York Times by political scientists (note the name!) Kevin A. Clarke and David M. Primo ((2012, March 30). Overcoming 'Physics Envy'. Available at http://www.nytimes.com/2012/04/01/opinion/sunday/the-social-sciences-physics-envy.html). Interestingly enough, this phrase, so commonly used regarding the social sciences, was actually coined by Joel E. Cohen with reference to biology. Cohen wrote a book review of a book on dynamical systems in biology (Cohen, J.E. (1971, May 14). Mathematics as Metaphor. *Science 172*, 674–675), which begins, "Everyone likes to discover general and unifying principles in biology" (p. 674) and then goes on to say, creating the famous phrase, "Physics-envy is the curse of biology" (p. 675)! So, even within the natural sciences, one should be careful to recognize that there may not be uniformity in appropriateness of methods and approaches.

[5]The English translation contained in the collection edited by Gertrud Lenzer was produced in Comte's day and, as Lenzer notes, was "enthusiastically approved" by Comte himself. The original French text can be found in Comte (1975b, leçon 46, p. 66).

This "positive," or as we might say, "scientific," knowledge was, for him, the final stage in an evolution of knowledge itself, beginning with what he termed the "theological stage" and then the "metaphysical stage" (see pp. 71–72).[6] Comte notes that social thinking will only bear fruit when it finds its way out of the metaphysical stage and fully enters the positive stage, which, he admits, has not yet been accomplished.

It is Comte's voice, his faith in progress through science, that one hears in the American *No Child Left Behind* policy. There we are told that we must aim for "Scientifically Based Research"[7] in order to bring about true educational improvement. This means research, according to *No Child Left Behind*, that:

(1) Employs systematic, empirical methods that draw on observation or experiment
(2) Involves rigorous data analyses that are adequate to test the stated hypothesis and justify the general conclusion
(3) Relies on measurement or observational methods that provide valid data across evaluators and observers, and across multiple measurements and observations
(4) Is accepted by a peer-reviewed or a panel of independent experts through comparatively rigorous, objective and scientific review (US Department of Education 2002a)

The implication, completely consistent with Comte's doctrine, is that research more philosophical, less empirical and experimental, even if it is "the best one can do now," is ultimately *to be replaced* by this "scientifically based" knowledge.

Interestingly enough, the opposing view, namely, that there are distinct modes of pursuing knowledge dependent on the object, that what might be appropriate for physics is not appropriate for the humanities, or, for that matter, educational studies, was recognized before *and* after Comte. Before Comte, one could point, say, to Aristotle, whose introduction to the *Nichomachean Ethics* begins with a discussion of just this point, saying, for example, that one should not expect probable arguments from a mathematician as one should not expect strict proofs from an rhetorician (Book I, 1095b:25–26). But a better example—one whose cogency remains unabated—is Pascal's distinction between two the different kinds of minds, "l'esprit de géométrie," the geometric mind, which proceeds by drawing conclusions from a few first principles, and "l'esprit de finesse," the intuitive mind, which proceeds with a kind of intuitive understanding of things whose principles are so numerous they cannot be grasped one-by-one but must be seen somehow all at once (tout d'un coup) (Pascal 1962, Lafuma 512). The importance for us is that where

---

[6]Comte claimed that education was, in his day, motivated by thinking of the theological, metaphysical and literary types. One of his hopes in laying out the positive philosophy was that education would turn in the positive direction: in effect, Comte was, in effect, pressing for education based more on the sciences and mathematics than on the traditional literary curriculum. This theme, now ubiquitous, was taken up often in the 19th century, for example, by the great biologist Thomas Huxley who suggested that liberal education should be science education.

[7]Comte's sense that progress is impeded by less-than-scientific research can be felt the discussion of mathematics education and "Scientifically Based Research" recorded at the US Department of Education Website (US Department of Education 2002b).

we have this complexity of principles (such as with, say, "learning" whose very definition is hard to frame) it is not enough to modify the analytical approach of the "l'esprit de géométrie": an entirely different approach is required. Pascal, makes it clear in this famous pensée, moreover, that one looks ridiculous, as he puts it, when applying the geometric mind to things that demand the intuitive mind, and the contrary; one cannot be reduced to the other.

After Comte, at the end of the 19th century and the beginning of the 20th, in the work of such figures as Wilhelm Dilthey and Wilhelm Windelband one finds an acute awareness of the difference between what was commonly called the "human sciences" and the "exact sciences." Dilthey (1989), for example, made it clear that in the human sciences, one is engaged in an activity of interpretation rather than deduction; one is driven by a kind of "understanding" (Verstehen), as he called it, of one's human subjects and what they produce. Windelband, a figure less known than Dilthey and perhaps less profound, made a pointed distinction between what he called *nomothetic* and the *idiographic* approaches to knowledge (Windelband 1894/1980); the one concerned phenomena that governed by universal law (*nomos* means "law" in Greek), while the other concern phenomena connected with individuals and their own perspectives.

Both Dilthey and Windelband (though their main object was historical inquiry) touch clearly on the kind of inquiry mathematics education research engages in, namely, studying the learning of mathematics and the place of mathematics in a student's life, as opposed to studying mathematics itself; what does it mean for a student to encounter and begin to assimilate a new mathematical idea, for a student to face and overcome a difficulty, to discern a difficulty? These questions involve exploring the understanding of a student from the inside, as it were. Windelband's distinction between *nomothetic* and *idiographic* inquiries is extremely important in this regard, for although mathematics education research often uses statistics and large populations, some of its most enlightening work is the result of case studies involving sometimes two or three students. The way in which one draws insight from an individual student is difficult, if not impossible, to grasp from the nomothetic perspective: how can a universal law be deduced from an individual case?

But perhaps more than figures such as Dilthey, it is Max Weber whom we must take account of in the post-Comtean world. For its in Weber one comes face to face with problem of values in a decisive way. Weber was deeply concerned about the scientific character of his sociological work. This led him to assert forcefully and repeatedly that values, or more precisely, value judgments, must be removed from social science (see the three long essays in Weber 1949).[8] What makes this fascinating and problematic is that work in the human sciences, Weber's work in particular,

---

[8]This point is also made in Weber's well-known address, "Science as Vocation" (English translation in the collection Gerth and Mills (1958, pp. 129–156) where he also, as in this place, refers to what university professors in science—specifically social science—can see as part of their vocation and what they cannot—and what they cannot includes pronouncements of value among their students.

refers constantly to values and value-systems.[9] There is no overt contradiction in this of course since it is conceivable to speak about a value-laden subject, say religion, in a value-free way. But there are already lines of tension, especially since the inner understanding of such subjects (and Weber has views here that are not inconsistent with Dilthey) presupposes what it is like to be committed to values.[10] These lines are stretched even further by Weber's acceptance of what he calls "value-relevance" in inquiry: the choice of subject matter for investigation may be related to the values of the investigator, even while the investigation itself is value-free (e.g. Weber 1949, pp. 21ff).

The notion of value-relevance applies of course even to the purest of sciences and to mathematics: it is at work in deciding whether a mathematical theory or problem is interesting and worth pursuing or whether a particular solution to a problem or proof deserves our praise (see, for example, Corry 1989 and Elkana 1981). In mathematics education research, however, as in other forms of educational research, not only must we speak about value-relevance, but, beyond that, we must speak about a role of values in a more direct way: here, by the very nature of the subject, engaging in "evaluation" is unavoidable (this will be discussed further in *Dialogue on a Dialogue*). For mathematics education research has ultimately the practical aim of *improving* mathematics education, of making it *better*, of saying how we *ought* to teach and how students *ought* to learn. This engagement with values together with its attention to individuals, its idiographic character, and its need to interpret rather than only to describe behavior, sets off mathematics education research from the kinds of research typically pursued in faculties of exact sciences and engineering. One cannot assume, as Comte did, that, in principle, there could be consistency in the general methodologies and general outlooks of these different forms of research.

The first step, therefore, in ameliorating the cooperation between researchers associated with the exact sciences and those in involved in research like mathematics education research is to recognize these radically different ways of pursuing research and to recognize the necessity of those differences. Mathematics education research must be understood as something apart from mathematics and mathematics from mathematics education research: one cannot be subsumed under the other or replaced by the other. It should not be our mission to "convert" mathematicians to what they cannot be as it should not be theirs to determine what mathematics education researchers should research. And yet, to reiterate what has been said in different ways throughout this introduction, this cannot be a formula to go in separate ways: the common focus on mathematics, one way or another, will not allow for that. Cooperation begins when there is at the same time the recognition that each side is looking in the same direction but with very different, complementary eyes.

---

[9]Weber's famous 1905 work *Die protestantische Ethik und der Geist des Kapitalismus* (*The Protestant Ethic and the Spirit of Capitalism*) is a case in point.

[10]In his chapter on the fact-value distinction in *Natural Right and History*, Leo Strauss (1953) makes a point along these lines.

The Structure of This Book

Almost as a demonstration of the possibility of cooperation, the authors of this book comprise mathematics education researchers and mathematicians. Some of the authors could wear both hats, and some do. But just putting the two communities together in one room is not enough to begin a dialogue. Indeed, the first part of this book raises the question of dialogue and is centered on a dialogue (Eisenberg and Fried 2009) written by one of the editors, Michael N. Fried, and Ted Eisenberg, to whom this book is dedicated. This dialogue, which concerned the state of mathematics education generally, was in fact a response to paper by Norma Presmeg (Presmeg 2009). In her paper, published in the same issue of *Zentralblatt für Didaktik der Mathematik* (ZDM), Presmeg had argued that since the purview of mathematics education includes more than mathematical content *per se*—that it concerns how students think about mathematics, how mathematics becomes part of students' inner and outer lives, how it is integrated into students' sociocultural world, for example—it is necessarily a multidisciplinary subject. Eisenberg, in particular, felt in the course of broadening mathematics education in this way, mathematical content was in fact becoming lost. The dialogue that he and Fried produced subsequently revolved around the question of mathematics education is truly about as a field, what are its true interests, and has it lost its identity by moving too far away from mathematics.

So *Dialogue on a Dialogue* revisits these two papers[11] and produces a new dialogue with the same players—Eisenberg, Fried, and Presmeg—providing thus three points of view. It sets the stage for the rest of the book by raising questions such as whether mathematics teaching has the same interests as mathematics education research and whether the latter should, as Presmeg originally claimed, be multidisciplinary. It also suggests some of the themes of commonality and difference joining and dividing the communities of mathematics and mathematics education—for example, visualization, proof, policy, problem-solving.

The remaining parts of the book treat eight of these themes. With two exceptions—*Mutual Expectations* and *Problem-Solving*—each part has a similar overall structure: a position paper followed by a chapter containing a series of short responses or reflections on the same subject. In each case, the latter also contains an introduction and synthesis of the main points and problems. To provide the reader with a kind of map for the book, we now summarize these eight parts and set out the players involved.

[11] The papers by Eisenberg and Fried and Presmeg were joined by a third written by David Pimm (Pimm 2009), who also discussed the relationships and provinces of the different disciplines contributing to mathematics education.

Mutual Expectations Between Mathematicians and Mathematics Educators

This part is one of two related to the preconditions for mathematics educators and mathematicians' working together. On the assumption that mathematicians and mathematics education researchers do wish to work together, what do they expect to receive from one another? What kinds of problems do they expect one another to focus upon? This, perhaps more than any other part, addresses the question of how each community is defined in light of the other, for what they expect from one another clearly reflects how they see one another. It is also the part in which one can see the tensions between the two communities, albeit sometimes between the lines.

The introduction and synthesis of the issues involved is written by Tommy Dreyfus; the other contributors include Stephen Lerman, Ioanna Mamona, and Uri Onn. The contributors were chosen carefully so that they would represent a spectrum of views from that of mathematics educator whose work is generally distant from mathematical content to a pure mathematician whose educational interests are closely tied to his university teaching.

History of Mathematics, Mathematics Education, and Mathematics

In a way, this is the oddest of the parts in this book. In contrast to a subject such as "proof," the history of mathematics is neither at the center of mathematics as a discipline nor at the center mathematics education as a discipline. Yet, it is of great interest to both even if it is often misunderstood by both. At the same time it is unavoidable in any effort to see mathematics as a part of general mathematical culture, as Felix Klein put it, and therefore goes far to address the difficulties of mathematical literacy and the meaning of being mathematically educated. A proper understanding of the place of history of mathematics in mathematics and mathematics education may end up being genuine common ground seeming, at present, foreign to both.

The introduction and synthesis of the issues involved is written by Luis Radford; the other contributors include Alain Bernard, Michael N. Fried, Fulvia Furinghetti, and Nathalie Sinclair. Luis Radford was chosen to produce the synthesis, not only because he himself has done some historical work in mathematics and is himself an eminent mathematics educator, but also because of his particular cultural-historical understanding of mathematics. This cultural-historical understanding places mathematics in a grey area between the "two cultures" (using C.P. Snow's famous phrase) and, therefore, shows more clearly the relationship between them.

The part opens with a paper by Hans-Niels Jahnke concerning the hermeneutic approach to history of mathematics, an approach that appreciates the historical character of mathematics of the past while taking into account modern mathematical notions.

Problem-Solving: A Problem for Both Mathematics and Mathematics Education

While history of mathematics may be foreign to both mainstream mathematics and mathematics education, this is certainly not the case with problem-solving. The centrality of problem-solving in mathematicians' own work and in their teaching, is incontrovertible. Problem-solving is also a central topic for mathematics educators, who have developed conceptual frameworks to formulate general ideas about problem-solving (as opposed to the specific ideas needed for solving specific problems). Both mathematics educators and mathematicians have given thought to problems helping students understand ideas, and both have given thought to the process of solving problems: George Pólya, of course, reflected deeply about this, and Pólya's work figures strongly in this chapter. This is, one hastens to add, not only because of the importance of Pólya's work regarding problem-solving, but also because Pólya himself represents a bridge between mathematics and mathematics education: he was an eminent mathematician and also a deep influence on mathematics education.

The introduction and synthesis of the questions raised by problem solving is written by Boris Koichu, who has done extensive work on problem solving especially among talented mathematics students, those most likely later to join the community of mathematicians. The other contributors to this part are Gerald Goldin, Roza Leikin, Shlomo Vinner, and Izzy Weinzweig.

Mathematical Literacy: What Is It and How Is It Determined?

One might say that the guiding question for this part on mathematical literacy is simply what does it mean to say someone is "mathematically educated"? In this light, its subject has a theoretical character. However, it also has a practical side with real consequences for teaching and curriculum development; a notion of literacy is, in this way, also a guide to the design of a mathematics policy, the subject of Part "Policy: What Should We Do, and Who Decides?". Moreover, literacy, precisely because it concerns the ends of policy, is connected to the practical problem of assessing educational policy and achievement. For this reason, operational definitions for literacy have been produced in conjunction with international assessment, notably the PISA program.

The synthesis here is written by Anna Sfard; other contributors include Abraham Arcavi, Iddo Gal, Ron Livné, and Hannah Perl. Sfard's own contribution clarifies the notion of literacy by connecting it to another theme of equal importance to mathematics educators and mathematicians, namely the idea of communication.

The part opens with a paper by Paul Goldenberg, who emphasizes what he calls habits of mind.

Visualization in Mathematics and Mathematics Education

The subject of visualization is important to both mathematics and mathematics education since it characterizes the way both students and mathematicians commonly think about mathematical ideas and solve mathematical problems. For this reason, Hadamard studied visual thinking in his famous work on the psychology of mathematical invention (Hadamard 1945), and mathematicians, such as Stanislav Ulam, writing about their own mathematical thinking attests to the importance of visualization (e.g. Ulam 1976, p. 183). Visualization is also related to the representation of mathematical objects with the aid of computers: the ability of computers to produce and manipulate pictures has allowed new ways for students to study and explore ideas in geometry and analysis. This part, then, takes into account mathematicians use visualization in their teaching, mathematics educators' proposals for employing and developing visual thinking in computer and non-computer environments, as well as research results from mathematics education.

The introduction and synthesis of this topic is written by Elena Nardi; the other contributors include, Rina Hershkowitz, Raz Kupferman, Norma Presmeg, and Michal Yerushalmy.

The part opens with a paper by Ken Clements that discusses, among other things, Clements work with the then young Terence Tao, later Field Prize medalist—a rare view into the ways, often visual ways, a young developing mathematician thinks.

Justification and Proof in Mathematics and Mathematics Education

Common ground here would at first sight seem unproblematic, since "justification," interpreted as "proof," is a subject is crucially important for both mathematics and mathematics education. Yet, there are in fact strong divisions. For in mathematics "proof" and "justification" are identified, whereas in mathematics education much attention is given to forms of justification that fall short of proof but nevertheless are deeply connected with processes of learning. The idea that an incomplete or even incorrect explanation may yield more insight for the mathematics educator than a rigorous proof runs counter to the way of thinking in a discipline that gives little credit to a justification which is not a proof. On the other hand, mathematicians do give weight to proof, even a heuristic argument, that actually persuades them of the truth of mathematical claims. Proofs must have in some sense pedagogical value.

The introduction and synthesis here is written by Keith Weber; the other contributors include, Gila Hanna, Guershon Harel, Ivy Kidron, and Annie and John Selden.

A paper by David Tall concerning research on mathematical reasoning and thinking generally is the opening paper for this part.

Policy: What Should We Do, and Who Decides?

The central concerns of "policy" are the principles and agents of decision-making and the program—the policy—actually decided. To the extent "policy" concerns the agents of decision-making, it is closely related to the subject of collaboration; to the extent it concerns the policy decided, including the curriculum, it must consider the ends the policy tries to achieve and is thus closely related to the subject of literacy. Naturally, beyond the curriculum, the policy decided takes in elements of teaching practice, assessment, and modalities for further decision-making.

The introduction and synthesis of the issues in this part is taken up by Nitsa Movshovitz-Hadar; the other contributors are Jonas Emanuelson, Davida Fischman, Azriel Levy, and Zalman Usiskin.

The part opens with a paper by Mogens Niss who was the architect of the competencies framework used in Denmark. Niss makes it particularly clear how broad the subject of policy is, involving not only decision making but also views about the nature of mathematics teaching and learning and even mathematics itself.

Collaboration Between Mathematics and Mathematics Education

This final part contains accounts of genuine instances of collaboration between mathematicians and mathematics educators or of scholars who have managed to work in both fields. These instances serve as existence proofs for the possibility of collaboration, but not uniqueness proofs. There may be different kinds of models for joint work between mathematicians and mathematics educators.

The introduction and synthesis is written by Pat Thompson; the other contributions are by Michèle Artigue, Ehud de Shalit, and Günter Törner.

The part opens with an account of collaboration written by Hyman Bass and Deborah Ball. They themselves are a superb example of the kind of collaboration that is possible.

One Final Word

Since the symposium from which this book emerged was held in the Negev, it is fitting keep in mind a Bedouin custom. When one comes to a Bedouin tent, the host offers coffee. It is always very strong, almost bitter. Then there is talk and food and talk. Finally, tea is served. It is always very sweet. The meaning of this, at least according to one account, is that a visit begins with a little unease, a little uncertainty—thus the bitter coffee. But after conversation, turning things over and exchanging thoughts, the visit ends sweet.

In a way, this is an image of how we hope readers of this book move from the beginning to the end. It is, we think, a hopeful book. So while the expectations discussed in the second part focus some of the uneasiness and friction existing between mathematics education and mathematics, the final part, *Collaboration*, shows signs that cooperation is possible. In between, many issues and questions are raised. These are not resolved completely, even at the end. However, the instances of cooperation and collaboration described in *Collaboration* and, perhaps more trenchantly, the very fact mentioned above that mathematicians and mathematics educators participated in the writing of this book show there is no inevitability in the growing distance between our two communities and that together we can work out these questions which are of mutual concern.

References

Althusser, L. (1971). *Lenin and philosophy and other essays*. New York: Monthly Review Press (trans: B. Brewster).

Begle, E. G. (1972). *Teacher knowledge and student achievement in algebra*. School Mathematics study group reports, number 9.

Boaler, J. (2012). When academic disagreement becomes harassment and persecution. http://www.stanford.edu/~joboaler/. Accessed 10 Dec 2012.

Comte, A. (1975a). In *Auguste Comte and positivism: the essential writings*. G. Lenzer (Ed.). New York: Harper & Row.

Comte, A. (1975b). *Physique social: cours de philosophie positive, leçons 46 à 60*. Paris: Hermann. Presentation and notes by Jean-Paul Enthoven.

Corry, L. (1989). Linearity and reflexivity in the growth of mathematical knowledge. *Science in Context*, *3*, 409–440.

Dilthey, W. (1989). In R. A. Makkreel & F. Rodi (Eds.). *Selected works: Vol. I. Introduction to the human sciences*. Princeton: Princeton University Press.

Eisenberg, T. (1977). Begle revisited: teacher knowledge and student achievement in algebra. *Journal for Research in Mathematics Education*, *8*, 216–222.

Eisenberg, T., & Fried, M. N. (2009). Dialogue on mathematics education: two points of views on the state of the art. *ZDM. The International Journal on Mathematics Education*, *41*(1&2), 143–150.

Elkana, Y. (1981). A programmatic attempt at an anthropology of knowledge. In E. Mendelsohn & Y. Elkana (Eds.), *Sciences and cultures: anthropological and historical studies of the sciences* (Vol. 5, pp. 1–68). Dordrecht: Reidel.

Even, R. (2011). The relevance of advanced mathematics to teaching secondary school mathematics. *ZDM. The International Journal on Mathematics Education*, *43*(6&7), 941–950.

Gerth, H. H. & Mills, C. W. (Eds.) (1958). *From Max Weber: essays in sociology*. New York: Oxford University Press.

Hadamard, J. (1945). *The psychology of invention in the mathematical field*. Princeton: Princeton University Press.

Holton, D. (Ed.) (2002). *The teaching and learning of mathematics at university level: an ICMI study*. New York: Kluwer Academic.

Howe, R. (1999). Knowing and teaching elementary mathematics. *Notices of the American Mathematical Society*, *46*(8), 881–887.

Kilpatrick, J. (1992). History of research in mathematics education. In D. A. Grouws (Ed.), *Handbook of research on mathematics teaching and learning* (pp. 3–37). New York: Macmillan Co.

National Science Foundation (2012). Science and engineering indicators. http://www.nsf.gov/statistics/seind12/start.htm. Accessed Dec 2012.

Pascal, B. (1962). *Pensées*. Paris: Seuil.

Pimm, D. (2009). Method, certainty and trust across disciplinary boundaries. *ZDM. The International Journal on Mathematics Education*, *41*(1&2), 155–159.

Presmeg, N. (2009). Mathematics education research embracing arts and sciences. *ZDM. The International Journal on Mathematics Education*, *41*(1&2), 131–142.

Shulman, L. S. (1986). Those who understand: knowledge growth in teaching. *Educational Researcher*, *15*(2), 4–31.

Singh, S. (1997). *Fermat's enigma*. New York: Walker and Company.

Snow, C. P. (1963). *The two cultures: and a second look*. New York: Mentor.

Sriraman, B., Roscoe, M., & English, L. (2010). Politicizing mathematics education: has politics gone too far? Or not far enough? In B. Srirarman & L. English (Eds.), *Theories of mathematics education: seeking new frontiers* (pp. 621–638). New York: Springer.

Steiner, G. (1997). *Errata*. London: Weidenfeld & Nicolson.

Strauss, L. (1953). *Natural right and history*. Chicago: University of Chicago Press.

Taylor, H., & Taylor, L. (1993). *George Pólya: master of discovery*. Palo Alto: Dale Seymour Publications.

Ulam, S. (1976). *Adventures of a mathematician*. New York: Scribner's.

US Department of Education (2002a). Comprehensive school reform program overview. http://www2.ed.gov/programs/compreform/csrdoverview/edlite-sld011.html. Accessed 24 Dec 2012.

US Department of Education (2002b). Scientifically based research. http://www2.ed.gov/nclb/methods/whatworks/research/page_pg6.html. Accessed 24 Dec 2012.

Weber, M. (1949). In E. A. Shils & H. A. Finch (Eds.), *The methodology of the social sciences*, Glencoe: Free Press (trans: Shils, E. A., & Finch, H. A.).

Windelband, W. (1894/1980). *Rectorial address, Strasbourg*. Reprinted and translated in *History and Theory*, *19*(2), 169–185.

Wu, H. (2011). The mis-education of mathematics teachers. *Notices of the American Mathematical Society*, *58*(3), 372–384.

Yasukawa, K. (2010). Commentary on politicizing mathematics education: has politics gone too far? Or not far enough? In B. Srirarman & L. English (Eds.), *Theories of mathematics education: seeking new frontiers* (pp. 639–643). New York: Springer.

Dialogue on a Dialogue

Chapter 2
Mathematics and Mathematics Education: Beginning a Dialogue in an Atmosphere of Increasing Estrangement

Michael N. Fried

Abstract In 2009, Norma Presmeg wrote a piece for a special issue of the ZDM on interdisciplinarity. Presmeg's paper presented her view of the general spirit of and possibilities for mathematics education research. This prompted a dialogue on the state of mathematics education by Ted Eisenberg and Michael Fried, published in the same issue of ZDM. This paper gives an account of that dialogue and the symposium in honor of Ted that arose out of it; in doing so, it also further elaborates on the themes that motivated this book.

Keywords Human sciences vs exact sciences · Mathematical content · Mathematicians · Mathematics education researchers · Mathematics education · Values

My Dialogue with Ted

When Dani Berend broached the idea of a conference in honor of Ted Eisenberg, it was immediately clear to me what the subject of the conference should be. It should concern the relationship between mathematics and mathematics education as disciplines. The thin mathematical backgrounds of many researchers in mathematics education, and worse, their apparent lack of interest in mathematics, had become one of Ted and my constant conversation topics. Ted often lamented to me how out of place he felt in a field more and more dominated by sociology, psychology, politics, anthropology, and philosophy, and less and less by mathematics. His feelings were understandable. For almost his entire academic career, Ted sat in a mathematics department, and, besides his own mathematics education research, he taught regular courses in the mathematics department, working hard to introduce students to calculus and linear algebra. Teaching mathematics, and, therefore, knowing mathematics, has always been for him at the center of mathematics education, and he has always maintained it should be. Not only is a mathematics department

M.N. Fried (✉)
Science and Technology Education, Ben-Gurion University of the Negev, Beer Sheva, Israel
e-mail: mfried@bgu.ac.il

M.N. Fried, T. Dreyfus (eds.), *Mathematics & Mathematics Education: Searching for Common Ground*, Advances in Mathematics Education,
DOI 10.1007/978-94-007-7473-5_2,

the right place for the study of mathematics education, in his view, but also a solid mathematics background was requisite for fruitful mathematics education *research.*

For my part, when we spoke about these things, I often took the position of the devil's advocate and defended the usefulness of non-mathematical mathematics education research. But I was not always consistent in that role, not wholly the devil's advocate. On the one hand, I could not help often commiserating with Ted about the limited mathematical understanding of many researchers who consider mathematics their specialty and about research that tries to solve the problems of the world rather than those of mathematics teaching and learning. On the other hand, I could deny some genuine leanings towards the other side, an inevitable outcome, perhaps, of my training as a historian of mathematics, where one learns to see mathematical thought as contingent and embedded in culture.

In any case, these conversations came to head in 2008 when Ted was asked to review a paper by Norma Presmeg entitled "Mathematics Education Research Embracing Arts and Sciences" (Presmeg 2009). In this position paper, Norma argued that since the purview of mathematics education research includes more than mathematical content *per se*—that it concerns how students think about mathematics, how mathematics becomes part of students' inner and outer lives, how it is integrated into students' sociocultural world, for example—it is necessarily a multidisciplinary affair. And the introduction of multidisciplinary considerations brings with it also the introduction of different kinds of methodologies. Research in the field, for this reason, takes on a character often closer to the human sciences than the exact sciences, even though the focus of the field is still mathematics.

Indeed, Norma never discounted the importance of mathematical content in her piece: she was explicit about that. "The subject matter," she said, "of research mathematicians is the content of mathematics, and without this content there would be no mathematics education" (p. 132). On the other hand, it was central to her argument that mathematics education research is not mathematics, not even applied mathematics! She recalled Millroy's 1992 study of carpenters in Cape Town which pointed to distinctions between the mathematics implicit in students' out-of-school culture and the explicit mathematics in their in-school culture. But bridging these two cultures requires the kind of ethnographic approach typical of anthropological research. Students' mathematical understanding in these cultural contexts requires understanding the language conditioned by such cultures, the systems of signs one uses to construct and communicate ideas, including mathematical ideas, and this brings one to the semiotic research which has become prevalent in mathematics education research. It becomes apparent by such examples how one can be led in a very natural way into extra-mathematical disciplines.

For Norma, these borrowings from other disciplines were not only a necessary widening of the field, but also a refreshing and welcome one. For Ted, ethnographic research, semiotics, and so on were dragging the field too far afield. Yet, Ted also understood that Norma's paper was an accurate picture of the state of the art, and while for her that meant finally "coming home" (p. 134), as she put it, for Ted it meant alienation. The situation was particularly painful for Ted because of the enormous respect he has for Norma, unquestionably and rightfully a leading figure in mathematics education. It was in that uneasy spirit Ted wrote his review of Norma's paper. The gist of his review was that while the paper went against his most basic beliefs about the nature of mathematics education and direction it should be going, he saw that according to criteria nearly universally accepted in the field he could hardly criticize the paper, let alone reject it.

Norma's paper was intended for a special issue on "interdisciplinarity in mathematics education" for ZDM—The International Journal on Mathematics Education. The editor of the special issue, Bharath Sriraman, saw in Ted's review an opportunity for some interesting counterpoint and suggested that Ted write up his criticism as a reaction to Norma's paper. He suggested, moreover, that Ted and I do it jointly. Perhaps because I had played the devil's advocate a bit too often, Ted responded that he thought I was more in Norma's camp than his. So Bharath, who does not give up easily, suggested that Ted and I write a dialogue on the issues Ted raised in his critique. When Ted finally told me about Bharath's proposal, my first inclination was to say that all this was so much at the center of Ted's concerns he should really take up the project himself. But, since I always have trouble uttering a simple "no," I said I would think about it.

Thus it stood until March 2008, when I went to Rome to attend the centenary of the ICMI, the International Commission on Mathematics Instruction, the oldest and most prominent international organization dedicated to mathematics education. As is well known, the ICMI was established at the Fourth International Congress of Mathematicians held in Rome in 1908. The ICMI still belongs to the IMU, the International Mathematics Union, and its connection with the greater mathematical world has deep roots. Felix Klein, for example, was the first president of the ICMI, and he was followed by other eminent mathematicians holding the presidency or other high posts in the commission, figures such as Jacques Hadamard, Marshall Stone, Sergei Sobolev, Saunders Mac Lane, Hans Freudenthal, and Hyman Bass. Yet, at this celebration of the first hundred years of the ICMI, years in which the commission survived tensions from nationalistic fervor and the violence of two world wars, and years of triumph in which it saw great changes in mathematics and mathematics education, the founding of international mathematics education journals and large scale international mathematics education conferences—it was at this happy occasion that some mathematics educators saw fit to ask for divorce.

The one that particularly stands out in my mind is Mamokgethi Setati. Setati did not want merely to broaden the scope of the field; she sought to reestablish its entire agenda—and in a way that left little room for mathematical content. For her, mathematics education should focus all of its energies on confronting the problems of the developing world, "the eradication of poverty, empowerment of women and gender equality" (Menghini et al. 2008, p. 182), no less. It was in this context that she also called for a reexamination of ICMI's relationship with the IMU (p. 184), a euphemistic way of saying, "End the marriage." To me, it was immediately clear that her position was untenable. As I wrote in my review of the proceedings of the ICMI centenary (Menghini et al. 2008):

> Following the implications of Setati's position, it seems difficult to avoid two equally dubious conclusions: (1) mathematics is not at the heart of mathematics education or at least

> must be subordinated to more general social issues, or, at the other extreme, (2) mathematics has a privileged position in dealing with global social problems such as poverty and gender inequality. (Fried 2009, p. 524)

As human beings, of course, we must be concerned with social justice. But this is not the question: the question is whether we must be concerned with social justice *as* mathematics educators and, more, whether social justice should trump all else relevant to mathematics education. I do not think Setati was a lone voice, though I hardly believe her view as to where mathematics education should be going reflected a consensus at the Rome meeting. That said, it was evident, whether one liked it or not, the field *had* broadened far beyond teaching and learning mathematical content: I could not help feeling we had reached a watershed and a real possibility that mathematical content might be swept away altogether. It was then and there I decided I had thought about it enough: I would tell Ted I am ready to work on the dialogue.

Although Norma's paper was the pretense, Ted and I wanted our piece as much as possible to be like the conversations he and I had so often. I think it was Ted's idea that the paper should take the form of an exchange of letters. He wrote about his vision and his discontent, and I responded. Although it gave to me, in effect, the last word, still it allowed us both to write a more or less connected account of our take on the state of mathematics education research. The format also allowed a certain informality appropriate to airing views rather than presenting findings. But that should not detract from the seriousness of the exchange. Where mathematics education ought to be going and what mathematics education research ought to be are not empirical questions that *findings* could ever settle. These are matters that require continual sober discussion. In fact, the place of empirical research in mathematics education was one issue we raised in our letters. There were many issues we put on the table.

Mathematical content in mathematics education and mathematics education research was only one of these issues; however, then and in our own off-the-page conversations it was a focal one. In Ted's way of thinking about it, it could be discussed in terms of university geography, that is, where on campus should a unit on mathematics education be located? Should it be where the exact sciences and engineering faculties are or where the humanities and social sciences faculties are? Even taken so literally, where one sees oneself is unavoidably a question of identity or self-definition, and ours *was* a discussion about self-definition. Certainly Norma's paper was about how mathematics education should be defined—an art? a science? both together? Setati's view too was surely a statement of self-definition, albeit one I could not swallow.

Mathematics and Mathematics Education: Difference and Confluence

Dani Berend's idea to have a conference in honor of Ted ultimately became, therefore, a symposium concerning the very identity of mathematics education as a field, specifically, its identity as it relates to mathematics as a field. But for this reason, it could not a question about mathematics education alone, for since it is asked *with respect* to the discipline of mathematics, whatever identity mathematics education crystalizes for itself will also leave a mark on the identity of mathematics as a discipline: the two really are wed. Indeed, having originally entitled this symposium "Is there still room for mathematics in mathematics education?" it quickly became apparent that this could not be asked without its complement, "Is there still room for mathematics education in mathematics?" Thus the symposium was renamed, "searching for common ground." As a title, it expressed, first of all, Ted's profound belief that mathematics education and mathematics do have a common ground that must be taken seriously and never ignored. But it also suggested we have to some extent lost hold of that common ground and must search together to regain it.

Naturally, with this in the background, we should want to concentrate on commonalities; yet, this cannot be done without, at the same time, being cognizant of how the mathematics education community is set off from the mathematics community. Failing to discern the separateness these communities only invites claims that mathematics education is populated by poor mathematicians and mathematics by poor educational thinkers. (And here, I should emphasize that by the mathematics education community I mean chiefly the academic community of mathematics educators rather than teachers in the field.)

Besides the more obvious differences between the two communities—that mathematics education researchers do not prove theorems as a matter of course and that mathematicians do not consider theories of learning and thinking, for example—I should mention two ways in which these communities at least appear move in different directions, particularly, ways in which mathematics education as an academic field of research turns towards the social sciences for its sense of identity.

The first has to do with how mathematics educators approach their questions, their methodology. The methodological approaches mathematics education must apply to understand aspects of teaching and learning and often questions of curriculum do truly have more in common with social sciences than with the exact sciences or engineering, the theory of "didactic engineering," notwithstanding. This is evident in part by the sheer variety of methodological approaches in the field, that is, by the lack of a single paradigm for doing research. In fact, it might be argued that this methodological eclecticism is one reason why the question of identity is so much more prevalent in mathematics education and the social sciences than it is in the exact sciences (perhaps with the exception of biology, but for different reasons). But more importantly, like the social sciences, the methodological concerns of mathematics education research share a history, rooted in Comte and Weber, of aiming to be "value-free," an "objective" science like physics, and, yet, ever falling short of that ambition. We want a *science* of learning and teaching, but we cannot escape, in its baldest form, our own commitments as to what we think students *ought* to learn and how they *ought* to learn it and how we *ought* to teach them; our most basic questions always lead us to questions of values.

The second has to do with our aims as educators. For the separateness of mathematics education research from the discipline of mathematics can also be felt even where both are focused on education. This is so because, being typically associated with a university mathematics department, mathematicians are placed in the position of training new mathematicians or scientists. Mathematics education, by contrast, concerns the whole gamut of learning and teaching mathematics, including university level mathematics, but typically concentrating on learning and teaching school mathematics. The one must ask what constitutes a mathematically *trained* person, while the other, a mathematically *educated* person. These are not, to be sure, mutually exclusive categories; however, what it means to be mathematically trained and what it means to be mathematically educated are also not identical, and the difference is not just one of degree. To be fully mathematically trained is to be a mathematician; but one can be highly mathematically educated without being a mathematician; and, conversely, there are competent mathematicians who are surprisingly uninformed about the history of mathematics and matters connected to its philosophical foundations. A trained mathematician must produce mathematics; one who is mathematically educated must feel at home with mathematics, appreciate its power, and know it as a part of one's culture. What is crucial for the latter is not always crucial for the former, and the contrary. And from this it follows also that researchers in mathematics education must take into account considerations that are at least broader than those mathematicians *qua* educators must take into account and some ways are qualitatively different.

But the picture is hardly black and white with regards to either of these differences; the differences are real, certainly, but they are not such that we can sit contently each in our own separate bailiwicks. Consider the second. To feel at home with mathematics and appreciate its power one must *engage* in mathematics at some level. To become mathematically educated one must understand something about mathematics from within, and that means having a foot in mathematically training. Even to understand *understanding*, as researchers in mathematics education should want to do (as pursued, for example, in Sierpinska 1994), one must engage in mathematics. For example, if one learns that the derivative of a function at a point P is the slope of the tangent to the graph of the function at P, one will have a certain level of understanding of the derivative; if one stops there, however, one could easily believe one understands the derivative *tout court*. Learning a little more, one is faced with a new idea, the "gradient," which is still called the "derivative"; one might convince oneself that it is similar to the old idea since the gradient can be related to the tangent plane of a surface. But then one goes further and learns another idea, the "Jacobian matrix," and, again, this is called a "derivative." One's notion of the derivative as a "slope" no longer suffices to understand the "derivative"; one needs a more general idea of a linear operator approximating the function at a point, which, in odd way, brings one back to the slope of the tangent. The point is without having *experienced* such levels of meaning and circulation of ideas, a student's understanding of mathematical concepts, as well as an educator's understanding of understanding are bound to be rather one dimensional.

On the other side of the equation, well-trained mathematicians who lack a broader view of their subject—its history, its place in society, its philosophy—may still be able to do what they have to do very well, but they face the danger of being something like excellent technicians only. It is thus not by accident that in the very best mathematicians there tends to be a confluence of training and education: knowing history or philosophy of mathematics or the social implications of mathematics may not allow them to solve more problems or prove more theorems, but it makes them more worthy of the name "mathematician." I am certain that Felix Klein had this in mind when he urged teachers to learn history. As he wrote in his *Elementary Mathematics from an Advanced Standpoint*:

> ... I shall draw attention, more than is usually done... to the *historical development of the science*, to the accomplishments of its great pioneers. I hope, by discussions of this sort, to further, as I like to say, your general *mathematical culture*: alongside of knowledge of details, as these are supplied by the special lectures, there should be a grasp of subject-matter and of historical relationship [emphases in the original]. (Klein 1908/1939, II, p. 2)

As for values, while it is true, as I described above, that mathematics education, like all education, is value-laden, it has been one of the leitmotifs of modern philosophy and history of science that even the most exact sciences themselves are not exactly *Wertfrei*! One begins to see this by considering how a certain question or idea or approach in mathematics is deemed interesting or important or beautiful. It is not just because it is correct, or even clever. One might say it is because it is useful. But what makes something useful? This has its own set of values attached to it. And the priority of utility as a measure of importance is itself a matter of values: one recalls Hardy's pride in never having done anything useful in mathematics!

The social background of values is also evident in mathematics and science: a mathematician or scientist's winning a prize or obtaining a speedy promotion depends on whether the community of mathematicians or scientists values the person's work—and that evaluation is not so much a determination as a collective judgment. Aesthetics plays an important part here too for the work is likely to be judged by the number of beautiful results it contains. Of course it may be that beauty only *appears* to be related to values, that it is actually a completely determined thing itself. However, if agreement is any measure of that, it worth recalling that when David Wells asked readers of the *Mathematical Intelligencer* to judge theorems according to their beauty, he had to conclude finally that "... the idea that mathematicians largely agree in their aesthetic judgments is at best grossly oversimplified" (Wells 1990, p. 40).

And there can be real clashes of values in mathematics. Such a clash was described in Siobhan Roberts' biography of H.S.M. Coxeter (Roberts 2006). Coxeter, being a classical geometer, represented a position favoring a visual and intuitive approach to geometry. Standing opposed to Coxeter—this was chiefly during the 1940s and 50s—was the more fashionable Bourbaki, who, according Pierre Cartier, considered that "... [geometry] was based on pure logical reasoning, as little visual insight as possible. Visual insight [in the view of Bourbaki] was considered a concession to human weakness" (quoted in Roberts 2006, p. 122) (a statement of values if ever there was one!).

What is interesting for us is that this difference of values within the mathematics community was played out in discussions about mathematics education. Recall it was in the context of debating reforms in French mathematics education (at Royaumont in 1959) that Dieudonné, one of the founding members of Bourbaki, cried out, famously, "Down with Euclid! Death to Triangles!" Coxeter, for his part, participating in activities and producing publications "...went on a crusade to bring his passion for the visual and intuitive methods to any and all willing spectators," as Roberts puts it (Roberts 2006, p. 163). Thus we see that far from a value-free existence, the mathematical world has its own biases and preferences and these bring it directly into regions of common ground with mathematics education. More precisely, it was as questions about mathematics education that these differences in mathematical values—those of the Bourbaki camp towards the formal, non-visual and those of the Coxeter camp towards the intuitive, visual approach—found a natural means of expression.

As a last word, I should say that when Ted and I wrote our dialogue, it was clear to both of us that this was only one round of a greater dialogue. We had no intention giving a final statement about any of the issues we raised. It was only a beginning. The question is where it should continue, where should its locus be? The remark above in connection to the Royaumont conference suggests, perhaps, this may be the role of mathematics education itself, even if, as I have already argued, mathematics has a stake in the dialogue as well. Indeed, the fact that our dialogue was published by a leading journal for mathematics education may not have been an anomaly, a departure from the main issues of mathematics education research, but an indication of a new issue of emerging importance in the field itself.

References

Eisenberg, T., & Fried, M. N. (2009). Dialogue on mathematics education: two points of view on the state of the art. *ZDM. The International Journal on Mathematics Education*, *41*, 143–150.

Fried, M. N. (2009). Review of M. Menghini, F. Furinghetti, L. Giacardi, F. Azarello (Eds.), *The first century of the International Commission on Mathematical Instruction (1908–2008): reflecting and shaping the world of mathematics education*. Instituto della Enciclopedia Italiana, 2008. *ZDM. The International Journal on Mathematics Education*, *41*, 521–524.

Hadamard, J. (1949). *The psychology of invention in the mathematical field*. Princeton: Princeton University Press.

Hanna, G. (2000). Proof, explanation and exploration: an overview. *Educational Studies in Mathematics*, *44*, 5–23.

Klein, F. (1908/1939). *Elementary mathematics from an advanced standpoint. Part I: arithmetic, algebra, analysis. Part II: geometry*. New York: Dover. Translated by E. R. Hedrick & C. A. Noble.

Menghini, M., Furinghetti, F., Giacardi, L., & Arzarello, F. (Eds.) (2008). *The first century of the International Commission on Mathematical Instruction (1908–2008): reflecting and shaping the world of mathematics education*. Rome: Instituto della Enciclopedia Italiana.

Millroy, W. L. (1992). *An ethnographic study of the mathematics of a group of carpenters*. Reston: National Council of Teachers of Mathematics. Monograph 5.

Presmeg, N. (2009). Mathematics education research embracing arts and sciences. *ZDM. The International Journal on Mathematics Education*, *41*, 131–142.
Roberts, S. (2006). *King of infinite space: Donald Coxeter, the man who saved geometry*. New York: Walker & Company.
Sierpinska, A. (1994). *Understanding in mathematics*. London: Falmer Press.
Ulam, S. (1976). *Adventures of a mathematician*. New York: Scribner's.
Wells, D. (1990). Are these the most beautiful? *The Mathematical Intelligencer*, *12*, 38–41.

Chapter 3
Some of My Pet-Peeves with Mathematics Education

Theodore Eisenberg

Abstract For nearly half-a-century I have been a mathematics-educator, and recently retired because of a mandatory retirement age for state workers in my country. As I think back over the years as to how the profession has changed, I am simultaneously proud and disillusioned. I am proud that there are so many different facets to our discipline, but at the same time I am disillusioned that there are so many different facets to our discipline, because we have seemingly lost sight of what our profession should be all about. Whereas many of us used to have appointments in departments of mathematics, the majority of us are now in departments of education, science teaching, cognitive science, and educational technology, where the teaching and learning of mathematics per se are attended to peripherally, if at all. Some colleagues claim we are discipline that has matured from it roots in mathematics; others however say we are a discipline that has lost its way. I am very much a member of this latter camp, a group that is shrinking in size daily. In an effort to inform the larger mathematical community of this state of affairs, I would like to put forth some of my pet-peeves on mathematics-education today.

Keywords Mathematics-education · Subtopic of mathematics · Curriculum for mathematics-educators · Future of classroom mathematics

I have sometimes been described as being a malcontent—but I like to think of myself as being more of a critical observer than a disgruntled ingrate. Whatever, in keeping with these monikers, I would like start with a cri-de-coeur, and put onto the table a few of my current beefs.

Where Is the "Math" in "Mathematics Education" These Days?

Please notice that I said "these days", because I feel that at one time there was mathematics in mathematics education.

T. Eisenberg (✉)
Department of Mathematics, Ben-Gurion University of the Negev, Beer-Sheva, Israel
e-mail: eisenbt@013.net

M.N. Fried, T. Dreyfus (eds.), *Mathematics & Mathematics Education: Searching for Common Ground*, Advances in Mathematics Education,
DOI 10.1007/978-94-007-7473-5_3,

As we all know, the sea of mathematics education is a large one, but most of us try to swim in only parts of it. There are those of us who worry about:

- Dispelling math-phobia; and there are those who worry about
- How to give students non-verbal reinforcement; and those who worry about
- Verbal and social interactions in the classroom; and those of us who worry about
- Teaching techniques in inner-cities, and still others who worry about teaching techniques in rural areas via long-distance education; and there are those of us who worry about
- Trying to understand how the brain works with respect to the sequencing and processing of information; and there are those of us who worry about
- Understanding if there are gender differences in learning styles; while others worry about
- The role of symbols in mathematics.

The list of our interests go on and on and on, *but what is strikingly absent these days in mathematics education research journals is the presence of mathematics itself*. And I am sure that you all know what I am talking about—just open any of the premier journals or the conference proceedings of our meetings. One only has to glance at these to see the dearth of anything that is directly related to mathematics. And this absence is not just in the journals, it permeates our graduate programs and the personae of the next generation in our profession. *We have moved away from our historical roots of being a subcategory of mathematics*, and I think that this move has been a bad one, and perhaps even an irreversible one.

It is not my intention to insult or denigrate anyone's work; quite the opposite. Many of our colleagues are absolutely brilliant; but many of their efforts are missing the mark as to what I think main-stream mathematics-education should be all about. Let me elaborate upon this a bit.

Defining Mathematics Education

My definition of mathematics education is that it is the discipline that worries about the teaching and learning of mathematics—and although the sprinkling of topics mentioned above with which some of us are concerned are important, they do not constitute the heart of our discipline. In the past, the heart of our discipline was concerned with mathematics—and not with topics peripheral to it. I recall as a student being amazed when I read Plato's accounting of Socrates' interaction with Meno's slave boy (*Meno*, 82b–85e). Through a simple series of questions Socrates showed that the Slave boy was able to construct a square having twice the area of a given square. I just couldn't get over the cleverness of the questions—and the sequencing in which Socrates asked them. Admittedly, Socrates' intention was on a much lofter philosophical plane than to teach the slave boy this piece of mathematics. His intent was to prove to Meno that certain human knowledge is innate, and that it resides within each and everyone of us, and that all we need is a little help in bringing that knowledge to the surface. But that philosophical goal didn't move me at all, I accepted it immediately; it was the manner in which he explicated this belief, that much of human knowledge is already within us, and that with a little help, it can be drawn to the surface, that made me say to myself "wow!" and I went on to incorporate that idea into my own personal creed. This notion I thought should stand as the philosophical base of all teaching.

In the 1960's, when I went to school, this notion became the mantra of the era: namely, that we can teach any topic to any person without compromising the topic's intellectual integrity. I, and many others swallowed this mantra, hook, line, and sinker. This mantra became the base of my personal philosophy, not just specifically for mathematics education, but for all of education. All one needed, I thought, was an open mind and a kind-hearted teacher.

My heroes in those days were mathematicians who had a sincere interest in teaching. Luckily I had many models in my undergraduate days at Illinois State, but on the national scene I kept coming across the personae of Peter Hilton, Bob Davis, Morris Kline, George Pólya, Lillian Lieber, Martin Gardner, and R.L. Moore who had this new (to me) way of teaching by pitting student against student in competitive situations. (I didn't like the competitive part of Moore's teaching, but one certainly couldn't argue with his success. And so in mathematics education classes we talked about how to humanize Moore's method.) All of this was in the late 1960's. Sputnik was already history, the new-math movement in the States was already being criticized by nearly everyone, the French school of mathematics education was following in the philosophical foot-steps of the Bourbaki approach to mathematics, the Nuffield Program in England was coming under attack, as well as most new programs that advocated discovery learning, inquiry teaching, math labs, open classrooms, and other programs that were centered on the students, and not on chalk and talk teachers. Innovative programs were under attack, and they provided fertile ground for the birth of the various back to basics movements.

I was entering a profession that was fraught with problems and in a state of disarray with nothing seemingly being coordinated between the many projects which were heavily funded. Some say that these were the golden years of our profession, but the motivation for them was fear: fear that spurred forth a decade of modern-day school mathematics. In the early 1960's the mood in the United States was one of desperation that the Western world was behind the USSR in science and mathematics; and that was simply unacceptable. The way to reverse that situation was to pour money into revamping the school curriculum. And most of the Western world did just that. And at the base of those reform movements was the belief that teachers have to know more math; informed teachers will put out informed students.

As some of you well remember, that hypothesis did not hold up under examination, but it certainly seemed to make sense. My point is that individuals of my generation who went into mathematics education at the secondary and higher levels, had to take core courses in mathematics per se—and those courses were offered by Departments of Mathematics.

The profession has moved away from the notion that mathematics educators should know mathematics at a level that is quite a bit higher than the level they would be teaching. An easy analogy for this is that there are two ways to learn the lay of a new terrain; one way is to go from tree to tree, and rock to rock, and from hill to hill; eventually one will learn the terrain and the constituent parts will all fit together. Another method is to climb the highest tree and look around. With respect to mathematics teaching, training programs can be constructed from the books the teachers will eventually use; or the other extreme is to take the pre-teachers into the lofts of higher mathematics; they will see the mathematics they will eventually be teaching, but they will see it from a different viewpoint. (Some say that these two methods represent a basic difference between teachers trained in teacher seminars, as opposed to those being trained in universities.)

You might be thinking that this is exactly how we train teachers today—we take students into the sky to see the terrain on the ground. And for most going into secondary teaching you are correct, at the undergraduate level. Even those in elementary school teaching have to take a sequence in mathematics, and this sequence is usually offered by mathematics departments. (Guidelines have been developed by the MAA and the NCTM that specify specific courses and skills teachers at various levels must know.) But this specification stops at the undergraduate level—and this, I believe, is wrong.

I believe that all master and doctoral programs that are designed for mathematics educators should have a set of required core courses in mathematics and that these courses should be offered by departments of mathematics. Moreover, I believe that mathematics educators and their students, and those in our profession who worry about the teaching of mathematics at the secondary and higher levels, should not sit in departments of education nor in departments of science teaching. They should sit in departments of mathematics. Let me explain why.

Atmospheres of Learning

I have always believed that environments affect learning. If one wants to learn how to ask clever questions and increase their problem solving skills, they should associate with people who are asking the kinds of questions that they wish they had asked; and they should associate with problem solvers who are solving the problems that they themselves want to solve. Pretty soon these individuals will be asking more penetrating questions, and they will be gaining the skills to answer them in innovative ways. If one wants to be a better problem solver, associate with others who are good problem solvers—listen to how they approach their solutions, and believe me, pretty soon these individuals themselves will be better problem solvers. The environment affects us in every conceivable way and this is why mathematics educators should sit in departments of mathematics. We are a sub-area of mathematics. For our own personal development we belong there. But we, in mathematics education, can also contribute the academic atmosphere of the department, and yes, even in the realm of mathematics itself.

So why do I prefer that mathematics educators who are working with higher level students not sit in departments of education and science teaching? For science teaching, the answer is simple. Science teaching units are just too small; they don't have the man-power to stimulate one another in their discipline. Complacency sets in, and it is a constant fight for most to keep it at bay. We, the teachers of teachers, learn through social-interactions with others. And if there are only one or two educators on staff, the numbers aren't there to foster a healthy learning environment.

And what's the problem with us sitting inside education departments? Well, most faculty members in education departments are too far away from what our intellectual interests should be—and that is the teaching and learning of mathematics. For example, I know of a department of education in which most on staff have trained themselves to not even see tables and numbers and statistical tests in research and journal papers. They simply don't see them—their eyes jump right over them—and they pooh-pooh the use of statistics and quantitative data. That culture is not conducive to mathematics educators as I see them. So the bottom line is very simple—for those of us in secondary and higher education we belong in departments of mathematics. But this isn't a one way street.

Mathematics departments have a moral obligation to take an interest in all aspects of mathematics teaching, particularly at the secondary and higher levels. Many mathematicians are critical of what is going on in mathematics classrooms in the schools—they complain bitterly that the level of knowledge of beginning students has been declining for years. But it is not enough to complain—they have a moral obligation to do something about it. And this means, at least to me, interacting with the mathematics educators—who hopefully will be, in my worldview, sitting down the hall from them.

So the question then becomes: Can we teach the mathematicians anything about mathematics? And my answer to that is that you will be surprised.

I doubt if there are many mathematicians on the staffs in your universities who can explain how one can find the harmonic conjugate of a given point on a given line segment; but most mathematics educators who have taught a course in the history of math can do it. And I doubt if there are many mathematicians on staff at your universities who can tell you under what conditions the roots of a cubic equation can be represented with Euclidean tools in the complex plane. I can continue on and on with this list, and they, in turn, could counter with lists ten times longer than mine, of basic notions that we (and probably most of their colleagues who are in different fields than they) don't know. But my point is that, "yes", we know some mathematics that many mathematicians don't know, and that having us on staff would contribute to the to mathematical reservoir of knowledge in the department, and to many other reservoirs in it too.

George Pólya had a list of commandments for teachers. The commandments that stick out for me are that teachers should know mathematics, they should like mathematics, and they should develop a healthy attitude toward problem solving (see Polya 1968, p. 116, or type "Pólya's commandments for teachers" into Google). These commandments have more of a probability of being inculcated in mathematics education students at every level, if they are sitting in departments of mathematics, and not in other departments on campus.

OK, so you know where I stand on this topic—mathematics educators who deal with secondary and more mature students, belong in departments of mathematics. As I see it, having mathematicians and mathematics educators sitting in the same department, enriches the department and its offerings; it is a win-win situation.

Some Comments on Teaching

Let me make some comments on teaching. I am absolutely frustrated and saddened when I see intelligent high school and university students who are unable to do the simplest of calculations. They don't know their times tables, they can't work with fractions, they don't understand percents, etc., etc., etc. All of you know what I am speaking about. Our profession has given this problem and all that emanates from it, the moniker of Numerical Literacy, and this lacking-of-skills phenomena seems to exist for many reasons. Poor teaching, lack of drill and practice, over-dependency on technology, it's material from an older age, etc., etc. And this lack of understanding of basic notions and of not having basic skills can be found everywhere amongst our students, and I find this terribly upsetting. Why? Why does this bother me so? *Because I believe that there are certain skills and belief-systems that should be handed down from generation to generation.*

I am sure that all of us have skills that we want to see in our children and grandchildren. The ability to read and write, the ability to be a rational and critical thinker, to generalize notions, etc., etc. The list goes on and on, and I am not speaking about the affective domain, where we want the school to instill in our children the ability to respect others who are different than we are in appearance and creed. What I am speaking about are cognitive skills. I find it particularly upsetting when students reach for their calculators to do basic computations (like 7×9); when they solve simple equations with algorithmic procedures (like using the quadratic formula to find x such that $x^2 + 5x = 0$. Nearly 50 % of my students in a beginning calculus class actually used the quadratic formula on this problem.)

One of my colleagues at this University (who is not in the mathematics department) believes that none of the above matters. As long as the students "know the underlying concept of the notion" the above-mentioned skills do not matter in today's world. She believes that every generation sets its own standards as to what is important and what isn't. Well, to a point she is correct, but in general, I beg to differ. I don't want to get into universal beliefs like the Magna Carta, the ten commandments and all of that, but when I see an 8th grader list the numeral 7 nine different times on a piece of paper and proceed to add them when faced with the problem of 7×9, I cry. (The person I am speaking about got the wrong answer 5 different times—before he put his head on the table and "tuned out".) And whose fault is this that such skills have not been internalized to the point of automaticity? Well, we, as a profession, are not completely blameless in this, and that is because we, as a profession, have down-played drill-and-practice to the point where it is ridiculous.

We have put our faith into computers and calculators, and we have built generations of adults who lack the basic numerical skills that their parents and grandparents possessed. And when I say basic, I mean basic!

Certain things should be automatic, and to my way of thinking, basic arithmetical operations fall under this rubric. And so should a million other things. I have had mathematics teachers tell me that when I write the base ten number, 5, in base 3 notation, the number has gone from being and odd and prime number, to a number that is now even and composite! Where does this nonsense come from? Well, once again, we are not completely blameless.

Concept Images

I think that one of the biggest innovations in our field has been the notion of "concept images." This notion was introduced by Shlomo Vinner and Rina Hershkowitz (1980), and in its early stages it was refined by Shlomo and David Tall (Tall and Vinner 1981), and many others (Bingolbali and Monaghan 2008; Harel 2004; Li and Tall 1993; Tall 1989). I feel that this notion should stand at the heart of mathematics teaching. When I see or hear the words logarithm, or integral, or differentiable functions, or the zeros of a function, or irreducible polynomials over the rationals, or, or, or, ... I immediately see pictures, and these pictures capture the concept for me. Admittedly, there are many ways to view these notions, but the sum total of these representations is my concept image of the notion. And I am sure that each of you has developed your own idiosyncratic way to think about these notions. My point is that this way of thinking about mathematical topics is important, and the development of these concept images should stand at the heart of what we are trying to do in the classroom. But they don't.

Why? Because mathematics education has evolved into something very different than helping teachers become more effective in the classroom. And most of you know exactly what I am speaking about.

On the Education of Mathematics Teachers and Educators for Higher Degrees

In many countries these days the content-level of a math teacher's background in mathematics ends with the awarding of their first degree. The assumption that is made in many of these Masters and Doctoral programs is that the teachers know their content, now let's teach them "how to teach" (and all that implies).

This is a fundamental error. Most of the teachers I know have tremendous holes in their content knowledge, and they have not mastered many of the topics we think they have learned. They then go on for a Masters Degree in departments of education and science teaching—but these degrees are often void of taking even a single course in mathematics that is offered by the mathematics department. And as such, the teachers enter the peripheral topics I mentioned above. Nonverbal reinforcement, alternative methods of grading, higher level questioning, etc. There is no end to it, but this all comes at the expense of making the teachers more subject matter competent. Taking more mathematics courses should be part of every graduate-level program offered for furthering the education of mathematics teachers. I am well aware that this is easier said than done, but as a profession, we have moved too far away from our historical roots.

I do not mean that every math educator should be a research mathematician in addition to everything else, but the mathematical training of mathematics educators should not stop with a first degree in mathematics. There are many tremendously brilliant people in all facets of our profession, and I believe that their graduate programs should take them a few rungs higher on the ladder of mathematical content. *When they finish a program in mathematics education, they should know more mathematics at the end of it, than they did at the beginning of it.*

OK, so how can all of this be changed? Well, the first step we should take is to change the atmosphere in which we have set our programs. Join math departments; and if that is not possible, make liaisons with your math department. Seek out those interested in your research interests and try to collaborate with them on a regular basis—even if this only means having coffee with them once a week. And even if you only chit-chat with them, sooner, rather than later, your conversations will turn to the business of doing mathematics.

Start doing mathematics again; work your way through a book or even an article with a colleague; both of you will benefit from this sort of activity. Even if you find the mathematics hard, don't give up. Many times when I get stuck on what seems to others to be a trivial piece of mathematics, I remind myself of a comment made by Martin Gardner. He always used to introduce himself as a journalist and not as a mathematician. His many books and columns in *Scientific American* certainly can be used to reverse his self-perception; but he did not consider himself to be a mathematician. And he was often asked how it is then that he can write so clearly about high level and often difficult and intricate topics—and his reply went something like this: "I can write about them because I myself have to work so hard, so very hard, to understand them in the first place." And his articles would often take readers down the paths that he built for himself to understand the notions. So, just because we ourselves have to work hard to understand something, it often makes us better teachers and more attuned to pitfalls in learning.

Glimpsing the Future

There is a popular movement in the US that is being heralded as the future of education. Essentially, it is a library of tapes on just about any subject one wants. It is called the Khan Academy and here is what Wikipedia says about it:

> Khan chose to avoid the standard format of a person standing by a whiteboard, deciding instead to present the learning concepts as if "popping out of a darkened universe and into

> one's mind with a voice out of nowhere" in a way akin to sitting next to someone and working out a problem on a sheet of paper: "If you're watching a guy do a problem [while] thinking out loud, I think people find that more valuable and not as daunting". Offline versions of the videos have been distributed by not-for-profit groups to rural areas in Asia, Latin America, and Africa. While the current content is mainly concerned with pre-college mathematics and physics, Khan's long-term goal is to provide "tens of thousands of videos in pretty much every subject" and to create "the world's first free, world-class virtual school where anyone can learn anything".
> Khan Academy also provides a web-based exercise system that generates problems for students based on skill level and performance. The exercise software is available as open source under the MIT license. Khan believes his academy points an opportunity to overhaul the traditional classroom by using software to create tests, grade assignments, highlight the challenges of certain students, and encourage those doing well to help struggling classmates. The tutorials are touted as helpful because, among other factors, they can be paused by students, while a classroom lecture cannot be.
> The success of his low-tech, conversational tutorials—Khan's face never appears, and viewers see only his unadorned step-by-step doodles and diagrams on an electronic blackboard—suggests an educational transformation that de-emphasizes lecture-based classroom interactions. (Wikipedia, Khan Institute, 2013)

I think that technology is a wonderful thing, but the above description reminds me of a newspaper cartoon I recently saw. Two individuals sitting in the same living room, each hunched over their computer, and they were corresponding with one another by e-mail! I certainly hope that the cartoon will not become a reality for the school of the future.

Frank Quinn, is a mathematician at Virginia Tech in the US. He was lamenting on the disconnect between school mathematics and university level mathematics, and he wrote:

> School mathematics is still firmly located in the nineteenth century, so student success rates in modern [university level] courses has been very low. There is a great deal of pressure to improve this situation, but recent changes, such as use of calculators and emphasis on vague understanding over skills, have actually worsened the disconnect. Something has to change. (Quinn 2012, p. 37)

I agree, something does have to change. Maybe, just maybe someone will remember Socrates' dialog with Meno's slave boy, and realize that this is what teaching should all about; a give and take between a student and a real teacher—not a virtual one. To me, the Socratic model of teaching is just as relevant now as it was 2000 years ago. As Einstein said about teachers: "Setting an example is not the main means of influencing another, it is the only means." If we sincerely believe that we can teach any subject to anyone, then we can.

References

Bingolbali, E., & Monaghan, J. (2008). Concept image revisited. *Educational Studies in Mathematics, 68*, 19–35.

Harel, G. (2004). A perspective on concept image and concept definition in mathematics with particular reference to limits and continuity. In T. P. Carpenter, J. A. Dossey, & J. L. Koehler (Eds.), *Classics in mathematics education research* (p. 98). Reston: NCTM.

Li, L., & Tall, D. (1993). Constructing different concept images of sequences and limits by programming. In I. Hirabayashi, N. Nohda, K. Shigematsu, & F.-L. Lin (Eds.), *Proceedings of the 17th international conference on the psychology of mathematics education*, Tsukuba, Japan, PME (Vol. 2, pp. 41–48).

Polya, G. (1968). *Mathematical discovery* (Vol. 2). New York: Wiley.

Quinn, F. (2012). A revolution in mathematics? What really happened a century ago and why it matters today. *Notices of the American Mathematical Society*, *59*, 31–37.

Tall, D. (1989). Concept images, generic organizers, computers and curriculum change. *For the Learning of Mathematics*, *9*, 37–42.

Tall, D., & Vinner, S. (1981). Concept image and concept definition in mathematics with particular reference to limit and continuity. *Educational Studies in Mathematics*, *12*, 151–169.

Vinner, S., & Hershkowitz, R. (1980). Concept images and common cognitive paths in the development of some simple geometrical concepts. In R. Karplus (Ed.), *Proceedings of the 4th international conference on the psychology of mathematics education*, Berkeley, CA, PME (pp. 177–184).

Wikipedia (2013). Khan Academy. http://en.wikipedia.org/wiki/Khan_Academy. Accessed 31 May 2013.

Chapter 4
Mathematics at the Center of Distinct Fields: A Response to Michael and Ted

Norma Presmeg

Abstract This response to Ted and Michael points out that we are unified in considering mathematics to be central in the work of mathematicians, mathematics educators, and mathematics education researchers. However, there are distinctions between the fields of pure mathematics research, the teaching and learning of mathematics, and research in mathematics education, and unless these differences are honored it is possible for researchers to talk past one another. The case of Swedish mathematics education research is examined to exemplify the distinctions. Another distinction is that between "training" and "education". To further characterize mathematics education research, submission of manuscripts to Educational Studies in Mathematics is explored. Values and aesthetics in various relevant fields are touched upon. Finally, an example is given of the mutual enhancement that exists when mathematicians and mathematics education researchers work together in university mathematics departments.

Keywords Distinct fields · Pure mathematics research · Mathematics education · Mathematics education research · Sweden · Training · Research manuscript submissions · Aesthetics · Mutual enhancement

What Are We Talking About?

I want to say at the outset that Ted, Michael and I have much in common in our sentiments about mathematics. There is no doubt in my mind that mathematics is central to the endeavors of mathematicians, mathematics educators, and mathematics education researchers alike.

However, it appears at times as if we are talking past each other, because there are distinct fields in question. Thus I have constructed a diagram to portray the distinctness of mathematics itself, mathematics education, and research in pure mathematics and in mathematics education respectively. Each of the subsets in the ellipses

Professor Emerita N. Presmeg (✉)
Illinois State University
Maryville, TN, USA
e-mail: npresmeg@msn.com

M.N. Fried, T. Dreyfus (eds.), *Mathematics & Mathematics Education: Searching for Common Ground*, Advances in Mathematics Education,
DOI 10.1007/978-94-007-7473-5_4,

may be regarded as the *topic* of the wider set in which it rests. Mathematics, mathematics education, and mathematics education research may be conceptualized as being nested, like Russian dolls, with mathematics at the center, nested in mathematics education, which is nested in turn in mathematics education research. The topic addressed in mathematics research is mathematics; the topic of mathematics education is also mathematics; and the topic of mathematics education research in not mathematics *per se*, but mathematics education.

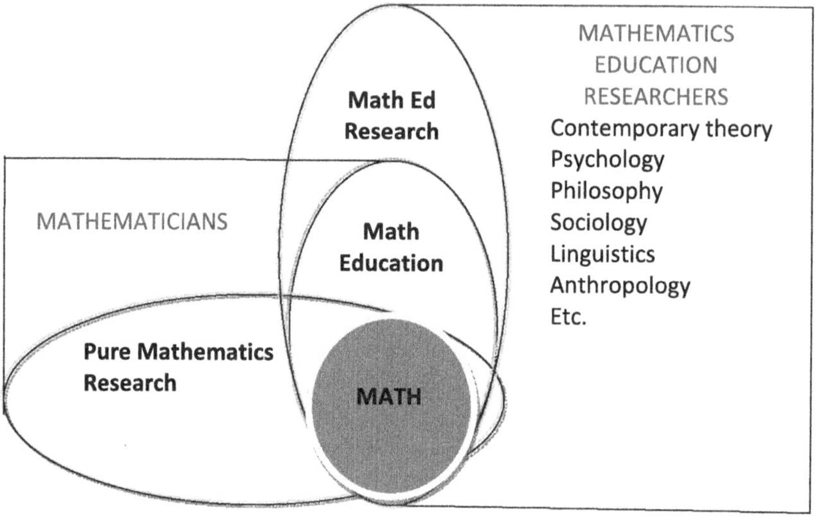

On the one hand, mathematicians are engaged in research in various fields of mathematics, and in mathematics education insofar as they teach mathematics. On the other hand, mathematics education researchers are not engaged primarily with research in pure mathematics, except insofar as they may be also mathematicians. Their research embraces the nested model of various aspects of the "complex human worlds" (Presmeg 1998, with hints of Bruner's *Actual Minds, Possible Worlds*, 1986) involved in the teaching and learning of mathematics at all levels. It is clear from this conceptualization that mathematics education research cannot simply be a branch of *applied mathematics*. Unlike, for instance, *business calculus*—which could be regarded as a branch of applied mathematics—in mathematics education research we are not applying the principles of mathematics to the topic of the research, namely mathematics education.

As I have pointed out many times, mathematics as a field is thousands of years old, and mathematicians have taught mathematics for thousands of years, but mathematics education *research* as a field in its own right is less than a century old, notwithstanding the important 1908 meeting in Rome at which the International Commission on Mathematics Instruction (ICMI) was founded, to which Michael referred. The International Mathematics Union (IMU) was founded in 1920 as an international community of mathematicians (Furinghetti and Giacardi 2010) and

ICMI is an affiliate of the IMU. I, too, attended the bright celebration of the ICMI centenary meeting in Rome in 2008, and I, too, appreciated the founding of ICMI, and the founders who were all eminent mathematicians. I, too, deplored the suggestion (and I do believe it was a minority position) that ICMI should break away from the IMU. But mathematics education research has matured as a field in the last half-century, as evidenced by a proliferation of journals, conferences, and kinds of research that it embraces. Yes, despite the title of Sierpinska and Kilpatrick's (1998) edited book that followed a high-level conference on mathematics education research's *search for identity*, we are still trying to find out who we are! But it is clear that mathematics education research is now an established field, and many universities internationally are acknowledging this fact, e.g., by establishing mathematics education professorships. I elaborate on this point in the next section.

In my paper (Presmeg 2009) that made Ted unhappy, I was not writing directly about mathematics. Nor was I writing directly about mathematics education. The topic of the paper was *Mathematics education* **research** *embracing arts and sciences*. When Ted writes in his response to my paper that "Research in mathematics education should be about the teaching and learning of mathematics" (Eisenberg and Fried 2009, p. 144), I agree fully—it is central, as in my diagram. And I believe that, mostly, this is the case. So, with mathematics education as its *subject of investigation*, why do we need psychology, sociology, and so on—even linguistics? I am convinced that the lenses of research methods used in these fields of the humanities are essential tools for mathematics education researchers to have at their disposal. The reason is simply because mathematics education involves *people*, with all their complexities. It is significant that the title of Ted and Michael's paper in response to mine is *Dialogue on* **mathematics education**: *Two points of view on the state of the art*. This title, involving mathematics education, not mathematics education research, implies that we were not addressing the same topic in our interchanges—hence the need for more clarity in what we are talking about.

With regard to the point raised by Ted that mathematics education should be housed in Mathematics Departments of universities and not in humanities departments such as Curriculum and Instruction—I fully agree with Ted. In my professional life, I have had the privilege of working in both scenarios. I was in the Department of Curriculum and Instruction for ten years, at The Florida State University. After that for ten years (until 2010), I belonged, as a mathematics education researcher, to the Mathematics Department of Illinois State University. The latter housing was exceptionally fruitful. ISU was originally the first teachers' college in Illinois (as the name of its town, Normal, attests), founded in 1859. With a tradition of education as a focus, even though mathematics education was housed in the Mathematics Department, education was appreciated and promulgated: Our Department held—with mutual respect—equal numbers of mathematicians and mathematics education researchers, which was and I believe still is an unusual phenomenon. (This Department at ISU will soon be the affiliation of the editors-in-chief of two top mathematics education research journals in our field: Cindy Langrall of ISU Mathematics Department will soon assume the editorship of *Journal for Research in Mathematics Education*, and as editor-in-chief of *Educational Studies in Mathematics* ISU is still my affiliation.) The cross-pollination that happens at the interface of mathematicians and mathematics education researchers is indeed fruitful to both groups, which affirms the importance of this symposium in honor of Ted Eisenberg.

A Deeper Look at Some of the Issues

The Case of Sweden

In order to clarify the emerging status of mathematics education research as a field in its own right, let me examine briefly the case of Sweden in this regard. In the 1990s, as in many other countries, there were no mathematics education professorships in Swedish universities. I was asked to serve as the co-advisor for the doctoral research of Andrejs Dunkels in Luleåin the north of Sweden, because there was nobody locally who could serve in this capacity, advising him on the qualitative research that his study demanded, on the teaching and learning of calculus in his class of engineering students. Before he could embark on his outstanding study in mathematics education, Andrejs had to prove his qualifications as a mathematician, by including in his dissertation his published papers on pure mathematics topics such as Dirichlet spaces and the Riesz kernel (Dunkels 1996).

It is to the credit of Andrejs's principal advisor, mathematician Lars-Erik Persson, that he saw the value of research such as that carried out by Andrejs. Supported by influential scholars such as Gerd Brandell, he and others were successful in establishing the Swedish National Program of doctoral studies in mathematics education, and over the following decade a new cadre of scholars who are expert in mathematics education research was established. This program embraced all the Swedish universities, and made use of international expertise by inviting distinguished scholars such as Jeremy Kilpatrick and Anna Sierpinska, along with many others, to conduct workshops and co-supervise the research of Swedish doctoral students. There was also funding for such students to visit overseas universities (e.g., Magnus Österholm from Linköping came to study with me at Illinois State University).

In 2002 I was privileged to serve on the three-person committee that served as the search team for the first *mathematics education* professor in Sweden, at Luleå University. It is significant for the topic of this symposium that many of the applicants for this position were mathematicians (several of them from Russia) who, from the evidence of the material in their application files, were unaware that mathematics education research had become a field in its own right. They believed that because they taught mathematics and were eminent in the field of *mathematics* research, they had the credentials to become the first mathematics *education* professor in Sweden. At that point in early 2002, there was nobody amongst the Swedish applicants whom the majority of the search team members considered to be a qualified candidate. Andrejs Dunkels, sadly, had died, and the new cadre of mathematics education researchers from the national doctoral program was still being formed. Thus Rudolf Strässer from Germany was appointed as the first mathematics education professor in Sweden. Now, in 2012, there are many scholars in Sweden qualified for such positions.

Rigorous Research Methods

One aspect that is pertinent to this account of the case of Sweden is the role that rigorous research methods have in the emergence of mathematics education research as a field in its own right. When Andrejs Dunkels started his research, his intention was to investigate his own teaching methods in his engineering calculus course. He initially did not realize that he would require evidence for every statement that he made in reporting his results: Anecdotal evidence would be insufficient to validate his claims. Validity and reliability were well known and accepted constructs in quantitative research, and Andrejs did use the numerical methods of exploratory data analysis as part of his methodology. However, it was becoming well accepted at that point in the 1990s that qualitative methods, too, have their forms of quality control. Andrejs learned to use triangulation of data collection methods, through audiotaped interviews, fieldnotes from observations, and artifacts such as the students' written work. There are various forms of triangulation (Stake 1995), and evidence for interpretation of data sources was also collated from the notes of an outside observer (in which capacity I served), from transcripts of the students' audiotaped interchanges in class, and from the notes of the researcher himself. And, importantly, in such research there is the opportunity for respondent validation ("member checks"): The participants are given the chance to examine the report of the researcher's interpretations, and their reactions become part of the data pool. We take these methods of quality control for granted in qualitative mathematics education research now, but in the 1990s these issues were still being explored. Andrejs used both quantitative and qualitative methods in his research. Each serves a different purpose, and now, more than a decade later, such "mixed methodologies" are becoming more common than they were in the 1990s.

I hope that the influence and value of methodologies adapted from other disciplines in the humanities is clear from this account. The quality control that I have described is prevalent in case studies in the humanities. But there are other forms that are more suited to different methodologies, e.g., the "key informant" notion adapted from ethnographic research in anthropology. The research question determines suitable methodology, and these concerns go beyond the bounds of both mathematics and mathematics education, while embracing both as topics.

Training Versus Education

One might say that the process of becoming a researcher in mathematics education (such as in Andrejs's case) is a form of "training". Michael uses the distinction between training and education to characterize the learning of pure mathematics majors as opposed to the more diffused learning of those who will become familiar with mathematics without becoming mathematicians. While recognizing the distinction he makes, I nonetheless want to suggest that "training" may have a rote connotation that is unsuitable for efficient learning of prospective mathematics education researchers and mathematicians alike. Familiarity with methods is important in both fields, but the capacity to judge suitability and to adapt to the context goes beyond mere training, and I would prefer to call learning in both cases education.

Another point I would like to make, somewhat provocatively, concerns Plato's *Meno*, which Ted cited as a paradigm case of good mathematics education. It seems to me that despite Socrates' excellent logic and choice of questions, very little *agency* was accorded to the slave boy, most of whose responses to the questions were of the form "Yes, Socrates" and "No, Socrates". I know that Michael has a deeper view of the concerns in the *Meno*: In fact, for a slave, the boy was exercising considerable agency—and Socrates had the purpose, not of teaching, but of convincing Meno of the innate quality of knowledge as manifested through *anamnesis*. In the more limited point I am making, I am not meaning to suggest that no role should be given to the teacher in students' mathematical constructions; on the contrary, I believe that teaching is highly important. However, there is a "dance" between instruction and construction in mathematics education (Presmeg 2012), as was debated at a lively conference in Frankfurt am Main, Germany, earlier this year.

Experiences with Submissions to ESM

With regard to the point I made in describing the pool of applicants for the first mathematics education professorship in Sweden, I would like to enlarge this issue by considering the range of manuscripts submitted to *Educational Studies in Mathematics* (ESM). As Tommy Dreyfus, as a former editor-in-chief of this journal, can confirm, we receive a range of submissions not all of which are suitable for publication in this venue. The mission statement appears in each hard-copy issue of the journal:

> *Educational Studies in Mathematics* presents new ideas and developments which are considered to be of major importance to those working in the field of mathematics education. It seeks to reflect both the variety of research concerns within this field and the range of methods used to study them. It deals with didactical, methodological and pedagogical subjects rather than with specific programmes for teaching mathematics. All papers are strictly refereed and the emphasis is on high-level articles which are of more than local or national interest. All contributions to this journal are peer reviewed.

As the statement suggests, research manuscripts on both theoretical and empirical issues concerning mathematics education are welcomed. However, all such manuscripts need to include all three of the initials in the acronym ESM to be suitable: Education, a research Study, and Mathematics. This requirement is sometimes violated in manuscripts that we receive. On the one hand some authors report on an analysis of topics that are of general educational interest, leaving out the essential component *Mathematics*. This aspect was of particular concern in the most recent double issue (Vol. 80, issues 1 & 2), a guest-edited special issue on *Contemporary theory in mathematics education*, which broadens the purview of theoretical considerations that may be useful as lenses for mathematics education researchers (Brown and Walshaw 2012). Several of the authors had to be reminded that the focus on *mathematics* education is central. On the other hand, on a regular basis we receive manuscripts written by mathematicians reporting their research results in pure mathematics: the Education component of the acronym is missing. These authors have chosen the wrong journal for the dissemination of their results, and an immediate decision letter to this effect follows.

These points emphasize the importance of distinguishing between the various components involved in mathematics research and mathematics education research, as sketched in my diagram.

Values and Aesthetics

In contemplating a suitable title for my response to Ted and Michael, I initially played with the possibility of using "Different kinds of beauty, in mathematics, mathematics education, and research in these two domains." It is appropriate that there is a session on *Visualization in mathematics and mathematics education* at this symposium, because as Michael pointed out, visualization and the broader categories of aesthetics and values may unearth potential common ground for mathematicians and mathematics education researchers. When I started my research on visualization in teaching and learning mathematics in the late 1970s, there were just a few scholars working in this area—and several of them are here at this symposium. Ken Clements (1982) had done important early research (some of it with Glen Lean: Lean and Clements 1981), and later Tommy and Ted published their influential research results (Dreyfus 1991; Dreyfus and Eisenberg 1990; Eisenberg 1994). There are more aspects to this topic than I can mention here (and some of them are explored in the session on visualization)—but let me just say that the complex issue concerning whether visual thinking is *valued* in the teaching of mathematics influences the quality of the learning that takes place (Presmeg 2006). There are some people who prefer to think visually in making sense of mathematical ideas. My test on *preference for visualization* in mathematics (Presmeg 1985) showed clearly that amongst the general population this preference follows a normal distribution—a Gaussian curve—hence the designation of "visualizers" to those who consistently prefer to think visually. A surprising result of my early research was that for the 54 visualizers in my study, it was the pedagogy of teachers in the *middle* group, not the visual or nonvisual groups (with regard to their mathematics teaching) that was optimal for these visualizers in their penultimate year of learning mathematics in high school. Issues of abstraction and generality are involved. The question of "reluctance to visualize" put forward by Ted and Tommy is a complicated one that embraces values as well as individual preferences and their interaction with instruction. These issues clearly go beyond the nature of mathematics itself, and implicate disciplines such as psychology, sociology, and even semiotics because all mathematics involves representations of various kinds.

One final point that I want to make regarding aesthetics has to do with my feeling (uncomfortable for Ted) that I was "coming home" in doing mathematics education research (Presmeg 2009). From the outset I enjoyed the aesthetic elements involved in doing mathematics, including mathematical proof—a cold, austere beauty. But I also felt drawn to the arts, especially poetry and music, with the different forms of soul-sense and beauty that they embodied. As a teenager, I set out to become a nuclear physicist, heavily involved in the hard sciences in the early specialization of the system, based on the British system, in which I found myself. I thought that the beauty of the arts could take care of itself in my life. But the need to work with people rather than with the objects of theoretical physics asserted itself, and I became a high school mathematics teacher, after one year of post-graduate mathematics study (involving 5 year-long courses in Functional Analysis, Measure Theory, Topology, Modern Algebra; and Projective Geometry in German because we were required to be able to read our subject in another language!) and a one-year university teaching diploma (UED). After 12 years of teaching high school I felt the need to study again, and after a Master's degree in Educational Psychology (which immersed me in Albert Einstein's incomparable capacity to visualize), and a Ph.D. on the topic of *The role of visually mediated processes in high school mathematics: A classroom investigation*, I became a mathematics education researcher. The rigors of the hard sciences as well as the beauty of the arts were involved in my research—which is why it felt like coming home.

What Mathematicians and Mathematics Education Researchers Can Contribute to Each Other's Fields

I appreciate and agree with Michael's point that the direction goes both ways in considering what mathematicians and mathematics education researchers can contribute to each other's fields. My experiences in the Mathematics Department at ISU validate this point. The 15 mathematics education researchers interacted freely with the 15 research mathematicians in our Department, and research collaborations were frequent. I attended their weekly colloquium on Discrete Mathematics, and we all taught mathematics courses in the Department. (I particularly enjoyed teaching the College Geometry course, which included advanced Euclidean geometry as well as non-Euclidean geometry.)

One vivid memory I have, which is pertinent to this topic, concerns the interview process when I applied for the professorship that was open in the Mathematics Department at ISU in 2000. During the two days of interviews, one of the meetings was with the mathematicians. They asked me what I, as a mathematics education researcher, could offer them in their work. I thought quickly, and then described the theoretical model of levels of learning geometry put forward by the van Hieles as a result of their research in The Netherlands in the 1970s. The mathematicians could see the value of such research in teaching mathematics. I was hired!

In the light of the damage that can be caused by misunderstandings between mathematicians and mathematics education researchers (Latterell 2005), I applaud the selection of the topic of this symposium, and look forward to the exchanges of ideas that can promote mutual appreciation and interaction.

References

Brown, T., & Walshaw, M. (Eds.) (2012). Contemporary theory in mathematics education. Special double issue. *Educational Studies in Mathematics*, *80*(1&2).

Bruner, J. (1986). *Actual minds, possible worlds*. Cambridge: Harvard University Press.

Clements, M. A. (Ken) (1982). Visual imagery and school mathematics. *For the Learning of Mathematics*, *2*, 2–9; *3*, 33–39.

Dreyfus, T. (1991). On the status of visual reasoning in mathematics and mathematics education. In F. Furinghetti (Ed.), *Proceedings of the 15th conference of the international group for the psychology of mathematics education* (Vol. 1, pp. 33–48).

Dreyfus, T., & Eisenberg, T. (1990). On difficulties with diagrams: theoretical issues. In G. Booker, P. Cobb, & T. Mendicuti (Eds.), *Proceedings of the 14th conference of the international group for the psychology of mathematics education* (Vol. 1, pp. 27–36).

Dunkels, A. (1996). *Contributions to mathematical knowledge and its acquisition*. Unpublished Ph.D. dissertation, Department of Mathematics, Luleå University, Sweden.

Eisenberg, T. (1994). On understanding the reluctance to visualize. *ZDM. Zentralblatt für Didaktik der Mathematik*, *26*, 109–113.

Eisenberg, T., & Fried, M. (2009). Dialogue on mathematics education: two points of view on the state of the art. *ZDM. The International Journal on Mathematics Education*, *41*, 143–149.

Furinghetti, F., & Giacardi, L. (2010). People, events, and documents of ICMI's first century. *Actes d'Historia de la Ciéncia de la Técnica*, *3*, 11–50.

Latterell, C. M. (2005). *Math wars: a guide for parents and teachers*. Westport: Praeger.

Lean, G. A., & Clements, M. A. (1981). Spatial ability, visual imagery, and mathematical performance. *Educational Studies in Mathematics*, *12*, 1–33.

Presmeg, N. C. (1985). *The role of visually mediated processes in high school mathematics: a classroom investigation*. Unpublished Ph.D. dissertation, University of Cambridge, England.

Presmeg, N. C. (1998). Balancing complex human worlds: mathematics education as an emergent discipline in its own right. In A. Sierpinska & J. Kilpatrick (Eds.), *Mathematics education as a research domain: a search for identity* (pp. 57–70). Dordrecht: Kluwer Academic.

Presmeg, N. C. (2006). Research on visualization in learning and teaching mathematics: emergence from psychology. In A. Gutierrez & P. Boero (Eds.), *Handbook of research on the psychology of mathematics education* (pp. 205–235). Rotterdam: Sense Publishers.

Presmeg, N. C. (2009). Mathematics education research embracing arts and sciences. *ZDM. The International Journal on Mathematics Education*, *41*, 131–141.

Presmeg, N. C. (2012). A dance of construction with instruction in mathematics education. In: *Plenary presentation, conference on a mathematics education perspective on early mathematics: learning between the poles of instruction and construction (POEM)*, Goethe University, Frankfurt am Main, Germany, 27–29 February 2012. http://cermat.org/poem2012/main/proceedings_files/Presmeg-POEM2012.pdf. Accessed 1 June 2013.

Sierpinska, A., & Kilpatrick, J. (Eds.) (1998). *Mathematics education as a research domain: a search for identity*. Dordrecht: Kluwer Academic.

Stake, R. E. (1995). *The art of case study research*. Thousand Oaks: Sage.

Mutual Expectations Between Mathematics and Mathematics Education

Chapter 5
Mutual Expectations Between Mathematicians and Mathematics Educators

Tommy Dreyfus

Abstract Four authors whose education and current position place them at different locations on the continuum from mathematician to mathematics educator use their experience to consider what contribution towards mathematics education mathematics educators might expect from mathematicians and what mathematicians might expect from mathematics educators and from mathematics education as a domain.

Keywords Expectations · Mutual expectations · Mathematics · Mathematics education · Mathematicians · Mathematics educators

Introduction

The purpose of this chapter is to make a start in asking what mathematics education should be about and what it should not be about in the eyes of mathematicians and mathematics educators: what do mathematicians as opposed to mathematics educators think the concerns of mathematics education should or should not be, and what do mathematics educators think mathematicians could, should, or should not contribute to mathematics education? What, in short, do these two groups expect from one another? Of course the divide between whom we should call a mathematician and whom we should call a mathematics educator is not sharp; there is almost a continuum in the degrees of involvement in mathematics education and mathematics pure and simple. The authors of this chapter occupy different positions along that continuum and represent different perspectives. From those different perspectives, some of the authors make claims about *mutual* expectations, some ask questions and some raise issues that suggest questions. All the authors, however, primarily intend

With contributions by
Uri Onn, Department of Mathematics, Ben-Gurion University, Beer Sheva, Israel
Joanna Mamona-Downs, Department of Mathematics, University of Patras, Patras, Greece
Stephen Lerman, Department of Education, London South Bank University, London, UK

T. Dreyfus (✉)
Department of Education in Mathematics, Science and Technology, Tel Aviv University, Tel Aviv, Israel
e-mail: tommyd@post.tau.ac.il

M.N. Fried, T. Dreyfus (eds.), *Mathematics & Mathematics Education: Searching for Common Ground*, Advances in Mathematics Education,
DOI 10.1007/978-94-007-7473-5_5,

that the reader become pointedly aware of questions about points of harmony and discord all along the continuum of mathematics and mathematics education. Many of these questions are taken up and partially answered later in the book in specific contexts.

Expectations of a Mathematician

Uri Onn

I was asked to sketch what are my expectations, as a mathematician, from researchers in math education. The following is drawn mainly from the teaching part of my work rather than my research.

What I See as a Teacher

Most of my students, engineers at all levels who will eventually be part of the technological backbone (high-tech or low-tech) of our country, have a completely damaged perception of mathematical objects.

Example 1 In a course on linear algebra for first-year bio-tech engineering majors, I gave the following example of a 2-by-2 rotation matrix:

$$\begin{bmatrix} \cos(a) & -\sin(a) \\ \sin(a) & \cos(a) \end{bmatrix}$$

First, the students were completely intimidated (there were voices asking, for example, "Why do you mention sine and cosine?—this is an algebra course"). Then, it turned out that although they would be able to solve any trigonometric identity I gave them, they did not understand that sine and cosine have a geometric meaning, that they represent ratios, that they are functions (they do not understand what a function is, and at best they identify it with its graph, let alone understand its inverse). Let me emphasize: out of a class of 70 students none knew what arcsine was; when I asked them "What is it?" there was a two minutes silence until one of them explained to the others: "Ah, I know what Uri means: it is shift-sine in the calculator," proudly pointing at his iPhone. Everybody nodded in agreement. Now they knew what it was: arcsine is an operation which involves pressing two specific buttons on a calculator. It is important to note that they nodded in agreement, that is, they *did see it in school.*

Example 2 In a final exam for a Calculus I course for computer science majors, I asked the students the following:

- Compute the Taylor polynomial $P_n(x_0; x)$ and the remainder $R_n(x_0; x)$ in Lagrange form of the function $f(x) = \sqrt{x}$ around the point $x_0 = 25$.
- Show that $|\sqrt{26} - P_2(25; 26)| < \frac{1}{50000}$.

Now, out of almost 300 students more than 150 did not even attempt the problem (there was no choice, they were supposed to solve all problems on the test to get full marks, and time was not the issue).

Most of those who did try to solve it could write the formula for the Taylor polynomial correctly:

$$P_n(x_0; x) = f(x_0) + f'(x_0)(x - x_0) + \frac{f''(x_0)}{2!}(x - x_0)^2 + \cdots + \frac{f^{(n)}(x_0)}{n!}(x - x_0)^n$$

and continued

$$f'(x_0) = \frac{1}{2}x_0^{-\frac{1}{2}}, \qquad f''(x_0) = \frac{1}{2}\left(-\frac{1}{2}\right)x_0^{-\frac{3}{2}}, \qquad \text{etc.}$$

In fact they were able to differentiate complicated function (this is technical!). But less than 30 could get up to (*) and effectively compute the remainder:

$$(^*) \quad \sqrt{26} \approx 5 + \frac{1}{2} \cdot \frac{1}{\sqrt{25}} + \frac{1}{2}\frac{1}{2}\left(-\frac{1}{2}\right)\frac{1}{25 \cdot \sqrt{25}}$$

From these 30 even fewer got to this point:

$$(^{**}) \quad \sqrt{26} \approx 5 + \frac{1}{10} - \frac{1}{1000}$$

Many answers looked like this:

$$\sqrt{26} \approx 300 + \frac{1}{3} - \frac{1}{150}$$

Students who wrote the latter have, in my opinion, no perception of the magnitude of numbers, no understanding of the meaning of approximation and no mechanism of self-control to know whether or not an approximation is reasonable.

But this is not the end. Out of the students who got to (**), several transformed it wrongly to a decimal fraction and did not arrive safely at

$$5 + \frac{1}{10} - \frac{1}{1000} = 5.099$$

Among the remaining students, some did not attempt to move to decimal fractions, but we can allow them the benefit of the doubt that they could. Thus, *about 15 out of 300 students correctly solved an elementary calculus problem and understood what they were doing. Many of the students lacked an understanding of material taught in elementary school.*

There are dozens of other examples which show that, at large, students' perception of mathematical objects is as if they were discrete set of points in space with no connections.

When you think of it, this is not surprising: *Our educational system is product oriented and not process oriented.* High-schools are infrastructures designed to produce students who will pass matriculation exams. The questions in these exams are arranged in distinct drawers and encourage technical skills. They encourage teachers to teach material in a meaningless order without showing the motivation for and the relation between different topics. In fact, these exams are meaningless. A student who got 100 surely mastered the sporadic disconnected techniques. But by no means does this grade indicate that he or she has a solid grasp of the mathematical concepts. And if you think that this number does represent something, then what about a student who got 70? Does that mean 70 % of the bridges he or she will design will hold? Does it mean that the algorithms he or she will implement in life-saving-equipment will work with probability 0.7?

My conclusion is that in Israel we are in an emergency situation, that the dark picture we see in international comparison tests do not even begin to reveal the depth of the catastrophe. A missile launched from the Gaza strip will reach Beer Sheva in about 1 minute: it represents a violent immediate threat against which we will all immediately act to protect ourselves. A slow and steady process of a society stripping itself of thinkers is much more dangerous and much harder to fix. There is a clear tendency downhill: I miss the students I had five years ago when I began to teach and they were already suffering from these symptoms.

My Expectations from Those Involved in Math Education

My expectations are that mathematics educators:

1. Start asking the right questions before it is too late.
2. Focus on the process and not only on the product.
    - Mathematics is not taught, it is mediated: it is a process which makes a deep structural change in learners' perceptions, structure of thinking, and internal representations of concepts. Part of this process is to transform the learner himself into a mediator who can independently present the world to himself and be able to solve future unknown problems.
    - Don't be afraid of failure. A failure is a healthy and crucial mechanism for learning; it indicates that there is a problem. The point is not to avoid failure but rather to grow out of it. Learn to live with the fact that we do make mistakes and equip the learner with the mechanism to identify them and fix them.
3. Know (and love) mathematics, at least one level above that immediately relevant to your teaching/research/decision making.
4. Realize that there are no shortcuts or new inventions that will make math more accessible. Learning mathematics is a long continuous process with many small but important layers and much internal ordering.

Expectations According to a Mathematics Educator from a Mathematics Department

Joanna Mamona-Downs

The symposium dedicated to Ted Eisenberg was a rare opportunity for mathematics educators and mathematicians to meet and exchange opinions on the teaching and learning of mathematics. As a mathematics educator, I was particularly refreshed by the contingent of mathematicians from their practical-based perspective. Indeed, many of the points made here will be as if from the eyes of a mathematician.

My task is to record some of the most pertinent issues of common interest between mathematicians and mathematics educators that were brought up in the round table session. I allow myself some personal liberty as the reader has the opportunity to read the other authors' interventions. I will write in a free style as I think respects the spirit of this book. In such a short note, I feel the most efficient format is to write down a list of points/questions.

1. A high percentage of mathematicians teach. Most mathematicians enjoy teaching in the main. A substantial proportion is prepared to devote time not only to making the exposition lively and meaningful, but also to seek out different approaches in proving theorems, etc. However careful they are, though, there likely will be numerous students that still seem perplexed. It is useful then to think of identifying learning problems; mathematicians are not so well equipped to deal with this perspective. For this reason, they usually accept the need of mathematics education.
2. Even if they accept the need of mathematics education, the majority of mathematicians have doubts about the output of the 'community' of mathematics educators as it is today. There is a measure of prejudice, but they might well complain that at least they have difficulty navigating the corpus of mathematics education literature, and, more seriously, fitting this corpus to their own 'images' about working in mathematics. Is there an onus for educators to make their results more transparent?
3. Naturally university mathematicians are interested about what and how mathematics is taught at school to understand what they can assume about the knowledge and skills of their incoming students. They prefer their students to be 'ready' for tertiary mathematics courses, and in their eyes a system of teaching at school to accelerate the most gifted is desired. However, they also want to disseminate a sense of the nature and the aesthetics of mathematics to the general public; most would advocate mathematics as a compulsory subject until some time in the teens. This fits with the banner 'Mathematics for all' espoused much in the mathematics education literature. This banner can be interpreted in manifold ways, but it usually suggests a strong strand of motivating the mathematics done. The educator, though, tends to do this by introducing modeling, the mathematician by examples and applications. The difference is to do with a sensibility for keeping to a mathematical theory. Concerning the issue of integrated ability classes at school, the educator typically believes that intercourse

between a 'poor' and a 'strong' student gives advantages to the 'strong' as well as the 'weak'; however, perhaps, not in a way that a mathematician would immediately recognize.

4. Mathematicians can be interested in teacher training from two slightly different perspectives; one is how far their ex-students retain relevance of university mathematics over their training period, and the other how they utilize that knowledge when they become independent teachers. One ex-student of mine, who subsequently became a secondary school teacher, told me that it took four years for him to adapt how he was thinking at university to suit his communication with his students effectively. In this regard, I think the program described by Tahl Nowik from Bar-Ilan University at the symposium in honor of Ted Eisenberg is profitable to examine; Nowik's own account of the program and his place in it was as follows:

    > I am a mathematician at the Mathematics Department of Bar-Ilan University, and my field of interest is topology. In our department I am in charge of the program for training teachers, which is one of several available programs (others are pure mathematics, applied mathematics, and mathematics & statistics). In addition to all mandatory courses common to the pure and applied math programs, the program for future teachers includes several specialized courses: Euclidean geometry, history of mathematics, a course titled "high-school mathematics from an advanced viewpoint", and beginning next year we are adding a course on applications of mathematics.
    > Over the years that I have been fulfilling this duty, I have put much thought into how mathematics teacher training should be done, and indeed my views have undergone changes, which relate to this chapter's topic. Since we have introduced the course "high-school mathematics from an advanced viewpoint", I have been teaching the first semester of the course, and another member of our department has been teaching the second semester. The aim of this course, which is given in the third year, is to close the gap between the "high level" mathematics that the students have studied in our department, and the "down to earth" mathematics that they will need to be teaching. It is very hard for the prospective teachers to close this gap by themselves, and there is a danger for them to ignore all knowledge acquired in their university studies when arriving to the classroom, and to teach using their own notebook from high-school...
    > This course aims at avoiding the potential disaster described above, and it was self evident to me when we introduced this course that the most suitable people for teaching it are professional mathematicians because we understand the mathematics best, and will be able to clarify the elementary mathematics in view of the advanced mathematics. Over the years I have been trying to bring the course more and more down to earth, many times following feedback from experienced teachers who take the course during their sabbatical year. Eventually I have realized that it is not I who should be teaching it since what is most important for teaching this course is not extensive in-depth familiarity with advanced mathematics, but rather extensive in-depth familiarity with the potential difficulties of high-school students in understanding the elementary mathematics. Thus, beginning this year the two semesters of this course are being taught by two excellent and very experienced high-school teachers whom we have recruited. So far I believe this change has been a great success. (Tahl Nowik, symposium lecture, April 30, 2012)

    Hence it would involve a team consisting of both mathematicians and educators to organize a curriculum for training teachers and run it, a curriculum in which prominence would be given to courses under the title 'elementary mathematics from an advanced viewpoint', or such like.

5. Following on from the previous note, who usually trains future teachers? Probably for the mathematician this could well seem the most tangible role of the educator. Accordingly, can we give a job profile for a 'typical' mathematics educator? Perhaps: basic teaching duties concern teacher training, advanced teaching duties involve instructing educational principles on which basis the training of teachers is made, research is about how to enhance the effectiveness of the above said principles. I feel most mathematics educators would object to this description as being narrow, if not representative. What other description fits then? Would this match up to how mathematicians think the work of educators should/might be? If not, does it matter?
6. Note 4 concerns the issue of teacher training; to some extent a 'converse' issue is the so-called transition stage between school mathematics and that of university. Many students find this transition difficult, and much educational research is devoted to this. Some educators see the problem as being artificial; it is just adjusting to another institutional setting. Others think that there is a natural break in the level of thinking that has to be crossed if a student pursues mathematics at the tertiary level. An educator can suggest how to prepare the student for the 'jump', but cannot delete it. Perhaps the first opinion could be tenable for mathematics taught for students outside the mathematics department. But the insistence of working with axioms that dominates the mathematics treated in a mathematics department marks a definite character not found at school. Given the diverse directions of future lives of school students, it is indeed not suitable to give much attention to axiomatics at school.
7. So if we want to give support to students entering a mathematics department, any project offering such support has to be located at university, not at school. Many universities now offer courses that are not driven by specific mathematical content or theory but are aimed to help students gain general skills. For example, sometimes a 'problem-solving' course is offered, sometimes a course on 'proof'. But precisely because such courses do not deal with specific mathematical content or theory, their educational intentions have to be clearly defined and pursued. Is the educator or the mathematician better equipped to teach them? Are they useful in the end?
8. We broach then the question whether a mathematics department could profit from the services of a mathematics educator. Beyond teaching general courses mentioned in the previous note, he/she should be able to teach any first year course and to teach it with a special care to students' understanding without compromising rigor; be a person that students come to with their overall problems; advise teaching staff that has communication problems; to undertake fieldwork when a lecturer is not sure which argument is 'best' on an agreed basis involving pedagogical concerns; how to alter the way to teach re-takes of key courses; to be present in departmental meetings concerning teaching, especially on how to balance the amount and the importance of the material taught against slowing the pace for better understanding; to trace what undergraduates retain of what they have learned on leaving university; to choose and review textbooks for first time readers—for example, 'classic' texts often can seem to address the

lecturer's apprehension rather than the student's; to help the department how to react to students' evaluation of courses; to inject enthusiasm in teaching: it is so important for students to see their lecturers teach with some passion. I can spin out other things. Would the mathematician buy this advertisement, especially if the department is small?

9. From the last note, the reader might imagine, and correctly, that I am a mathematics educator with a full-time appointment in a mathematics department. In reality, the interaction with my colleagues is minimal, at least on educational issues. Despite this, my being within the department has advantages; I am in a position to interact with undergraduate students informally as well as formally, and by attending meetings I am fully aware of the politics guiding the changes in the whole working of the department. If I were placed in an education department and only visited the mathematics department to do fieldwork, I would feel that my work would be less effective because I would not be aware of essential aspects of the learning of mathematics as a scientific discipline.
10. Particularly over this last decade, there have been quite a few educational studies where the researcher interviews lecturers about their ways of thinking whilst doing mathematics, their teaching principles and practice, and other facets that impinge on their professional life. The output; usually a published paper in an educational journal, aimed mostly for a readership of educators. Hence the interaction between educators and mathematicians largely runs one way. Educators do have a legitimate case for studying mathematicians' practices and beliefs, but perhaps the ultimate test of the value of the collaboration occurs when the educator regards it as a normal regular activity expected of him/her with an active role without necessarily having a research motive in studying mathematicians. At the University level, lecturers have freedom to teach as they wish; the educator cannot influence their teaching without direct contact and persuasion. This introduces issues of dissemination; for example, it would be useful to deliberately set up national and international journals such that the articles published are accessible and of interest to both communities.

Expectations According to a Mathematics Educator

Steve Lerman

The title of the panel discussion was not couched in terms of people; it did not ask us to address: 'Mutual expectations between mathematicians and mathematics educators'. Nevertheless I think this is what was implied. In many places around the world relations between the mathematics and mathematics education research communities are typified by heated interactions, sometimes very heated indeed, leading to what are called 'math wars'.

I would like to believe that it is possible to take some of the heat out of those interactions and that may best be aided by an analysis of the nature of intellectual fields, and so I begin with a few comments on these two intellectual fields, mathematics and mathematics education, and the relations between them, from a sociological point of view. Indeed the question addressed by authors of this chapter can not be addressed within the field of mathematics, with mathematical insights, nor within mathematics education. It is a sociological issue, one that is about boundaries, status, and control; control of what is taught in school mathematics and of the content of mathematics teacher education both pre-service and in-service.

Mathematics has a long and highly valued history within the academy, covering hundreds of years. Mathematics education is a newcomer. It recently celebrated just 100 years of the establishment of the International Commission on Mathematical Instruction, a sub-committee of the International Mathematical Union. The problems of the teaching of mathematics have been of great interest to the mathematics community for much longer, but this event marked an important stage. Systematic research into the teaching and learning of mathematics has a much shorter history still. The International Group for the Psychology of Mathematics Education (PME), perhaps the leading research group in the world, held only its 37th annual meeting this year, 2013. However the community has a huge range of journals, conferences, doctoral students, research grants and so on. This is not surprising of course; it is the nature of intellectual fields in the academy (Said 1977).

Mathematics is a highly valued, hierarchical discourse, with concepts building on each other, expressing ideas in the most abstract terms. As Hilbert put it:

> But, in the further development of a branch of mathematics, the human mind, encouraged by the success of its solutions, becomes conscious of its independence. It evolves from itself alone, often without appreciable influence from without, by means of logical combination, generalization, specialization, by separating and collecting ideas in fortunate ways, new and fruitful problems, and appears then itself as the real questioner. (Hilbert 1902, quoted from the 2000-reprint, p. 409)

Mathematics education is very different (Lerman 2010). First, unlike most academic fields, including mathematics, it has a Janus-like character; a face towards practice as well as a face towards theory. Its first face leads to an essential demand of research that it is rooted in teaching-learning problems and should address them in some way, whether that be in an early years, mainstream school, university, or adult settings, including the workplace. These research problems are not mathematical. Whether to teach algebra at elementary school in parallel with arithmetic or afterwards, or why certain social groups fail mathematics disproportionally are not mathematical questions, they are educational ones. 'Truth' ends up being relative; some people are convinced by a piece of research, others disagree and criticize it. I suspect this is the case with all mathematics education research and readers' reactions to the two research issues above will illustrate the diversity of opinions and ideas.

Its second face is towards those bodies of theories that can assist in achieving the demands of the first face. Mathematics educators have traditionally drawn on psychology, but nowadays draw also on sociology, anthropology, philosophy, ethics and other fields. The central focus is, of course, the teaching and learning of mathematics, and thus the nature of mathematical activity and thinking are a crucial focus for study in the field (see Burton 2004). The nature of mathematical activity and thinking have to be studied using those fields, psychology, sociology, anthropology, philosophy, and so on.

Second, educational theories exhibit a horizontal knowledge structure; that is, mathematics education has developed multiple theories, each one a separate discourse. Researchers can examine the same data, a classroom transcript for example, and say quite different things in describing what is happening there. This phenomenon can be quite frustrating both for educational researchers and others looking in. In some senses mathematics has the same kind of structure. Mathematics grows both hierarchically, as knowledge grows within a theory, and horizontally as a new field of mathematics develops. But there is a further distinguishing feature which helps to explain the multiple theories in (mathematics) education: mathematics has a strong grammar, in contrast to mathematics education which has a weak grammar. One can readily agree that a particular structure has all the features of a Boolean algebra. But mathematics educators will argue about how understanding develops and how, indeed whether, teachers can identify that a student understands a mathematical concept or not. This exemplifies the different strengths of the grammars; one precise and singular, the other diverse and multiple.

In determining a curriculum a selection from the field of mathematics is made (as with history, art, etc.). For example, we might want to see school students acquiring the ways of thinking and acting of mathematicians and so will look for a curriculum modeled as an apprenticeship into mathematical thinking—perhaps what we might call a problem-solving curriculum. We might prefer a curriculum that focuses on enabling pupils to be able to act mathematically in the real world, and so might design a curriculum that harnesses real world problems, a modeling curriculum. We might, as a third option, be concerned that pupils acquire the skills, algebraic ones especially, that enable them to access some of the powerful ideas in mathematics, such as infinity and infinitesimals, in order to have access to higher study that requires a mathematical qualification. A fourth possibility is to mirror the ways that mathematicians often work, collaboratively, and build a curriculum around tasks that, together with pedagogic moves, enable students to be creative and innovative in their mathematical work with others. These choices, and further ones we could think of, are driven by ideology, by beliefs and values. Mathematicians will draw on different values amongst themselves in addressing what should be taught and come up with competing emphases; how much more contested will it be when the decision-making includes other interested groups.

Of course the way I have presented this discussion makes it appear that mathematicians and mathematics educators determine what is to be taught in all the phases of education. In fact this is to ignore policy makers. Governments determine what is to be taught. In some countries they will consult mathematicians, mathematics educators and other interested parties, including parents and teachers. In other places they will consult mathematicians only, the high status academic community. In other places still the consultation will be merely token; the decisions have been made in advance by politicians (see Bernstein 2000).

This is the underlying reality to the 'math wars'; competing ideologies regarding what school mathematics should consist of. At this conference we are celebrating the work of a mathematician whose heart has always been in helping students to learn, but especially to enjoy and be challenged by, mathematics. Hence I will respond to the panel question in a way which takes into account the understandings that sociological theories give us, but based also on the assumption that mathematicians and mathematics educators want the best for students, however they might interpret that aim.

Mathematicians can expect that mathematics educators should: recognize the necessity of taking account of mathematical knowledge and mathematical activity; engage in mathematical activity themselves from time to time; and take the views of mathematicians concerning what constitute significant areas of mathematics and of mathematical activity into account in curriculum planning, assessment etc.

Mathematics educators, in return, can expect that mathematicians should: respect the differences between the two fields; recognize the uniqueness of the research problems in mathematics education; recognize the ideological positions implied by any selection of mathematics that constitutes a school curriculum; and acknowledge the help that mathematicians can receive from mathematics educators' engagement with them on researching mathematics teaching and learning at University.

Concluding Comments

Tommy Dreyfus

Most of the comments in this concluding section have been influenced by the above contributions; however, rather than repeating or summarizing them, I have made some stronger, others weaker, and given a different perspective to still others, adding further questions.

Everything I say will be true about some and nothing about all mathematicians/mathematics educators. The reader is asked to interpret the indefinite "mathematicians" as "some mathematicians" rather than as "the mathematicians" or "all mathematicians". This will relieve me from adding "some" to every statement. It also illustrates that no statements in mathematics education are universally true.

Expectations by Mathematicians

Mathematicians might see the role of mathematics educators as having to do mainly with teacher training, or mainly with making sure school graduates have appropriate mathematical knowledge and ability (whatever that might mean for different groups of graduates, university bound ones or not). Even a quick look at the mathematics education literature will show that mathematics educators see the teacher training and the mathematical knowledge of high school graduates as some of their concerns but also deal with many others, some of which might be considered unimportant by mathematicians, among them students' learning difficulties; the role of technology in learning mathematics; students' and teachers' beliefs and attitudes toward mathematics and ways of teaching and learning it; social aspects of learning mathematics such as whether students should learn in small groups within the classroom, whether learning mathematics should happen in homogeneous or heterogeneous groups and at what ages; ethnographic aspects of learning mathematics; epistemological issues related to learning mathematics; theories about how people learn mathematics; and many others.

Mathematicians may see mathematics education as a practical, applied rather than a theoretical domain. They may believe that results in mathematics education should be practical, applicable. Mathematics educators might ask whether mathematicians also expect results in mathematics to be applicable. Is mathematics a fundamental science, where theoretical as well as applied results are acceptable whereas mathematics education is an applied domain where only applicable results are acceptable? If so, how about "applied mathematics"? Is it an applied domain in this sense? Are only applicable results acceptable in applied mathematics? And is applied mathematics (the applied mathematician) closer to mathematics education than pure mathematics as Törner seems to imply in his contribution to Chap. 18 of this volume.

As a consequence of their beliefs about mathematics education, mathematicians may not understand what 'research in mathematics education' could mean. And mathematicians who do accept research in mathematics education as a valuable undertaking may have an image of mathematics education as of a scientific discipline with a (single) theory and a methodology, rather than the more accurate one of dozens of theories and at least as many methodologies. Indeed, as Fried discussed at length in the introduction to this volume, the methodological outlook of the exact sciences, with which mathematics is aligned, and that of the social sciences and humanities have profoundly different foundations, the misunderstanding of which can lead to false or unrealistic expectations. On the basis of such misunderstanding, mathematicians may well expect mathematics education research to provide "clear apodeictic answers" as to whether, say, one teaching method is more effective than another. They may grant that "effective" has to be defined, but it might not be easy to convince them that the adequacy of such a definition is limited since, for example, the same teacher teaching the same content with the same textbook in two classes may provide rather different experiences of algebra to the students in the two classes (Eisenmann and Even 2009).

Mathematicians may expect mathematics education to provide simple, clear-cut solid findings—results that are robust and can be reproduced. While there are such findings in mathematics education (Education Committee of the European Mathematical Society 2011–2013), they are not the norm, and mathematicians might find the reasons for this hard to fathom. One such reason is that it is impossible to repeat an experiment under the "same" conditions because the same educational conditions never return. As a consequence, mathematicians may respect mathematics educators as people, as expert teachers or curriculum designers, but many will not respect mathematics education as a research domain.

Mathematics teachers at university may well blame themselves if the students in their classes don't grasp what they teach. They may feel that they haven't done their job well. And in many cases, they will expect that explaining things better a second time, will make a big difference, something mathematics educators might doubt. How to explain the mathematics, mathematicians assume, they know better than mathematics educators since they have a deeper understanding of the content domain. So they have the expectation that mathematics educators cannot help them. They may not be interested in mathematics educators getting involved in what is happening in their university mathematics classes, even if these mathematics educators have studied similar experiences at other universities and can point to literature relevant to the issues.

On the other hand, university mathematicians may expect mathematics educators to be responsible for what is learned in elementary and high school. They may not have a clear grasp of factors and bodies that have a far stronger influence than mathematics educators on mathematics in high school, namely education systems; they may also fail to clearly distinguish between mathematics teachers and mathematics education researchers (see Fried, Chap. 2; Presmeg, Chap. 4). Mathematicians may in fact blame mathematics educators for what's wrong or what they consider as being wrong (such as low achievement on high stakes tests like TIMSS or PISA) with school mathematics.

Expectations by Mathematics Educators

Mathematics educators may expect mathematicians to be interested in students' lack of understanding, misconceptions, and so on, at least as far as the mathematicians' own students are concerned.

Mathematics educators may expect mathematicians to understand that a curriculum consists of more than a syllabus, a list of topics to be taught. Many mathematicians will grant that a curriculum consists of more, that there are, for example, more or less formal approaches to the content. However, mathematics educators may expect more from mathematicians, namely that they also understand that a curriculum includes an underlying epistemology of mathematics, a (possibly theory based) view of how students learn, a position on how to design units and activities, and how to achieve an equilibrium between student and teacher activity, and so on (Dreyfus et al. 1987).

Mathematics educators, especially those who deal with the use of technology in teaching and learning mathematics, may expect mathematicians to understand that using technology actually changes the mathematics itself rather than being only an accessory to help "see" things better.

Mathematics educators may expect mathematicians to see research in mathematics education as relevant, even to teaching at the tertiary level. They may expect mathematicians to show some interest in theoretical constructs of mathematics education (after all, mathematics itself consists of theoretical constructs).

Mathematics educators may expect mathematicians to appreciate the complexity of the processes of teaching and learning mathematics (Davis 2011), and the consequent need for interdisciplinary work involving (at least) mathematics, didactics, psychology (cognitive and social), sociology, and possibly also anthropology and other domains. Mathematics educators may expect mathematicians to be amenable to the conviction that because of this complexity, quantitative research results are often not meaningful (Schoenfeld 2001).

Mathematics educators might expect mathematicians to consider mathematics education as more than a group of people responsible for what mathematical knowledge students have when they finish high school, and to appreciate that the education systems rather than mathematics educators control these outcomes to a large extent. The situation may be analogous to the one in economics: The people at the helm, the people making policy decisions may be more than happy not knowing what university economists know and what their theories predict because such knowledge would make their decision making far too complicated. Indeed, this is just the kind of complexity Niss discusses in his paper in Chap. 15 of this volume. Mathematics educators might not like matriculation exams and their influence on what is going on in school classrooms any better than mathematicians do; in fact, mathematics educators might be even more antagonistic to such exams because they might feel frustrated by years of failure in trying to effect some change. In fact, they might think (rightly or not) that the exams have been dictated by what university mathematicians have prescribed to the education ministry. Things are complex.

Given this complexity, where should mathematics educators, groups of mathematics educators, or departments of mathematics education be placed in a university? In the mathematics department? In the school of education? What would be their role in a mathematics department and how would it differ from their role in a school of education?

Closing Remark

Some expectations are mutual; after starting with teacher education, let me use mathematics teacher education again to illustrate this mutuality: Some perennial questions concern teacher education: Should it be under the responsibility of mathematicians or mathematics educators or both, or does the answer depend on the level at which the prospective teachers are intended to teach? And if both, mathematicians and educators, should the mathematics courses be simultaneous, consecutive or integrated with the mathematics education courses? Should mathematicians and educators work in collaboration with one another? And what form would such collaboration take? De Shalit discusses one option in his contribution to Chap. 18 of this volume. What would mathematicians expect from the mathematics educators and vice versa in such collaboration?

Bridges and collaboration, common ground, need as a prerequisite mutual respect. This is often missing, sometimes from both sides. The authors of the chapters in this book have demonstrated such mutual respect in their professional lives in general and when producing this book. With them, we hope that the book will contribute to such respect more generally.

References

Bernstein, B. (2000). *Pedagogy, symbolic control and identity: theory, research, critique*. New York: Rowman & Littlefield (rev. edn.).

Burton, L. (2004). *Mathematicians as inquirers: learning about learning mathematics*. Dordrecht: Kluwer Academic.

Davis, B. (2011). Mathematics teachers' subtle complex disciplinary knowledge. *Science*, *332*, 1506–1507.

Dreyfus, T., Hershkowitz, R., & Bruckheimer, M. (1987). Processes in the transition from syllabus to curriculum. *ZDM. Zentralblatt für Didaktik der Mathematik*, *19*, 19–25.

Education Committee of the European Mathematical Society (2011–2013). "Solid findings" in mathematics education. *Newsletter - European Mathematical Society 81*, 46–48, *82*, 50–53, *83*, 46–50, *84*, 53–55, *86*, 49–52, *87*, 42–44, *88*, 59–61, *89*, 51–53.

Eisenmann, T., & Even, R. (2009). Enacted types of algebraic activity in different classes taught by the same teacher. *International Journal of Science and Mathematical Education*, *9*, 867–891.

Hilbert, D. (1902). Mathematical problems. *Bulletin of the American Mathematical Society*, *8*, 437–479. Reprinted (2000) in *Bulletin (New Series) of the American Mathematical Society*, *37*, 407–436.

Lerman, S. (2010). Theories of mathematics education: is plurality a problem? In B. Sriraman & L. English (Eds.), *Theories of mathematics education* (pp. 99–110). New York: Springer.

Said, E. (1977). *Orientalism*. London: Penguin.

Schoenfeld, A. H. (2001). Purposes and methods of research in mathematics education. In D. Holton (Ed.), *The teaching and learning of mathematics at university level* (pp. 221–236). Dordrecht: Kluwer Academic.

History of Mathematics, Mathematics Education, and Mathematics

Chapter 6
History in Mathematics Education. A Hermeneutic Approach

Hans Niels Jahnke

Abstract The paper discusses the possibility of bringing history in the mathematics classroom by studying historical sources with students. A manuscript by Johann Bernoulli about the differential calculus which was brought to a grade 11 classroom serves as an example. Reading a source is fundamentally a hermeneutic activity and can be conceptualised by the term 'horizon merging'. In the so-called hermeneutic circle the horizons of the reader and the author of a text are supposed to merge by a repeated reading. In contrast to common ideas about the genetic principle the hermeneutic approach described in the present paper assumes that students have already some experience with and knowledge of the modern counter-part of the concepts treated in the source. Reading a source is an activity of applying mathematics in a way completely new to students. It provides opportunities for reflecting deeply about their images of the respective mathematical concepts.

Keywords Johann Bernoulli · Concept image · Differential · Genetic principle · Hermeneutics · Hermeneutic circle · Horizon merging · Historical source · Infinitely small quantity

Preliminary Remark

Before I entered the field which usually is called "History and pedagogy of mathematics" (HPM) I had met two rather different notions of what this could mean. One was the idea to consider history of mathematics as a collection of interesting mathematical problems some of which were suitable to be treated at school. This idea was mainly supported by teachers at schools respectively math educators who saw themselves predominantly as teachers. To be sure, some work in this direction is impressive but I asked myself where there was any substantial relation to the history of mathematics. All these problems were meaningful by themselves and could be treated without any reference to history.

H.N. Jahnke (✉)
Faculty of Mathematics, University of Duisburg-Essen, 45117 Essen, Germany
e-mail: njahnke@uni-due.de

M.N. Fried, T. Dreyfus (eds.), *Mathematics & Mathematics Education: Searching for Common Ground*, Advances in Mathematics Education,
DOI 10.1007/978-94-007-7473-5_6,

The other idea was in a vague sense related to what could be called the "genetic principle". Prominent mathematicians like Felix Klein and Otto Toeplitz believed that history of mathematics could contribute to the learner's understanding by making visible great lines of development and thereby connecting seemingly unconnected subjects. Klein had exemplified this by his magnificent book "Lectures on the development of mathematics in the 19th century" in which he had reconstructed the immediate pre-history of the mathematics of his time. One can imagine that students who attended these lectures got a sound idea of what was going on in mathematics. But these lectures were definitely intended for an advanced audience.

Otto Toeplitz, on the other hand, was more involved in the teaching of beginning university students of mathematics. He, too, intended to present for the case of analysis the great lines of thought. He justified his "genetic approach" by saying that if we go back to the roots of mathematical concepts the dust of time and the scratches of long use would be removed. Infinitesimal analysis would become attractive for students when they can see that its basic concepts had been objects of an exciting process of research at the time of their invention. But Toeplitz hastened to add the remark that this is completely different from a "historical method". This term, historical method, he says, "... brings to mind the idea, which we, on the contrary, would particularly like to eliminate, of the old and antiquated, the roundabout paths often followed by research, the subjective and haphazard nature of scientific discoveries. It is especially important to me to draw a dividing line in this direction" (Toeplitz 1927, 93, translation by the author and Michael N. Fried, to appear). As a consequence, in his "Genesis of the infinitesimal calculus" he never mentioned Newton's binomial theorem which Newton himself had considered as one of his most important discoveries nor did he made an attempt to discuss indivisibles or infinitesimals. This tension then between a historical and a mathematical perspective on history is a running theme in HPM and an important issue for a book about the common ground between mathematics and mathematics education. I shall return to this at the end of my paper.

A third experience in the beginnings of my involvement with HPM was a talk by Jan van Maanen, later co-editor of the ICMI Study on HPM, which he gave in Toronto in 1992. He reported about a teaching unit on a 17th century Dutch textbook on algebra which he did with pupils of grade 8 (see van Maanen 1997). Obviously, there was a strong contrast to Toeplitz' ideas. There was no great mathematician discovering something new, no exciting solution of new and deep problems. Instead of this, there were questions such as these: How did pupils three hundred years ago solve quadratic equations? Which symbols were used in the textbooks, which procedures, which applications? This means, instead of studying big ideas and great lines of thought pupils were invited to look for the specific and the context. They were invited to compare their own concept image of quadratic equations with that of pupils at their place two hundred years earlier. Listening to this talk it seemed to me obvious that these pupils had learnt a lot of substantial mathematics. Somehow the tension between the historical view of a concept and its modern counterpart itself provided the productive element which was so exciting to students. For similar approaches the reader might consult Arcavi et al. (1982), Laubenbacher and Pengelley (1999) as well as the survey of Tzanakis and Arcavi (2000).

Thus, I decided to continue in this direction. From the outset it was clear that the inclusion of history of mathematics into teaching cannot make things easier to students. Rather the opposite is the case. At all times, scientists have written for a narrow circle of specialists and not for pupils of later generations. Frequently, their ideas were still vague and insufficiently formulated in a clumsy language. Thus as a rule, when it comes to reading texts of eminent mathematicians of the past we have to expect considerable difficulties. This is even a frequent experience of working mathematicians when they consult papers in their field which have been written in a distant past.

In the following I will describe an experience with reading a historical source with students and then discuss general principles and difficulties underlying this enterprise.

Johann Bernoulli's Textbook on the Differential Calculus

First of all I should mention that at the time of this teaching experience I conducted regularly courses for practicing teachers in which we read historical sources. Every course comprised 5 sessions of 2 hours each. The participating teachers were asked to prepare every session by reading a source. Participation was completely voluntary, the teachers did not receive any reward or credit. Some of them intended to use historical material in their classroom, others came out of interest in the history of mathematics. With hindsight, it is still astonishing to me that this rather naïve procedure really worked. But it is a fact that after two years there was a group of ca. 100 interested teachers whom I invited and of which ca. 20 used to participate in these reading courses. It was in the context of this in-service teacher training that I did this teaching experiment, other teachers did similar experiments with their pupils.

Let us now turn to our source. Brothers Jacob Bernoulli (1655–1705) and Johann Bernoulli (1667–1748) were the most important mathematicians of the Leibniz school of analysis. It was due to their substantial work that not Newton's fluxions, but Leibniz' differentials became generally accepted on the European continent. After Leibniz and Newton had passed away Johann Bernoulli was for some 15 years the leading mathematician of Europe. It is an indication of the highly competitive mathematical climate at that time that he was involved in numerous controversies and quarrels with colleagues. He even managed to fall out with his son Daniel because the latter had won a prize of the Paris Academy of the Sciences for which he himself had applied, too. Thus, he earns a place of honour in Ted's list of idiosyncratic mathematicians (Eisenberg 2008).

Johann Bernoulli wrote the manuscript "De calculo differentialium" in 1690/91 when he was 23 years old. At that time, he conducted private lectures about Leibniz' new calculus to the Marquis de L'Hospital (1661–1704). Johann and L'Hospital had agreed on a secret contract saying that Johann would teach L'Hospital and would leave his mathematical discoveries to the latter's exclusive use. In return,

L'Hospital payed Johann a considerable salary until his death. For a long time Johann's manuscript "De calculo differentialium" was regarded as lost, some historians even believed that Johann never had written such a paper. Only in 1922 the manuscript was detected by Paul Schafheitlin in the library of the university of Basel. This, of course, was no accident since Johann had been a professor at that university. The more astonishing it is that it took such a long time after Bernoulli's death that historians became aware of this important manuscript.

Schafheitlin published the manuscript and translated it into German (Schafheitlin 1924). A comparison of l'Hospital's textbook "Analyse des infiniment petits" (1696) with Johann's manuscript shows that the latter was sort of a draft to the former. L'Hospital had considerably extended Johann's text, corrected some mathematical mistakes (see below) and dressed it up didactically. In the teaching sequence it was especially stressed that at the time when Bernoulli and L'Hospital wrote their books the meaning of the basic concepts of Leibniz' calculus was not at all clear. Thus, both books were efforts of interpretation and clarification, and not just reproductions of a given body of knowledge.

Bernoulli's "differential calculus" comprises 38 printed pages. It contains:

- 3 postulates,
- calculation rules for differentials,
- 11 problems on the determination of tangents to curves,
- 9 problems of maxima and minima,
- methods for determining points of inflection.

For the teaching unit I produced a montage of texts containing the postulates, the rules for differentials, some problems on determining tangents to curves and problems on minima and maxima.

In the following I will at first introduce some parts of the source and then describe how the students worked with it. First, the postulates:

Postulates

1. A quantity which is decreased or increased by an infinitely smaller quantity is neither decreased nor increased.
2. Every curved line consists of infinitely many segments which are infinitely small.
3. [omitted here: refers to the integral calculus]

A symbolization of the first postulate might be written as

$$x + e = x$$

where e is infinitely smaller than x. According to the ordinary rules of algebra this implies

$$e = 0.$$

But this was an implication not intended by the analysts of the Leibniz school—a fact which does not become clear from the postulate itself but from its later applications to differentials. To be sure, all of his life Leibniz was rather vague about what differentials and the infinitely small were. Thus it was young Johann Bernoulli who in plain language formulated the central rule underlying any calculation with differentials. This was a conscious act of interpretation and a determination of meaning which had not been done before and which later was taken over by the Marquis de L'Hospital.

The second postulate, too, made explicit what the analysts of the Leibniz school had tacitly assumed, namely to consider a curve as a polygon with infinitely many sides of infinitely small lengths.

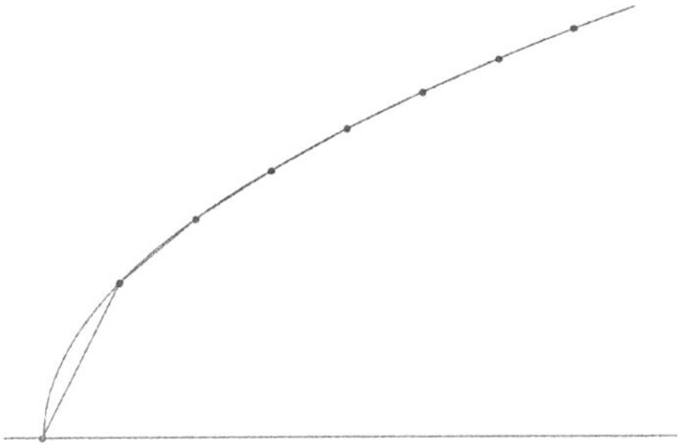

Thus, the second postulate shows the geometrical meaning of the calculus and determines its application to geometry. In regard to the idea to replace curved lines by straight lines we could anachronistically speak of a geometrical version of what we today call linearization.

What infinitely smaller quantities and infinitely small segments are is nowhere explained. Thus, we have to look at the applications of these concepts to learn more about them.

Thus, let us look at the calculation rules for differentials, e.g. the product rule.

> "The differential of xy is $x\mathrm{d}y + y\mathrm{d}x$. When $x + e$ is multiplied by $x + f$ (where $e = \mathrm{d}x$ and $f = \mathrm{d}y$) then the product is $xy + ey + fx + ef$. After subtracting xy we get $ey + fx + ef$ which is according to postulate 1 equal to $ey + fx = x\mathrm{d}y + y\mathrm{d}x$. q.e.d."

Bernoulli does not explicitly say what a differential is but from his calculation it becomes clear that it is a difference between two states of a quantity which are infinitely near to each other.

$$d(xy) = (x + dx)(y + dy) - xy$$

The calculation gives

$$d(xy) = xdy + ydx + dx \cdot dy = xdy + ydx$$

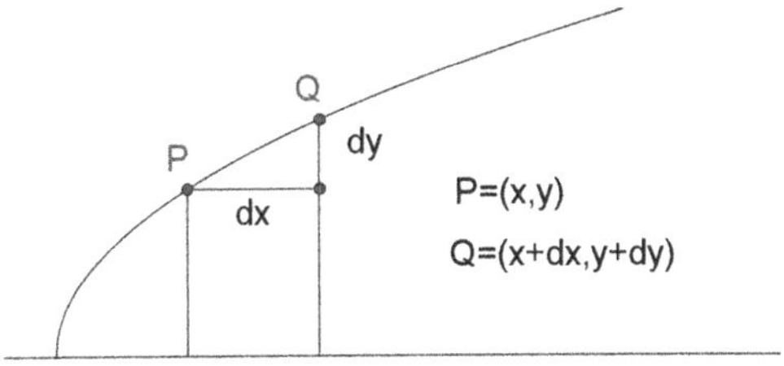

Thus $d(xy)$ is infinitely smaller than xy. The critical point in Bernoulli's proof is of course the argument that $dx \cdot dy$ is infinitely smaller than xdy or ydx which are for their part infinitely smaller than the quantities x and y. Thus like Russian dolls we have three nested universes of quantities each one infinitely smaller than its predecessor. I mention in passing that the discussion with the students why $dx \cdot dy$ is infinitely smaller than xdy was far from easy and we escaped to some plausibility arguments. (Please, do not misinterpret the figure. y should not be considered as function of x, rather to each point of the curve the quantities y (ordinate) and x (abscissa) are assigned.)

In a further step we consider how the calculus of differentials is applied to the study of curves.

To Find the Tangent to the Parabola

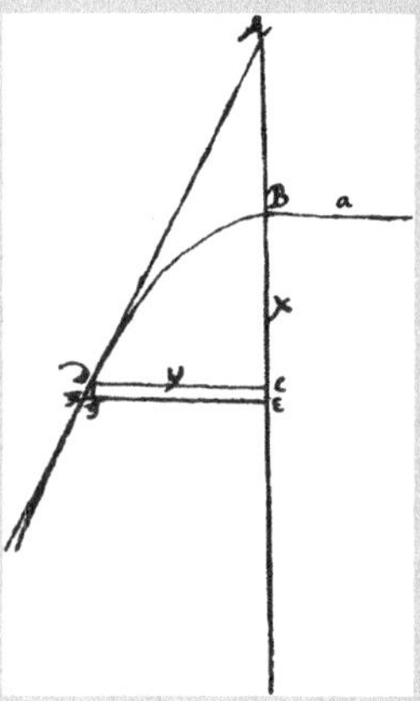

According to the definition of the parabola we have $ax = y^2$, thus $adx = 2ydy$ or $a : 2y = \frac{dy}{dx}$ and since according to postulate 2 it is supposed that every curve consists of infinitely many straight lines, the tangent AD and the infinitely small segment DF of the parabola BDF will be a straight line. Therefore, if one draws DG parallel to diameter AE, then triangle $\Delta DGF \sim \Delta ACD$. Thus we have $FG : GD = CD : AC$, and if s designates the subtangent, then $\frac{dy}{dx} = \frac{y}{s} = \frac{a}{2y}$. Consequently, $s = \frac{2y^2}{a} = \frac{2ax}{a} = 2x$. If, therefore, AC is taken twice as large as the abscissa BC of point D and if through A the straight line AD is drawn, then this is the sought tangent.

For the sake of clarity we replace Bernoulli's freehand sketch by a computer drawing.

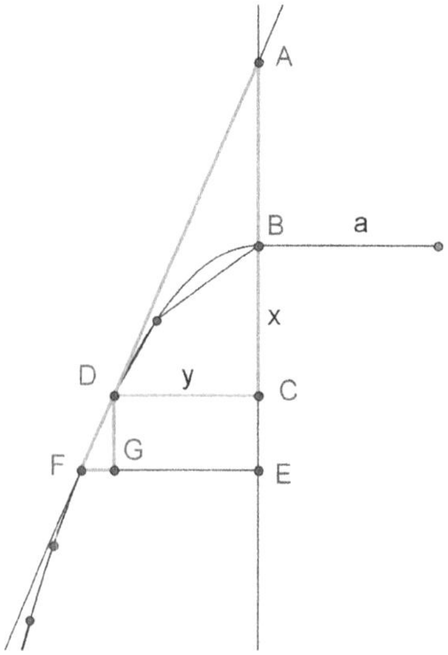

In short: Bernoulli put up the equation of the parabola $ax = y^2$ (in Bernoulli's words its 'nature') from which he derived a differential equation. Then he applied postulate 2 with two remarkable consequences. (1) A tangent line to a curve is simply the prolongation of the infinitely small segment adjacent to that point. (2) This creates two similar triangles, the infinitely small triangle FGD and the finite triangle DCA. This gives $\frac{dy}{dx} = \frac{y}{s} = \frac{a}{2y}$ where s is the subtangent $\overline{AC}$. From this follows

$$s = 2x$$

Three features of this text immediately strike us. (1) There is no coordinate system, instead the variables refer to the symmetry axis of the curve. (2) The equation of the parabola is of a geometrical nature. Any point of the parabola is constructed by transforming geometrically the rectangle ax into the square y^2. See the segment a attached to B. (3) For determining the tangent Bernoulli does not calculate its slope, but he calculates the x-coordinate of a second point of the tangent, namely A, that is the subtangent s. Then the tangent can be constructed as the straight line connecting A and D. (4) We pointed already at the remarkable definition of a tangent as an extension of an infinitely small side of the polygon representing the curve. All in all, we see here a basically geometrical conception of infinitesimal analysis in which the role of the algebraic symbolism is reduced to auxiliary calculations.

Students Read Bernoulli's Text

Sections of Bernoulli's manuscript were read with students of an advanced mathematical course ("Leistungskurs") in grade 11 at a Gymnasium near Bielefeld, Germany (see Jahnke 1995, for details). The students had already been introduced to the fundamentals of the differential calculus and they knew the concepts of limit and derivative and how to apply them in order to determine tangents, extremal values and points of inflection. In Bernoulli's manuscript they found a conceptual framework completely different from their own. The textbook on which their calculus lessons were based consistently avoided differentials, even as a notation. Derivatives were exclusively written as $f'(x)$. However, students had met differentials in their physics classes as very small but finite quantities. Usually, pupils worked in groups on a section of the source to which I had added an assignment with special questions. For example, I asked them to give an intuitive interpretation of Bernoulli's concept of "infinitely small quantity", or I asked them to calculate the differential of x^2 after they had studied Bernoulli's proof of the product rule for differentials or they had to find out from the text what a subtangent is.

The teaching started with a 'map of the history of mathematics.' Students were asked to give names and dates of mathematicians they had heard of. Since they were convinced that they didn't know anything about history of mathematics they were astonished that after half an hour the blackboard was full of names and dates. As it is often the case, as a group they were more successful and brighter than as individuals.

In a second step they read a short sheet of information about the early history of analysis, Johann Bernoulli's biography and the history of the source they were expected to read. Since studying the source took more time than predicted the treatment of extremal values was finally omitted.

Asked for an intuitive interpretation of "infinitely small quantity" students were very inventive. They offered:

- "a quantity is always positive"
- $x + e = x$, but $e \neq 0$?

- they set up the proportion:

$$\text{quantity} : \text{infinitely small quantity} \equiv \text{line} : \text{point} \equiv \text{area} : \text{line} \equiv \text{solid} : \text{area}$$

- an infinitely small quantity is like the difference between a rational approximation of π and the number π itself
- an infinitely small quantity is like the difference between 1 and 0.999...

The proportion line : point ≡ area : line suggested that students had heard something about Cavalieri's principle. However, some of them expressed doubts: Can one really say that a line is composed of points?

These questions found a remarkable continuation when the tangent to a parabola was studied. First of all, for the students it was very demanding to handle the unusual form of Bernoulli's equation of a parabola. And it was even more demanding to find out from the involved argument with similar triangles what a subtangent is. Everybody with teaching experience would expect these difficulties. However, after an understanding of the meaning of 'subtangent' had emerged students were excited to realize that in order to determine the tangent Bernoulli did not calculate its slope, but the coordinates of a second point of the tangent. To be sure, the proportion defining the slope was needed to get rid of the infinitely small quantities, but the target quantity was not the slope, but a second point. Two points being known one can construct the tangent by a ruler much easier than using its slope.

Bernoulli's definition of a tangent as a prolongation of an infinitesimal side of the polygon caused lengthy discussions. First of all, if one fixes a point which of the two adjacent sides of the polygon should be chosen? Second, and more important, if the intuition is correct that an infinitely small quantity can be considered as a point, how can such an entity define a direction as is supposed in the definition of a tangent. When it defines a direction an infinitesimal should at least contain two points since two points are needed to determine a direction. The conclusion of this discussion was that an infinitely small quantity must be more than a point, it cannot be 0-dimensional, but has to be 1-dimensional. Of course, this was also the conclusion of the analysts of the Leibniz school. Thus, we have an infinitely small universe in which we can do Euclidean geometry just as we can in the domain of the normally sized quantities.

We stop the analysis of the students' discussion at this point. Of course, the source contains a host of additional interesting problems. Half a year after this teaching unit, students invited me to come again to their class for a further study of Bernoulli's manuscript. We decided to investigate Bernoulli's methods for determining points of inflection. Curve sketching is a routine topic in German calculus classes and the criteria for maxima/minima and points of inflection are dead stuff. They are learnt and mechanically applied. Thus, it was exciting to see that besides the usual $ddy = 0$ Bernoulli hat a second criterion for points of inflection.

When one runs through a curve in the neighbourhood of a point of inflection P then the sketch seems to show that the points of intersection of the successive tangents move along the x-axis in a way that the tangent in P has an extremal position. In his commentary to Bernoulli's manuscript Paul Schafheitlin remarked that this criterion was not adopted by L'Hospital because it is equivalent to the usual

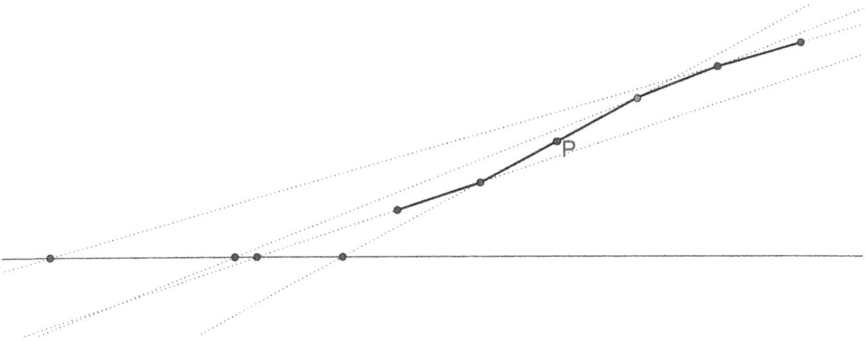

$ddy = 0$. But we found that this is not true. Bernoulli's criterion is neither necessary nor sufficient, it is in general not correct. Of course, the students were impressed to see that such an important mathematician like Bernoulli made such a mistake. To investigate this criterion is a wonderful exercise for good high school and even university students.

The Hermeneutic Approach

In the introduction of my paper I referred to the reconstruction of a great line of thought leading from past roots of a mathematical concept to its modern version. To me this idea (a typically 19th century idea!) is the essence of the genetic principle as it was understood by mathematicians of the late 19th and early 20th century and which is influential even today. For school teaching this is simply unrealistic. This is the case since school teaching is necessarily episodic and progresses in small steps. Thus, we have to be realistic and should not overload the enterprise "History and pedagogy of mathematics" by demands which necessarily must lead to failure. One can call this a *pragmatic categorical imperative*.

Thus, the hermeneutic approach grew first of all out of the idea that you should confine history to a local experience which is quite a modest approach compared to what you have in mind when you imagine a historically guided reconstruction of a mathematical concept. In the hermeneutic approach, students are asked to examine a source in close detail and explore its various contexts of historical, cultural and scientific nature. The hermeneutic approach will not give you an overview. Rather, it is a hope that some pupils will like history and develop a certain interest in it which might motivate them to search for further reading.

The basic guidelines of the hermeneutic procedure can be summarized in 6 principles.

(1) Students study a historical source *after* they have acquired a good understanding of the respective mathematical topic in a *modern* form and a *modern perspective*. The source is studied in a phase of teaching when the new subject-matter

is applied and technical competencies are trained. Reading a source in this context is another manner of applying new concepts, quite different from usual exercises.

(2) Students gather and study information about *context* and *biography* of the author.
(3) The historical *peculiarity* of the source is kept as far as possible.
(4) Students are encouraged to produce *free associations*.
(5) The teacher insists on *reasoned arguments*, but not on accepting an interpretation which has to be shared by everybody.
(6) The historical understanding of a concept is contrasted with the modern view, that is the source should encourage processes of reflection.

What then is hermeneutics? For the following the reader should compare Glaubitz (2010, 2011) and Jahnke (1994, 1996). Simply said it is the "art or the science of interpreting texts." It distinguishes systematically between the author and the reader of a text and their different perspectives. Thus, the strong tension between the historical perspective and the modern view on a mathematical topic is not something which should be smoothed out or eliminated, but is considered as the essential achievement a historical text might contribute to the intellectual development of a person. Thus, the whole enterprise of reading a source rests on experiences of "dépaysement" as the French say or "Verfremdung" ("alienation") as the German writer Bertold Brecht would have said. Sources introduce into teaching an unwieldy element. But how comes it that such unwieldy elements do not lead to failure? This is so only when they have anchor points. The student who deals with something that he already knows but that is presented in a radically different, unfamiliar way or an unknown guise, should be able to make connections to these anchor points. In hermeneutics you would say: His horizon merges with the horizon of the past. *Horizon merging* is a term that was coined by Hans Georg Gadamer (1900–2002). In the horizon merging the student may begin to wonder and to reflect upon what he possibly had never thought about before. In essence he begins to develop deeper awareness. This is in fact an instance of broadening one's horizon. And it does so by utilizing a strategy of dissonance. It is well known that this kind of incompatible information ensures greater retention and ease of retrieval from memory. But to do so, there must be a familiar reference frame. It is therefore applied only to subject-matters that students are already familiar with.

In hermeneutics the process by which the merging of horizons occurs is described by a spiral, the so-called 'hermeneutic circle' which points to the necessity of already possessing an interpretation of a text in order to gain a new interpretation. For us as mathematics educators this appears not so strange as it might be for other people since we are used to reflect about spiral processes, the most prominent being the process of modelling. I take a picture from the dissertation of my former Ph.D. student Michael Glaubitz (2011, 61).

You start with a certain image of the text reflecting your expectations about what it might be about. Then you read the text and realize that some aspects of your image do not agree with what is said in the source. Thus, you have to modify your image,

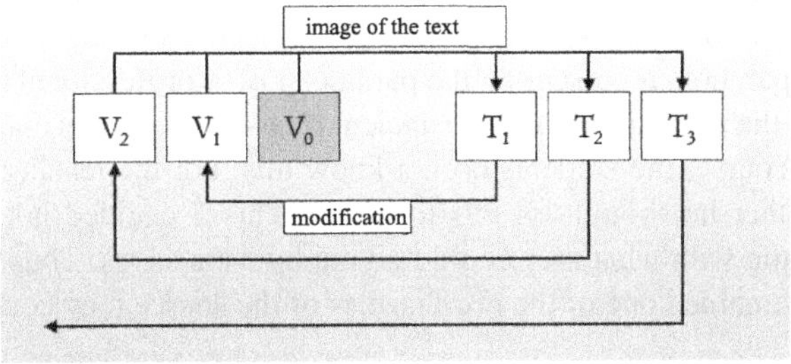

read again, modify and so on until you are satisfied with the result or simply do not like to continue.

In our case students started with the expectation that Bernoulli's text might be about determining tangents to and extremal values of curves, that the idea of a limit of an infinite process might be central to the subject, that concepts like derivative and slope of a tangent (a quotient) will frequently appear and that all is sort of algebra, with new rules, but symbolic in nature. After some windings of the hermeneutic spiral they had realized that the source was in fact about tangents to and extremal values of curves, but that there was no limit concept, instead there was the difficult concept of infinitely small quantity. Also, the slope of a tangent was less important than expected. It was used in the source to get rid of the infinitely small quantities, but the target object was a second point of the tangent. Thus, the status of a slope changed to that of an auxiliary object. And so on.

On a more basic level the hermeneutic circle can be considered as a process in which a *hypothesis* is put up, tested against the source, modified, tested again and so on until the reader arrives at a satisfying result. For example, the students were asked to infer from the source what a differential is. From their knowledge of calculus some students formed the idea that a differential is something similar to a derivative. With this hypothesis they studied Bernoulli's derivation of the product rule and realized that this cannot be true, since Bernoulli did not calculate a quotient. After some further attempts they saw that a differential is in fact a difference. In a similar manner they found out what a subtangent is.

Can one say then that students behave like historians of mathematics when they read a source? In principle, this is the case. When they entered the source they had questions similar to those a professional historian of mathematics would ask. Roughly spoken, these questions refer to the different meanings of concepts and the different conceptual structures at the time of Bernoulli and today. There are other natural questions they did not explicitly pose but which were obviously in their minds. These refer to what math educators are used to call concept images. Another question they asked was whether Bernoulli really believed that infinitely small quantities exist or whether he considered them as useful but meaningless tools. Of course, a professional historian would ask this question, too. There are other questions a historian would routinely study and which our students didn't ask, namely to compare Bernoulli's text with other writings of the Leibniz school.

The most important difference concerns the previous knowledge a historian and a student have at their disposal. For example, consider the segment in Bernoulli's sketch of the parabola representing the parameter a. A professional historian knows of course that the segment hints at the ancient ruler-and-compass construction of the parabola. Of course, the students do not know this. For the teacher it is a difficult question whether he should tell this to his students. I decided not to do that and preferred to stop with what they could find out by themselves. Thus, to the students the segment remained one of the peculiarities of the source they couldn't explain.

Discussion

In a recent paper Uffe Jankvist (2009) has distinguished between the use of *history as a tool* and that of *history as a goal*. The first concerns the use of history as an assisting means in the teaching and learning of mathematics (mathematical concepts, theories, methods, motivation and so on). In contrast to this, a use of history as a goal does not serve the primary purpose of being an aid, but rather that of being an aim in itself. For instance, it is considered a goal to show students that mathematics exists and evolves in time and space, that it is a discipline which has undergone an evolution over millennia, that human beings have taken part in the evolution. The distinction between tool and goal is quite useful for becoming conceptually clear about the "whys" and "hows" of history of mathematics in teaching. However, as Jankvist himself remarked both dimensions are intertwined, and I would say inseparably intertwined. In the hermeneutic approach, mathematics enters at least in two ways. First, there is the experience of dissonance or alienation. Students learn something about their own mathematics by experiencing and *reflecting on the contrast between modern concepts and their historical counterparts*. And the point of the "hermeneutic circle" as understood here is that the reflection is in both directions, so that the students deepen both their understanding of history and of their own set of modern conceptualizations. Second, and equally important, is the fact that in reading a source (modern) mathematics itself is *applied as a tool*. The task to think oneself into the situation of persons living at a time long ago requires to be able to argue from the assumptions of these persons, to use their symbols and methods of calculation. This poses completely new demands on the students' abilities to argue and to prove mathematically. The teaching unit I have described in this paper showed clearly that it operated on the upper limit of the students. It was a real stress test to the mathematics they had learnt. Thus, reading a source deepens the mathematical understanding on both levels, on that of *doing mathematics* and on that of *reflecting about mathematics*.

Michael Fried nicely distinguishes between different attitudes (of mathematicians) in regard to the mathematics of the past (2011), that of "colleagues", "treasure-hunters", "conquerors", "privileged observers" and "historical historians of mathematics". The latter "view the past as fundamentally different from the present and see the treatment of the past demanding more than present mathematical knowledge". Of course, this tacitly says that modern mathematical knowledge *is* necessary for understanding the past.

How then can we describe this "more", this extra component which goes beyond modern mathematics? This brings us back to the theme of this conference "the search for a common ground between mathematics and mathematics education". I would like to describe this "more" by the term "respect". At first sight "respect" is a category of human relations. When we communicate honestly with other people mutual respect is a necessary condition. In the present context I use this concept with an additional connotation. This is an *epistemological* one and says that we have to accept that there are different legitimate perspectives on the history of mathematics and none of them has the right to claim exclusive truth for itself. According to hermeneutics understanding a text consists in the merging of different horizons, the horizon of the reader and the horizon of the text/author. Different readers with their different backgrounds arrive at different interpretations. Thus, a history of ideas (produced by a leading mathematician) which might neglect many historical details is as legitimate as a social or cultural history of mathematics or the history produced by our students. However, respect as an epistemological category is injured when somebody neglects the difference between his reconstruction of the past and the past itself and takes the mathematical tools he applies as the matter itself.

References

Arcavi, A., Bruckheimer, M., & Ben-Zvi, R. (1982). Maybe a mathematics teacher can profit from the study of the history of mathematics. *For the Learning of Mathematics*, *3*(1), 30–37.

Barbin, É. (1994). Préface. In Commission Inter-Irem Épistémologie et Histoire des Mathématiques (Ed.), *Quatrième université d'été d'histoire des mathématiques* (pp. ii–iii). Lille: IREM.

de l'Hospital, G. M. (1696). *Analyse des infiniment petits pour l'intelligence de lignes courbes*. Paris: Imprimérie Royale.

Eisenberg, T. (2008). Flaws and idiosyncrasies in mathematicians: food for the classroom? *The Montana Mathematics Enthusiast*, *5*, 3–14.

Fauvel, J., & van Maanen, J. (Eds.) (2000). *History in mathematics education: the ICMI study*. Dordrecht: Kluwer Academic.

Fried, M. N. (2011). *Postures towards mathematics of the past: mathematicians, mathematician-historians, historians of mathematics*. Talk given at the Cohn Institute for the history and philosophy of science and ideas, Tel Aviv University, Tel Aviv, 17 January 2011.

Furinghetti, F., Jahnke, H. N., & van Maanen, J. (2006). *Studying original sources in mathematics education*. Report 22/2006, pp. 1313–1315, Mathematisches Forschungsinstitut Oberwolfach.

Glaubitz, M. R. (2010). The use of original sources in the classroom. Empirical research findings. In E. Barbin, M. Kronfellner, & C. Tzanakis (Eds.), *History and epistemology in mathematics education. Proceedings of the sixth European summer university ESU 6 (section 3.01)*. Vienna: Holzhausen.

Glaubitz, M. R. (2011). *Mathematikgeschichte lesen und verstehen. Eine theoretische und empirische Vergleichsstudie*. Dissertation, Universität Duisburg-Essen. http://duepublico.uni-duisburg-essen.de/servlets/DocumentServlet?id=25416. Accessed 31 May 2013.

Jahnke, H. N. (1994). The historical dimension of mathematical understanding—objectifying the subjective. In *Proceedings of the eighteenth international conference for the psychology of mathematics education*. University of Lisbon, Lisbon (Vol. I, pp. 139–156).

Jahnke, H. N. (1995). Historische Reflexion im Unterricht. Das erste Lehrbuch der Differentialrechnung (Bernoulli 1692) in einer elften Klasse. *Mathematica Didactica, 18*(2), 30–58.
Jahnke, H. N. (1996). Mathematikgeschichte für Lehrer – Gründe und Beispiele. *Mathematische Semesterberichte, 43*(1), 21–46.
Jankvist, U. Th. (2009). A categorization of the "whys" and "hows" of using history in mathematics education. *Educational Studies in Mathematics, 71*, 235–261.
Laubenbacher, R. C., & Pengelley, D. (1999). *Mathematical expeditions*. New York: Springer.
Schafheitlin, P. (1924). *Ostwalds Klassiker der exakten Wissenschaften: Vol. 211. Die Differentialrechnung von Johann Bernoulli aus dem Jahre 1691/92. Nach der in der Basler Universitätsbibliothek befindlichen Handschrift übersetzt, mit einem Vorwort und Anmerkungen versehen.* Leipzig: Akad. Verlagsgesellschaft.
Toeplitz, O. (1927). Das Problem der Universitätsvorlesungen über Infinitesimalrechnung und ihrer Abgrenzung gegenüber der Infinitesimalrechnung an den höheren Schulen. *Jahresbericht der Deutschen Mathematiker-Vereinigung, 36*, 88–100.
Tzanakis, C., & Arcavi, A. (2000). Integrating history of mathematics in the classroom: an analytic survey. In J. Fauvel & J. van Maanen (Eds.), *History in mathematics education: the ICMI study* (pp. 201–240). Dordrecht: Kluwer Academic.
van Maanen, J. A. (1997). New maths may profit from old methods. *For the Learning of Mathematics, 17*(2), 39–46.

Chapter 7
Reflections on History of Mathematics

History of Mathematics and Mathematics Education

Luis Radford

Abstract The specialization that mathematics education research has undergone in the past decades has led to a sense of division and disconnection between mathematicians and mathematics education researchers. This chapter deals with the possibilities that the history of mathematics may afford to reduce the divide. Although the recourse to the history of mathematics is an interesting prospect, it unavoidably induces new problems. A range of tensions becomes visible among the involved communities of teachers, historians of mathematics, mathematics education researchers, and mathematicians. Some of these tensions are investigated in this chapter, in particular in the case of a hermeneutic reading of original sources. The tensions that the history of mathematics induces, it is argued, may function as a way to foster a critical reflection and dialogue to contribute to a rich multi-layered understanding of mathematics, its history, and its teaching and learning.

Keywords History of mathematics · Hermeneutic approach · Teachers' beliefs about mathematics · Aesthetic in mathematical thinking · Mathematics and culture · Nature of mathematics

Introduction

This chapter presents a reflection on some of the contributions of the history of mathematics to mathematics education. It also explores the manner in which the history of mathematics may serve as a bridge across the intensifying divide between mathematics and mathematics education research. While the first aforementioned point has been a matter of extensive discussion (see, e.g., Barbin et al. 2008), the

With contributions by:
Alain Bernard, Centre Alexandre Koyré, Paris, France
Michael N. Fried, Ben Gurion University of the Negev, Beer Sheva, Israel
Fulvia Furinghetti, Universita di Genova, Genoa, Italy
Nathalie Sinclair, Simon Fraser University, Burnaby, Canada

L. Radford (✉)
École des sciences de l'éducation, Université Laurentienne, Sudbury, Ontario P3E 2C6, Canada
e-mail: lradford@laurentian.ca

M.N. Fried, T. Dreyfus (eds.), *Mathematics & Mathematics Education: Searching for Common Ground*, Advances in Mathematics Education,
DOI 10.1007/978-94-007-7473-5_7,

second point results from the increasing specialization of mathematics education as a research discipline. It is clear indeed that in the early 20th century, research in mathematics education revolved around curricular problems and international cooperation, as the epoch-making articles published in *L'Enseignement Mathématique* in the first decades of the 20th century make clear (see, e.g., Borel 1914; Bourlet 1910). During the 1970s and 1980s, research in mathematics education moved to new arenas: problems of a psychological nature moved to center stage, with an interest in understanding the students' thinking, and, more recently, a shift has become clear with the political concerns of today (see, e.g., the Educational Studies in Mathematics Special Issue edited by Brown and Walshaw 2012), where a need to go beyond the definition of mathematics education as the diffusion of the mathematical content (see, e.g., Brousseau 1997) is questioned, if not contested. Often, these shifts have led to a sense of disconnection between the work of professional mathematicians and mathematics education research. In the latter, new terminologies, concepts, and methods have been introduced and developed, often with the recourse to theories in the social sciences, such as linguistics, semiotics, sociology, and anthropology. Sometimes these new trends, which are clearly embedded within social science research, appear as excessive and even unnecessary (Eisenberg and Fried 2009).

The chapter has its origin in a symposium organized in Beer-Sheva in 2012 to honour the seminal work of Ted Eisenberg. The symposium was an extraordinary opportunity to reflect on these matters and to try to come up with possible actions to overcome the separation that seems to affect the communities of mathematicians and mathematics education researchers. By focusing on the history of mathematics, the contributors to this chapter present some of the potential and challenges that such an endeavour entails. The following sections still retain the aural dimension of the presentations made during the Symposium. We decided to keep them this way for reasons that will become apparent later.

In the opening section, Alain Bernard reflects on the challenges that a hermeneutic approach may present to the teachers of mathematics. An interpretative turn to historical texts, Bernard argues, entails some skills and knowledge that teachers may be lacking in order to ensure suitable hermeneutic classroom discussions. Michael Fried stresses the tension that a historical attitude induces in mathematics, mathematics teaching, and mathematics education research. He suggests that the recourse to history in mathematics education may bring a philosophical view that may provide historians and mathematicians with an opportunity for even philosophical discussion and historical understanding. Fulvia Furinghetti argues that the history of mathematics offers a unique window through which teachers' beliefs about mathematics can be made explicit and turned into possibilities of conceptual growth and development. Nathalie Sinclair puts forward an interesting conception of mathematics as temporal and material activity embodying diverse modes of thought and forms of subjectivity. In addition to exploring the past through the written dimension of historical texts, a hermeneutic approach could also encompass mathematics as performance and unveil the richness of mathematical narrative styles. In my own section, I ask three questions and make a remark. They intersect with Sinclair's arguments. The questions are a rhetorical device used to invite mathematicians and mathematics educators to rethink what we mean by mathematics. The thrust of the questions is my concern with the fact that mathematics has unfortunately become a technical domain under the influence of contemporary neo-liberal forms of production and its emphasis on marketing and consumption as the modern and postmodern predominant forms of life.

History Within Math and Science Teaching: A Historical Issue

Alain Bernard

My reaction to Jahnke's insightful proposal for a *hermeneutic* approach of the history of mathematics in mathematics teaching is guided by a few basic claims that are indicated in my title. The first is that there are difficult questions underlying the introduction or promotion of history of mathematics into math teaching. The second is that we are now in a situation in which the history of mathematics has come to be considered almost as a *necessary* component of mathematics education in many (though not all) countries. Finally my central claim is that reflecting on the above questions should be deeply informed and guided by a *historical* reflection about our current situation.

I do think, first of all, that "mathematical education" should be understood just in the general way that was underlined by Presmeg: namely as *the activity of teaching mathematics* or even, as was suggested by Niss' categories, as *the organization of the human and institutional framework for it.* By contrast, both Presmeg and Eisenberg have rightly reminded us that math education *research* is a much more recent academic field that did not exist at the beginning of the 20th century or before. From a historical perspective, recognizing these facts is crucial, for it enables one to say, without anachronism, that several major (and also lesser known) mathematicians have been constantly and deeply involved in math education, that is, into the reform of math curricula and pedagogical methods *from the 19th century to the present day.* For it must be recalled that the *level of interest* taken in math education by mathematicians in particular or by political people in general *has neither really decreased nor changed.* As Niss shows very clearly, there are still active mathematicians, as well as many other people, that pay much interest to mathematics education and try to influence it in deep and significant ways. Following Jahnke's reaction to Niss' talk, one should add the timeline to his picture: in other words, what is true today has already been true for a long period.

There are, in turn, deep and long lasting historical reasons, for which many people have been concerned by (if not involved in) the *constant reform* of math education since the 19th century. My remarks will elaborate on one important point made by Movshovitz-Hadar about the *ever evolving perimeter of mathematics* and its necessary consequences on mathematics education. Indeed, a recurrent argument seems to evoke a golden age in which the very notion of what math education is about was clearly defined and not a subject of contention. But I fear the Golden Age was already an Iron age: the eve of the 20th century was already characterized by dramatic and significant changes in orientation and subject matter. Thus, introducing the concept of function as a fully legitimate subject for mathematics education was precisely the move promoted by mathematicians like Borel, Klein or Poincaré, who were deeply involved in the reform of curricula in their respective countries: before then, the concept of function was not a central subject in math education, but more a subject for advanced studies.

Many other examples could be given. But the question is: why would those people have felt the urge for change? Generally speaking, we have to remind ourselves that mathematics itself, throughout the 19th and 20th century, and especially after WWII, has constantly undergone drastic and profound changes—more than in every other period. These changes have been quite directly related to the changes in industrial societies: new technologies, new industries, new sciences. These rapid and radical changes have naturally led many people, first of all mathematicians, to consider the unavoidable fact that questions concerning mathematics and their education on a large scale could *not* be any more considered in isolation of many others. Those other questions, which are so constantly present in the background, such that we tend to forget them, touch on the development of industry, experimental science, economy or politics—especially educational policies. For example, much of the concern about mathematics education was fostered by the launch of Sputnik in 1957 and by the kind of *shock* it produced in western countries, and first of all in the USA, by that time. To my view, this remark can and should be extended to much of the history of the 19th and 20th century mathematics and mathematics education, with an acceleration after WWII.

I thus come to the last historical fact that I would like to connect to the previous ones. Ever since there has been a question of changing (mathematics) education to accommodate the 'new' industrial world and its needs, namely from the 19th century onwards, *it has also been a question of introducing history of science in general (history of mathematics in particular) into science curricula.* This was explicitly done, at least in France in which this contextual question has been much studied, in order to *compensate* the inevitable split induced by the twofold curricula, separating science from literature tracks—see, among others, Hulin's (2011) synthesis. Very early, during the industrial age, therefore, a teaching of history of science was welcomed and called for—at least theoretically.

The situation of today has made this traditional wish all the more central than it has largely entered official curricula: the teaching of science in general, and of mathematics in particular, is now meant to be deeply 'cultural': this means, in particular, that it should include more history, more epistemology and more facts concerning society at large. We should be strongly aware of the fact that this is *not* a recent move, but already the continuation of long-lasting concern; and that this concern, in turn, should not be dissociated from the history of modern science, industry and philosophy (especially positivism). We should finally be aware of the practical implication of these moves: namely that science and mathematics teachers are requested, more than ever, to deal with science and mathematics as cultural subjects, however this might be interpreted.

With these historical observations and preliminaries in mind, I think we can address Jahnke's proposals and questions. First of all, I can only agree with Jahnke's suggestion that readers of historical sources of mathematics should be placed in the situation of *interpreting* these sources, and that *reading* should be considered a hermeneutical task in a strong sense. Indeed, this proposal addresses somewhat in depth the real difficulties encountered by teachers when they try to address the long-lasting demand signaled above.[1] But such proposals should be understood against the historical background that I have summarized before. For, to put it simply, we retrieve the bizarre situation I have spoken about:

> *Most students of mathematics* who are now asked, as teachers or future teachers, to develop the kind of pedagogical activities that are indispensable from interpretation, are not ready for it. From my experience, many such teachers were often not prepared to organize a discussion on a complex argument, implying the elaboration of a coherent interpretation that might be expressed orally or by written means.[2]
>
> *Teachers of literature*, philosophy or history, by contrast, would consider natural that students should develop a complex point of view on a given document; this idea is more or less alien to many students in science.

The split is not only a problem of being trained or not in the kind of pedagogical activities that foster interpretation, but also, of course, a question of epistemology of what science and literature studies are about: there is a kind of invisible limit here that forbids such cooperation. We are collectively far away from the old idea that mathematics and science are a legitimate part of literature in a strong sense.

On the other hand, we are perhaps not so far away from these classical ideas. For these questions and implicit frontiers are now quickly changing: the new technologies and especially the rise of "digital humanities" are deeply changing the way we read and write. This does not mean that printed matter is outdated, but we must nevertheless count on the new media and on the way in which they radically transform the access to historical and cultural information, as well as the way students and teachers might cooperate with each other. This recent change should probably be part of the discussion.

---

[1]Other proposals, like Michael Fried's notion of a "radical accommodation" between history of mathematics and mathematics teaching seem to come close to it—although I understood from Michael himself that Jahnke's proposal is not yet radical enough to correspond to the aforesaid category.

[2]For example, I am ready to bet that the kind of "open discussion" that Jahnke organized with his students on Bernoullis' conception of infinitesimals quantities would be dreadful to many mathematics teachers: both the spirit and the concrete organization of such debates is in many cases alien to them.

Mathematicians, Historians of Mathematics, Mathematics Teachers, and Mathematics Education Researchers: The Tense but Ineluctable Relations of Four Communities

Michael N. Fried

Rather than asking whether history of mathematics is good or bad for students or for teachers of mathematics—and I *do* think it is good!—I would like to focus on how our presuppositions about mathematics and about the history of mathematics play out in the relations among four communities, that is, to the extent each is concerned with history of mathematics in mathematics education. I have in mind mathematicians, historians of mathematics, mathematics teachers, and mathematics education researchers. These different communities cannot be assumed to speak in one voice. I would like to suggest that mathematics education as a whole must somehow situate itself within a web of tensions created by the interests and commitments of these four communities. Furthermore, I would suggest that mathematics education research, though it itself is a pole within this web, has a distinct role of creating a view of mathematics education in which these tensions can be productive for mathematics learning, not paralyzing.

To begin, I should remark that the relationship between the *history of mathematics* and mathematics education mirrors the overall problem of this book namely, *mathematics* and mathematics education. And continuing down this hall of mirrors, the relationship between mathematics and history of mathematics themselves reflects the problem as well. As with mathematics educators, historians of mathematics until the middle of the twentieth century were invariably well-trained mathematicians, most often working mathematicians. The awakening of history of mathematics as an independent historical discipline created a certain amount of tension within the mathematics community just as the crystallization of mathematics education as a separate academic discipline has. When, for example, Unguru wrote that the history of Greek mathematics needed to be revised (Unguru 1975) so that it would be based on sound history rather than sound (modern) mathematics, he was attacked by mathematicians as not understanding mathematical thinking, as he was attacking them for not understanding historiography (e.g. van der Waerden 1976).

Taking into account history of mathematics in mathematics education adds another level of complexity, however (it was this that was analyzed in Fried 2001, 2007). For history of mathematics as *history* tries to see the how mathematics of the past was *different* from the mathematics of today. With that, it treats mathematics as a product of culture and, as Judith Grabiner (1974) has put it, as "time dependent." This means one cannot assume a modern mathematician should be the final arbiter in judging what was said or done in the mathematical past. To the extent then that mathematics educators turn to a cultural view of mathematics, they align themselves with the history community. On the other hand, because working mathematics teachers must teach mathematics with an eye to its application in the sciences or its investigation in mathematics itself, it must give mathematics an unconditional objectivity that aligns them with the mathematics community or, more generally,

Fig. 7.1 A three-way tension between mathematics, mathematics teaching, and the history of mathematics

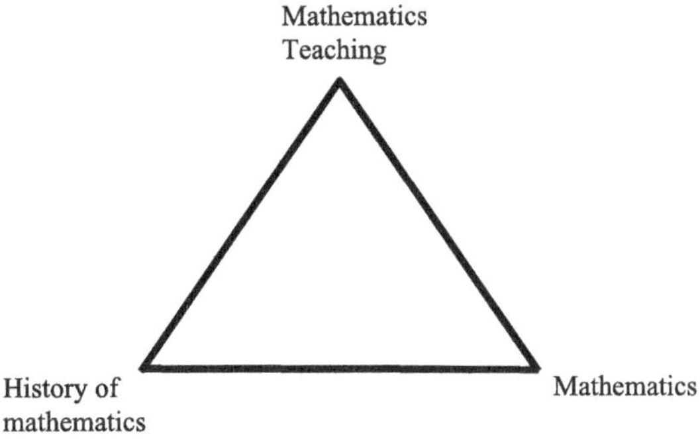

with those who see mathematics of the past as fundamentally the same as the mathematics of today. Pulled in this latter direction, educators are hardly obliged to take history into account in their teaching; that is, history can be added or subtracted at pleasure. I would point out that even Freudenthal, who supported the inclusion of history in mathematics education vigorously, still did not think history could help one understand better the subject matter of mathematics (Freudenthal 1981).

So there arises, to start, a kind of three-way tension (Fig. 7.1):

(1) Between mathematicians for whom mathematics exists outside of time waiting to be discovered, and historians of mathematics who see mathematics as a cultural product which by definition develops in time;
(2) Between mathematicians concerned with mathematical content and ideas, and mathematics educators concerned with the development of students' mathematical thinking and their situating their mathematical understanding within their culture and everyday experience;
(3) Between mathematical educators concerned with preparing students for work in science and engineering and the concomitant need to teach them modern mathematical procedures and concepts, and historians of mathematics who keep the present at a distance in order to understand the past and who we are as beings possessing that past.

But despite the symmetry of the triangle, there is in fact a clear asymmetry between the mathematicians and historians of mathematics on the one hand and the mathematics educators on the other. For while mathematicians and historians of mathematics (taken as types, of course, not individuals) are fairly consistent in their respective positions, mathematics educators are somewhat chameleon-like in theirs. This, however, is the nature of the subject: mathematics education has a diverse set of ends that are themselves not completely consistent. It is for that reason that mathematics education can by itself mirror the tension between mathematics as a discipline and the history of mathematics.

The fourth pole is mathematics education research, which Norma Presmeg was careful to distinguish from mathematics teaching in Chap. 4. And with this fourth pole, we have, in fact, a tetrahedral web of relations. This is represented in the tetrahedron below (Fig. 7.2), where each face corresponds to a different simplicial set

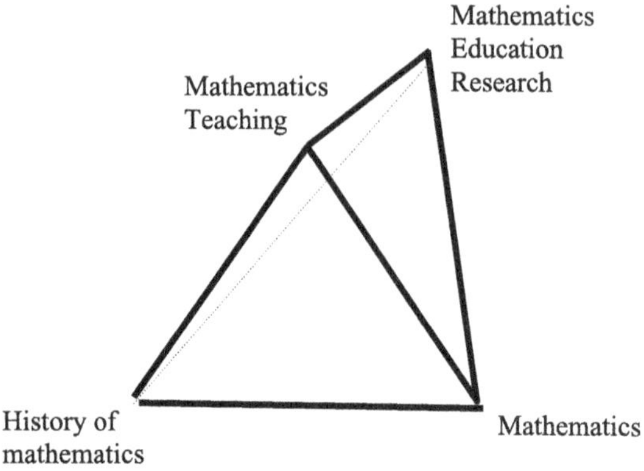

Fig. 7.2 A four-way tension between mathematics, mathematics teaching, mathematics education research, and the history of mathematics

of tensions (of course the face whose vertices are mathematics education research, mathematics, and mathematics teaching, respectively, is that which is the main focus of this book).

The relationship of mathematics education research to history of mathematics and mathematics is in many ways similar to that of the mathematics teacher: the disciplines of mathematics and history of mathematics make conflicting demands as to what the teacher is supposed to teach, as well as to what the researcher is supposed to investigate. Does mathematics education research waste time if it does not try to find ways of teaching functions or how to solve a system of linear equations? Does it neglect the "true hard-core" of the subject, if it looks at mathematics as a semiotic-cultural system?

However, mathematics education research, as an academic community, has considerably more freedom to define itself and shape a view of mathematics than does the community of mathematics teachers. Mathematics education research possesses, accordingly, a greater potential to bridge the divide between mathematics and history of mathematics by defining a view of mathematics education which can accommodate both. As I emphasized in Chap. 2, this means defining what exactly it is to be mathematically educated. Mathematics education research may take on an important function in enriching mathematics itself in this regard, while also defining a role for mathematics teachers in such a way that situates them neatly between mathematics and history of mathematics. A tetrahedron such as that in Fig. 7.3 might, therefore, better show the relation between mathematics education research and the communities of teachers, mathematicians, and historians, namely, that of a kind of orchestrator poised above the plane containing mathematics teaching, history of mathematics, and disciplinary mathematics.

In accepting this bridging role, both with regards to mathematics and history of mathematics and between mathematics teaching and both of these disciplines, mathematics education research must contend with questions both of a practical and theoretical nature.

On the theoretical side, we ought to ask questions such as the following: Does the introduction of history of mathematics into mathematics education, where this is not a trivial introduction for the sake of motivation or "spicing up" mathematics lessons,

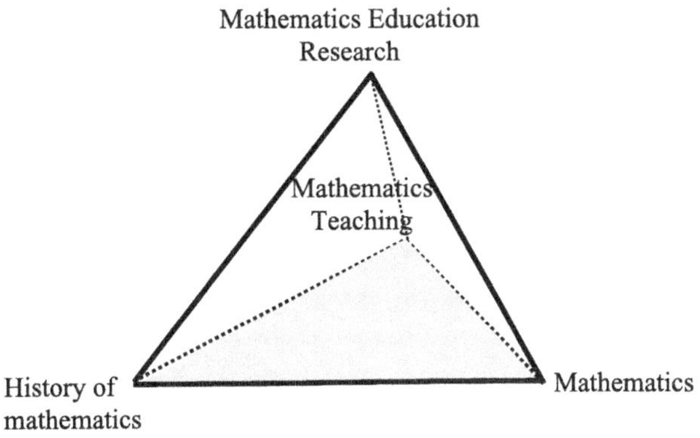

Fig. 7.3 The relation between mathematics education research and the communities of teachers, mathematicians, and historians

imply a philosophical position with respect to mathematics? Does this position distance mathematicians from mathematics education, for example, by weakening the claim that modern mathematical understanding is the key to historical understanding? Can mathematics education act as a context for a philosophical discussion of mathematics where mathematicians and historians can have an equal say (a kind of amplification of the classroom setting Lakatos chose for his historico-philosophical discussion of proof)?

On the practical side, I would suggest that bringing history of mathematics into mathematics education in such a way that it is both mathematics and also truly history of mathematics consists, first of all, in the study of original texts—really looking at Euclid's *Elements* or working through Descartes' *Geometrie*. How to introduce original texts, however, is not obvious. In its most uncompromising form, learning by way of original texts falls under the category of what I called in the past (Fried 2001), "radical accommodation" (as opposed to the other radical alternative, "radical separation"). It is radical because the texts become *primary in every sense of the word*: the study of mathematics in "radical accommodation" becomes precisely the study of mathematical texts, just as literature is the study of great works of prose and poetry. I had claimed that this would indeed also be mathematics:

> ...in the study of mathematical texts, one is not only engaged in solving problems and developing ideas with a great mathematician, and therefore becoming deeply acquainted with the human activity of mathematical work, but one is also engaged in a kind of reflective thinking or inquiry that ultimately is of the highest importance for one who deals with *technical* scientific and mathematical work. (Fried 2001, p. 402)

But original texts can be introduced without making them the exclusive source of learning. Indeed, one can gain much historical understanding by bringing mathematics classroom learning to original texts *if* one is made ever cognizant that one is indeed an interpreter with a point of view. This is the core of what Jahnke calls the "hermeneutic" approach (e.g. Jahnke 2000). In taking up a theme such as the hermeneutic approach, mathematics education research provides an example of how it can investigate a practical approach to learning that defines a set of mutual relationships between students and teachers of mathematics, modern mathematical knowledge, and historical understanding.

History in the Mathematics Classroom

Fulvia Furinghetti

One of the struggles inside the community of HPM (International Study Group on the Relations between History and Pedagogy of Mathematics, affiliated to ICMI) is to make clear the relation between history and pedagogy of mathematics. Some papers written by Ted Eisenberg provide hints for pointing out the links between these two domains. In the following I will refer in particular to teachers' beliefs.

Let me start with the claim by Eisenberg (1977) that there is close to a zero correlation between teacher knowledge and student achievement, and that other factors appear to be responsible for student achievement. Among the factors responsible of this failure in teaching I consider important teachers' beliefs about mathematics and its teaching. These beliefs, for example, make teachers neglect what they learnt at university and reproduce in their classroom what they have been taught in secondary school. I am not the first to take this point of view. At the beginning of the twentieth century two important mathematicians engaged in mathematics education and focused on the problem of teachers' knowledge by pointing out this fact (Borel 1907; Klein 1924).

Then, a main aim of teacher educators is to challenge prospective teachers' beliefs. As I have discussed in some papers, history may be a good tool for attaining this aim, since it provides an unknown landscape where people are obliged to look at things from a different perspective and to grasp aspects that previously escaped their attention (Furinghetti 2007). As an example, I mention the case of algebra. Usually teachers tend to consider algebra as an extension of arithmetic, generalization, abstraction, and use of symbolization. I challenge this view by asking them to solve medieval problems such as the following Problem 47 taken from *Trattato d'Aritmetica* by Paolo Dell'Abbaco:

> A gentleman asked his servant to bring him seven apples from the garden. He said: "You will meet three doorkeepers and each of them will ask you for half of all apples plus two taken from the remaining apples." How many apples must the servant pick if he wishes to have seven apples left?

In solving this kind of problem two paths may be followed, which may be put in relation with the analytic and synthetic methods:

- arithmetic path: from the known (left apples) to the unknown (apples to be picked)
- algebraic path: from the unknown (apples to be picked) to the known (left apples)

Reflecting on these paths leads the teachers to focus on the fact that algebra is not only generalization, not only abstraction, not only using symbols, not only an extension of arithmetic: algebra is a method and the analytic method is its core. Then François Viète's introduction of parameters and variables is not perceived as something coming out of the blue, but as a consequence of this way of looking at algebra.

A further belief challenged by history is the view of the role of intuition and rigor in mathematics teaching and learning (Dreyfus and Eisenberg 1982). Mathematicians such as Poincaré (1899) and Klein (1896) considered the history of mathematics a suitable context for bringing intuition back into the teaching process against the excesses of rigor advocated by some of their contemporary colleagues.

The main idea expressed by Klein is that students need to approach a topic at an "intuitive" level and later on to pass to the formal level. History may be useful in this regard because it brings back the polished concepts as are presented in the modern textbooks to their origin. History recovers the cognitive roots, described by Tall (2003) as concepts which are (potentially) meaningful to the student at the time, yet contain the seeds of cognitive expansion to formal definitions and later theoretical development. The historian Gino Loria, who was a convinced supporter of the use of history in mathematics teaching, epitomized this idea about cognitive roots by using a sentence found in Victor Hugo's novel *Les travailleurs de la mer*, which says that any embryo of sciences presents this double aspect: monster as a fetus; marvel as a germ (Loria 1914, p. vii).

The focus on formal approaches has the consequence that a vast majority of students do not like thinking in terms of pictures (Eisenberg and Dreyfus 1991). This way of thinking may be promoted by history, since the early stages in the development of concepts often reside in the visual domain. This aspect has been exploited in significant experiments concerning the teaching of calculus.

Another challenge to teachers' beliefs is to make them shift the focus of their teaching from product to process. This shift may be fostered by history, since reading original sources directs the attention to processes, which leads to the genesis of concepts. The engagement promoted by the contact with an author's thinking obtained by means of historical passages is an aesthetic value introduced into mathematics teaching. In the words of Hawkins (as cited in Featherstone and Featherstone 2002), aesthetics "is a mode of behavior in which the distinction between ends and means collapses; it is its own end and it is its own reinforcement" (p. 25). Then history of mathematics may be a means for introducing a form of aesthetics into the mathematical discourse in the classroom, as advocated by Dreyfus and Eisenberg (1986).

The few aspects outlined above suggest ways in which the history of mathematics challenges teachers' imagination in finding new modes of dealing with mathematical discourse in the classroom.

Aesthetic Considerations

Nathalie Sinclair

This chapter offers some reactions to Jahnke's chapter, which outlines the possibility of using historical texts in the mathematics classroom to enable students to (1) develop insights into the development of mathematics, (2) develop an understanding of the role of mathematics in our society and (3) encourage the perception of the subjective dimension of mathematics.

I would like to contribute to this discussion by way of Eisenberg's 1986 paper (with Dreyfus) on the role of the aesthetic in mathematical thinking, which has had a strong influence on my own work in mathematics education. They suggested that explicit attention to aesthetics in the mathematics classroom could help improve students' problem-solving abilities. One challenge that mathematics educators must reckon with is not only finding ways of welcoming—and even eliciting—aesthetic values in the classroom, but also accepting that the values students have in the classroom, with respect to mathematics, do not always align with those of mathematicians (see Sinclair 2001).

Of course, judgments of aesthetic values are not only subjective, but also strongly influenced by socio-historical factors. Yes, it is true that adjectives such as symmetry, order and precision re-occur across different historical time periods and in diverse cultures. But these are very broad descriptors that do not just operate in mathematics. Moreover, there are many examples in the history of mathematics where asymmetry, chaos and fuzziness also vied as aesthetic values.

So, one way of thinking about the historical project proposed by Jahnke is to focus not only on the changes in mathematical content, but also changes in the aesthetic values that constitute the discipline. These values become evident when one looks, for example, at the dominant activities of the day—perhaps focusing on problems related to the foundations of mathematics or on solving specific open problems. They also become evident when considering the techniques and strategies that are used to solve these problems—be they algebraic means or experimental ones. Finally, they become evident when questioning the reasons for focusing on certain problems or techniques over others—because they are more beautiful, right, useful, ideal or true. In other words, aesthetic considerations of the historical variety would concern *what* was attended to, *how* it was attended to and *why* it was attended to. These questions could easily be raised in the context of historical activity in the classroom and the answers, I think, would certainly support the threefold aims of Jahnke's proposal.

From an aesthetic point of view, much can be learned about mathematics and the people doing mathematics by reading historical texts. One can focus, for example, on the styles of writing that are used. Netz (2009) has argued that the Archimedean so-called "ludic" style of writing, which he characterises as involving narrative surprise, mosaic structure and generic experiment, and a certain "carnivalesque" atmosphere, evokes very different aesthetic qualities than, for example, the Euclidean style. Archimedes writes mathematics to delight and inspire; Euclid does so to organise and convince through strictly logical means. Clearly, these considerations are strongly related to the role of mathematics in society (to entertain? To be useful? To be pure?). It also suggests that the issue of subjectivity is not just of epistemological interest, but also of aesthetic order.

I have been arguing that the study of historical texts can (and should) have a strong aesthetic component to it. But now I'd like to raise questions about the limitations of this approach, still within the context of aesthetics. I will discuss three concerns:

The Historical Text Can Reveal Only the "Body" of Mathematics The first relates to the historian of mathematics Corry's (2006) distinction between the *body* and the *image* of mathematics. In contrast to the body of mathematics, which includes "questions directly related to the subject matter of any given mathematical discipline: theorems, proofs, techniques, open problem," the images of mathematics "refer to, and help elucidating, questions arising from the body of knowledge but which in general are not part of, and cannot be settled within, the body of knowledge itself".

Thus, while the body of mathematics might concern itself with describing a technique used in the course of a proof, the images of mathematics refer to the motivations, choices and values related to the use of certain techniques. While the body of mathematics concerns itself with defining objects, the image of mathematics questions which objects are defined and which are not. As Corry points out, mathematicians do not customarily write about their images. But images of mathematics constitute a layer of mathematical knowledge, one that centrally involves aesthetic concerns—and one that will not easily be revealed in a study of historical texts. Indeed, Netz's study of Archimedes' style required the use of sophisticated and specialised analytic tools from disciplines such as archaeology and cognitive linguistics. I would argue that it is just as much, if not more, in the changing images of mathematics, that we can learn about the development of mathematics, the role of mathematics in our society and the subjective dimension of mathematics.

From Written Text to Performance A second point I would like to raise relates to the idea of mathematical writing style that I have already mentioned and the question of what kinds of historical texts we might choose for students to read. Consider some of the linguistic features of modern mathematical writing that attempt to render utterly transparent the 'logical structure' of the text. These include the prevalence of non-active verb forms, the lack of direct address and the frequent use of imperatives. One can also read a more covert agenda aimed at creating the very sense of decontextualised authority and certainty that is then claimed as the hallmark of mathematics (Pimm and Sinclair 2009).

In any case, Solomon and O'Neill (1998) have usefully identified two contrasting styles of writing using a variety of texts authored by the 19th-century mathematician William Rowan Hamilton (the narrative and the paradigmatic). They argued that the main difference between these contrasting styles lies precisely in this 'glue' of logical versus chronological structuring (and their surface manifestations in terms of verb tense, personal pronoun use, connectives between sentences and other lexical choices). Interestingly, in Hamilton's range of mathematical writing, the syntactic glue changes depending on whether he was writing diary notes to himself, letters to friends or journal articles or monographs (ostensibly addressed to his colleagues). Of course, in all the writing, quaternions are the central topic. But when choosing a text for students to read, we might ask which piece of writing would work better in the classroom? Might the diary notes provide greater scope for writing about images of mathematics? Might studying the transition from the diary notes to the journal article help make explicit the ways in which professional mathematical communication seeks to immanent, immaterial truth and obscures personal motivations, feelings and doubts?

Because published texts tend to be the endpoint of mathematical investigation, both of the problem-solving process and of the writing process, these texts give a limited sense of mathematical activity. They fail to convey the narrative modes of thought that characterize discovery and, hence, run the risk of distorting the development of mathematics and even maintaining its objectivity. At issue, I think, at least in part, is the technology of the written word. As Brian Rotman (2008) has argued, the sequential logic of the printed (and copied) word in and of itself, independent of style, has a character of immanence and immutability.

It is interesting to imagine alternatives to the text in the historical project that Jahnke proposes. Consider, for example, being able to study the live performance of Archimedes, drawing geometric figures in the sand, going back and forth from diagram to symbol to gesture to spoken word. Or a YouTube video of a Terrence Tao lecture. The perception of subjectivity would be inescapable. Mathematics would be a temporal, material activity. In our digital era, not only is performance gaining ground over textual forms of communication, but the ability to manipulate time (reverse it, repeat it, fast forward it) will change the mathematical discourse. I wonder how classroom activities centered on historical texts will have the effect of celebrating a static, alphabetic way of mathematical communication.

Whither Subjectivity, Agency and Materiality? My third point relates to this discussion of the authority of the written word. It stems from the ideas of the historical and philosopher of mathematics Gilles Châtelet (2000) whose interest laid in the subjectivity, materiality and embodiment of mathematics. He studied several inventive instances in the history of mathematics, such as Hamilton's quaternions, Grassman's theory of the extension and Cauchy's residue theorem. But he studied these examples by analysing the diagrams that these mathematicians used to create new objects and relationships. For Châtelet, diagrams transduce the mobility of the body; they are "concerned with experience and reveal themselves capable of appropriating and conveying 'all this talking with the hand'."

And thus, his analysis of these historical episodes is an analysis of the these two, intertwined pivotal sources of mathematical meaning, mutually presupposing each other, and sharing a similar mobility and potentiality. Diagramming and gesturing are embodied acts that constitute new relationships between the person doing the mathematics and the material world. For Châtelet, the study of mathematical texts is not just an epistemological undertaking but an ontological one—the points and lines in the diagram do not represent ways of thinking about mathematical objects or spaces; rather, they *are* those objects and spaces; they can move, extend, cut, meet.

Although his study is an historical one, Châtelet's aims are philosophical. They challenge received notions of mathematics, insisting on its materiality, seeking to close the gap Aristotle erected between the abstract immobile mathematics and the concrete, mobile physical. But in terms of this book section, and of Jahnke's proposal, many questions come to mind: At the most general level, and similar to my previous point, does the study of text run the risk of ignoring an important part of the development and subjectivity of mathematics? At a more specific level, might the study of texts also include a Châtelet-like study of diagrams, not so much for its philosophical implications, but as a way to excavate the embodied meanings that created the objects and relationships under study? Lastly, is there room not only for Jahnke's epistemological laboratory, but also an ontological one in the mathematics classroom?

Three Provocative Questions and One Remark

Luis Radford

I start with a general observation. When we try to convince people of the benefits of history in mathematics education we resort to several possibilities—for instance, that the history of mathematics may help our students to attain a better understanding of the mathematics that they are learning today or to make the students sensitive to the fact that mathematics is a cultural construction.

Although laudable, our reasons tend to leave some views unquestioned. We tend to talk as if there were *one* mathematics, *one* history, and *one* history of mathematics. Perhaps we should start by asking ourselves what we mean by mathematics; only then might we be able to deal with the question of its possible histories.

Rationalist epistemologies present us with a view according to which mathematics is a body of objective knowledge that predicates truths that were already true even before they were discovered. If this is so, what then is the role of culture in the construction of mathematics? Culture, it turns out in rationalist accounts, is something that can only constrain or accelerate the rhythm of mathematics evolution but can in no way modify its natural course.

Yet, studies such as those of Emmanuel Lizcano (2009) bring to light the fact that mathematics is immersed in cultural and historical symbolic systems on which it draws its basic concepts, like those of number and figure. These cultural symbolic systems function as a semiotic superstructure that endows with meaning mathematical ideas and activities. In the case of ancient Greece, the whole mathematical edifice was governed by epistemic and ontological beliefs organized around the distinction between Being and Non-Being, and the logical principle of the Excluded Third. In the case of ancient Chinese mathematics, by contrast, mathematical thinking was organized around the *yin-yang* opposition. Since within the oppositional context of the yin-yang ontology each number has to have an opposed counterpart, what we now call negative numbers were "natural" to Chinese mathematics and remained unthinkable to the Greek *episteme*. To come up with the idea of negative numbers in Western culture, it was necessary to wait for the creation of new forms of labour and production and in fact to invent capitalism and its mercantilist practices of debt. (This does not mean that debts did not exist before. They did, but not in the typically surplus capitalist sense.)[3]

This short example opens up a possibility to try to envision, in new non-rationalist terms, the question of the nature of mathematics and its relationship to culture (Radford 2008). Of course, this example is not an isolated one. Current research in ethnomathematics offers a multitude of examples of mathematics that are quite different from the one we grew up into—many of them practiced orally only, as the Pythagorean brotherhood did in its own time.

The Provocative Question Is: Should We Be Concerned with Those Mathematics? I am not referring only to mathematics in other cultures that have made substantial contributions to our mathematics (e.g., Arabic mathematics and its impressive development of algebra). What I have in mind is the mathematics of cultural formations such as the one of the Lobodan people of the Normanby Island in Papua New Guinea. Lobodan mathematics is very distinctive in that it remains a-numerical. Lobodan people think relationally in ways that are different from ours. Drawing on an epistemology that is different from mainstream Western epistemology, Lobodan people do not quantify as we do: they compare in contextual *ad hoc* ways (Radford 2008).

Art scholars seem to be more prone to navigate between cultural forms of art than mathematicians are to navigate between radically distinct cultural forms of mathematical thinking. The goodness of concerning oneself with other mathematics—mathematics of other ethnical formations, present and past—bears on the question—I think I can already hear it—of *why*? Why should we be concerned with the mathematics of other cultural formations?

From a utilitarian viewpoint perhaps there is no reason. The mathematics that has been developed in the West is precisely the one that responds the best to the

[3]One of the oldest examples of negative numbers in the Renaissance appears in Nicolas Chuquet's Triparty en la Science des Nombres (Marre (ed.) 1880). Chuquet tackles a problem dealing with a merchant who has bought two kinds of cloths of different price. The total amount of pieces of cloth and the total amount of money are known. Solving what we would now call a system of linear equations, Chuquet finds out that the amount of pieces of cloth of the first kind is 15; he infers that the amount of pieces of cloth of the second kind is equal to 15 minus $17\frac{1}{2}$ and concludes that the problem is impossible, unless one interprets the difference $(-2\frac{1}{2})$ as a debt: the merchant bought $2\frac{1}{2}$ of cloth on credit! ("creance"; see Spiesser 2006, p. 19). In the following centuries, when algebraists like Bombelli and others do calculations on negative numbers, they are drawing on a conceptualization that has its roots in a commercial practice that has offered the possibility to think of negative numbers in a specific way—a social practice that has provided algebraists like Bombelli and Cardano with the conceptual ground to carry out a formidable cultural abstraction.

needs of progress as it came to be understood in the early 19th century—progress in a technological sense, where mathematics became the right hand of massive industrialization. But this is precisely my point. We need to rethink the *nature of mathematics*. Is mathematics technical stuff only? I think that most mathematicians would agree that the answer is no. In fact, there is a long list of philosophers (among them Hegel and Heidegger) who perceived the danger of reducing mathematics to its technical aspect, to a science of computation, to a kind of sophisticated technology. This conception of mathematics, the philosophers argued, eradicates the individuals from the discipline. Their concern was the depreciation of the subject of historical-cultural action. The student of mathematics is put on one side; mathematical knowledge is put on the other. Their contact is in the technological point. No wonder that mathematics is often found to be an unappealing subject by so many.

This discussion brings me to my second two-fold question: If mathematics is not technical stuff only, what else is it? And how can we take into account this neglected albeit important dimension of mathematics in school mathematics?

We need to rethink the nature of mathematics in general and the nature of school mathematics in particular. The technification of modern societies from the 19th century on led to a technification of mathematics. The justification of mathematics shifted from a discipline dealing with truth to one dealing with the efficient mastering of nature and the search for an optimal mechanism of production (Radford 2004). In the course of this process truth became obsolete. Euclidean geometry, to give but one example, has now disappeared from many school curricula. And if some vestiges can still be noticed, they are the remnants of the past. In Ontario, where I come from, what remains of geometry is what is susceptible to be translated into calculations. Our students do not prove theorems. They calculate. We have analytic geometry now.

My aim here is not to plead for a return to Euclid—at least not with the idea of resurrecting the splendors of truth as the Greeks conceived it. I take an incommensurate pleasure in going back to Euclid's *Elements* not to find truth there, but to see how the Greeks conceived of it, much as I come back to Piero della Francesca's paintings to see how the Renaissance conceived of the transcendental realm and pictured the world. I think that we have come to understand that truth is no longer the adequacy of our representations with the objects they represent. Truth is not of the order of adequacy. Truth, as Cornelius Castoriadis argued, "is the constant effort of dismantling the fence in which we find ourselves and to think otherwise, and to think no longer quantitatively, but deeper, better" (Castoriadis 1999, p. 54). Truth would rather be an attitude, what the Greeks called an *ethos*.

Along these lines, let me suggest that maybe we can think of mathematics as a historically constituted social practice, a cultural form of reflection and action, much like music, poetry, or painting, something practiced not in a vacuum but *with* others and *for* others. Mathematics would be hence not something to acquire (as if mathematics were merchandise) but a practice in which we come to insert ourselves, where we step into the public space. It would be what Arendt (1958), following the Greeks, called the *polis*—a place where we come to hear others' voices and perspectives and to speak out. There may be some hope then that in doing so, our students will no longer find themselves in front of an impenetrable alienating discourse, but rather will grow up as subjects of mathematics, as critical cultural subjects.

Let me now move to the question of history and start with the following remark. Something that distinguishes the human species from other species is our historical nature. Indeed, while rats are still doing what they were doing five hundred years ago, individuals are not. We draw on what previous generations have accomplished.

This is why history cannot be merely a tool to make mathematics accessible to our students. History is a necessity. As Russian philosopher Eval Ilyenkov put the matter, history is a necessity because "A concrete understanding of reality cannot be attained without a historical approach to it." (Ilyenkov 1982, p. 212)

Reality, indeed, is not something that you can grasp by mere observation. Neither can it be grasped by the applications of concepts, regardless of how subtle your conceptual tools are. The current configurations of reality are tied, in a kind of continuous organic system, to those historic-conceptual strata that have made reality what it is. Reality is not a thing. It is a process which, without being perceived, discreetly goes back, every moment, to the thoughts and ideas of previous generations. History is embedded in reality and reality in history.

To confine history to a tool for cognitive improvement is certainly a good idea. But would we not be missing the most important point? This is my third provocative question.

History is something that can make us aware of who we are, and how we have come to be the individuals that we are.

Yet, as Brown (2011) reminds us, history is a problematic concept. Indeed, we can ask ourselves: Whose history? Told by whom? History is not something out there. History is not and cannot be an objective, neutral account of events. History is our spatial and temporal situated understanding of something that tells not only the story of some events but also our own story. One of the challenges that we have to face when using history... I am sorry, I do not like the utilitarian expression "using history"... Let me restart my phrase... One of the challenges that we have to face when resorting to the histories (as a plural noun) of mathematics (as a plural noun), is that we have to go beyond the rationalist, regulative view of history that sees it unfolding as naturally as the movement of a pendulum. There is no history, but histories. And histories are political in the sense that we cannot focus on everything and that our histories leave in the margins events, voices, and presences. Bringing in political histories of mathematics in teaching and learning may help us and our students understand that mathematics can only make sense within the context of a history of its own culture; it can help us see how mathematics operates within the centrifugal forces of society, how it accomplishes inclusion and exclusion, how it offers cognitive templates of development, and how it helps to shape the selves into which we evolve in our lives. This is why histories are not narratives of the past. Histories configure our present and make it possible to envision a future that, ironically, is already historical, even if it is unpredictable.

Reflective Summary

The various sections of this chapter point towards challenges and possibilities in resorting to the history of mathematics in mathematics education and mathematics education research. The possibilities are certainly substantial, but so are the challenges. The history of mathematics cannot be simply imported into the classroom, nor can it be used as a transparent mediating term between the poles of mathematics and mathematics education research. To be a meaningful mediator, the history of mathematics needs to appear as a problematic field—one where one can interrogate notions and ideas that we usually leave unthematized, such as mathematics, its development, and its relationship to culture. To be a meaningful mediator, history has to subject itself to an enquiry of its own meaning. Such a task, of course, is extremely difficult. We can only vaguely perceive its contours. At this point, it appears as an *abstract notion* in Hegel's sense: something that has to find determination in the event of its concrete activity. It might be the case that such an endeavor will lead us to a better and deeper appreciation of what mathematics is, and of what teaching and learning mathematics entail. It might be the case that we end up finding new forms of thinking mathematically that have remained buried underneath the thick layers of technicalities and calculations that characterize to a large extent the mathematics of today. Maybe we can still extract from the deepness of today's practice the aesthetic, intersubjective and embodied dimensions of mathematics and mathematical thinking.

References

Arendt, H. (1958). *The human condition*. Chicago: University of Chicago Press.

Barbin, E., Stehlíková, N., & Tzanakis, C. (2008). *History and epistemology in mathematics education*. Prague: Vydavatelský Servis.

Borel, É. (1907). À propos de l'enseignement de la géométrie élémentaire. *Revue Scientifique*, *5*(8), 656–658 (On the teaching of elementary geometry).

Borel, É. (1914). L'adaptation de l'enseignement secondaire aux progrès de la science. *L'Enseignement Mathématique*, *16*, 198–210 (The adaptation of high-school teaching to the progress of science).

Bourlet, C. (1910). La pénétration réciproque des mathématiques pures et des mathématiques appliquées dans l'enseignement secondaire. *L'Enseignement Mathématique*, *12*, 372–387 (The reciprocal penetration of pure and applied mathematics in high school teaching).

Brousseau, G. (1997). *Theory of didactical situations in mathematics*. Dordrecht: Kluwer Academic.

Brown, T. (2011). *Mathematics education and subjectivity: cultures and cultural renewal*. New York: Springer.

Brown, T., & Walshaw, M. (Guest editors) (2012). *Mathematics education and contemporary theory. Educational Studies in Mathematics*, *80*(1–2), 1–285.

Castoriadis, C. (1999). *Dialogue*. La Tour D'Aigues: Éditions de l'Aube.

Châtelet, G. (2000). *Figuring space: philosophy, mathematics, and physics*. Dordrecht: Kluwer Academic.

Corry, L. (2006). Axiomatics, empiricism, and Anschauung in Hilbert's conception of geometry: between arithmetic and general relativity. In J. Ferreirós & J. Gray (Eds.), *The architecture of modern mathematics: essays in history and philosophy* (pp. 133–156). Oxford: Oxford University Press.

Dreyfus, T., & Eisenberg, T. (1982). Intuitive functional concepts: a baseline study on intuitions. *Journal for Research in Mathematics Education*, *13*, 360–380.

Dreyfus, T., & Eisenberg, T. (1986). On the aesthetics of mathematical thought. *For the Learning of Mathematics*, *6*(1), 2–10.

Eisenberg, T. (1977). Begle revisited: teacher knowledge and student achievement in algebra. *Journal for Research in Mathematics Education*, *8*, 216–222.

Eisenberg, T., & Dreyfus, T. (1991). On the reluctance to visualize in mathematics. In W. Zimmermann & S. Cunningham (Eds.), *MAA notes: Vol. 19. Visualization in teaching and learning mathematics* (pp. 25–37). Washington: Math. Assoc. of America.

Eisenberg, T., & Fried, M. (2009). Dialogue on mathematics education: two points of view on the state of the art. *ZDM*, *41*, 143–149.

Featherstone, H., & Featherstone, J. (2002). "The word I would use is aesthetic": reading David Hawkins. *For the Learning of Mathematics*, *22*(2), 24–27.

Fried, M. N. (2001). Can mathematics education and history of mathematics coexist? *Science & Education*, *10*, 391–408.

Fried, M. N. (2007). Didactics and history of mathematics: knowledge and self-knowledge. *Educational Studies in Mathematics*, *66*, 203–223.

Freudenthal, H. (1981). Mathematics teacher know something about the history of mathematics? *For the Learning of Mathematics*, *2*(1), 30–33.

Furinghetti, F. (2007). Teacher education through the history of mathematics. *Educational Studies in Mathematics*, *66*, 131–143.

Grabiner, J. V. (1974). Is mathematical truth time-dependent? *The American Mathematical Monthly*, *81*(4), 354–365.

Hulin, N. (2011). *Culture scientifique et humanism*. Paris: L'Harmattan (Scientific culture and humanism).

Ilyenkov, E. V. (1982). *The dialectic of the abstract and the concrete in Marx's Capital*, Moscow: Progress Publishers.

Jahnke, H. N. (2000). The use of original sources in the mathematics classroom. In J. Fauvel & J. van Maanen (Eds.), *History in mathematics education: the ICMI study* (pp. 291–328). Dordrecht: Kluwer Academic.

Klein, F. (1896). The arithmetizing of mathematics. *Bulletin of the American Mathematical Society*, *2*(2), 241–249. An address delivered at the public meeting of the Royal Academy of Sciences of Göttingen, November 2, 1895 (Trans. I. Maddison)

Klein, F. (1924). *Elementarmathematik vom höheren Standpunkte aus* (Vol. 1). Berlin: Springer (Elementary mathematics from an advanced point of view).

Lizcano, E. (2009). *Imaginario colectivo y creación matemática*. Madrid: Gedisa (Collective imaginary and mathematical creation).

Loria, G. (1914). *Le scienze esatte nell'antica Grecia*. Milan: Hoepli (The exact sciences in ancient Greece).

Marre, A. (1880). *Bullettino de bibliografia e di storia delle scienze matematiche et fisiche: Vol. 13. Le triparty en la science des nombres de maistre Nicolas Chuquet Parisien* (The triparty of number science of master Nicolas Chuquet, Parisien) (Original work published 1484).

Netz, R. (2009). *Ludic proof: Greek mathematics and the Alexandrian aesthetic*. New York: Cambridge University Press.

Pimm, D., & Sinclair, N. (2009). Audience, style and criticism. *For the learning of mathematics*, *29*(2), 23–27.

Poincaré, H. (1899). La logique et l'intuition dans la science mathématique et dans l'enseignement. *L'Enseignement Mathématique*, *1*, 157–162 (Logic and intuition in the mathematics science and teaching).

Radford, L. (2004). From truth to efficiency: comments on some aspects of the development of mathematics education. *Canadian Journal of Science, Mathematics and Technology Educa-*

tion/Revue Canadienne de l'enseignement des Sciences, des Mathématiques et des Technologies, 4(4), 551–556.

Radford, L. (2008). Culture and cognition: towards an anthropology of mathematical thinking. In L. English (Ed.), *Handbook of international research in mathematics education* (2nd ed., pp. 439–464). New York: Routledge/Taylor & Francis.

Rotman, B. (2008). *Becoming beside ourselves: the alphabet, ghosts, and distributed human beings*. Durham: Duke University Press.

Sinclair, N. (2001). The aesthetic is relevant. *For the learning of mathematics, 21*(1), 25–33.

Solomon, Y., & O'Neill, J. (1998). Mathematics and narrative. *Language and Education, 12*(3), 210–221.

Spiesser, M. (2006). L'algèbre de Nicolas Chuquet dans le contexte français de l'arithmétique commercial. *Revue d'histoire des Mathématiques, 12*, 7–33 (Nicolas Chuquet's algebra in the French context of commercial arithmetic).

Tall, D. (2003). Using technology to support an embodied approach to learning concepts in mathematics. In L. M. Carvalho & L. C. Guimarães (Eds.), *Historia e tecnologia no ensino da matemática* (Vol. 1, pp. 1–28).

Unguru, S. (1975). On the need to rewrite the history of Greek mathematics. *Archive for History of Exact Sciences, 15*, 67–114.

van der Waerden B.L. (1976). Defence of a 'shocking' point of view. *Archive for History of Exact Science, 15*, 199–210.

Problem-Solving: A Problem for Both Mathematics and Mathematics Education

Chapter 8
Reflections on Problem-Solving

Problem Solving in Mathematics and in Mathematics Education

Boris Koichu

Abstract The chapter includes four contributions on different aspects of the relationship between problem solving in mathematics and in mathematics education. Gerald Goldin points out that besides the importance of teaching students how to solve certain classes of problems, problem solving is a means of achieving some more general purposes pertaining to mathematics learning. Israel Weinzweig develops the claim that certain sequences of mathematical questions can provide students with problem-solving experiences similar to those of research mathematicians, and that such experiences are beneficial for promoting students' conceptual understanding. Shlomo Vinner discusses the role of schemata and creativity in mathematical problem solving, and argues that the notions "problem solving in mathematics" and "problem solving in exam-oriented mathematics instruction" are incompatible. Roza Leikin presents a study aimed at identifying unique cognitive traits of intellectually gifted students who have the potential to become research mathematicians in the future. The chapter concludes with a reflective summary, in which the points made by the contributors are considered as parts of a longer-term debate on the relationships between problem solving in mathematics and in mathematics education, a conversation that has developed over the years according to a certain spiral pattern.

Keywords Conceptual understanding · Dimensions of mathematical giftedness: creativity · Cognition · Neuro-cognition · Insight-based and routine problems · Evolution of problem solving within mathematics education · Problem-solving expertise · Problem solving by mathematicians · Teaching for problem solving · Teaching through problem solving

With contributions by
Gerald A. Goldin, Rutgers University, Piscataway, NJ, USA
A. Israel Weinzweig, University of Illinois at Chicago, Chicago, IL, USA
Shlomo Vinner, Achva College of Education, Hebrew University, Ben-Gurion University of Negev, Jerusalem, Israel
Roza Leikin, University of Haifa, Haifa, Israel

B. Koichu (✉)
Technion—Israel Institute of Technology, Technion City, Haifa 32000, Israel
e-mail: bkoichu@technion.ac.il

M.N. Fried, T. Dreyfus (eds.), *Mathematics & Mathematics Education: Searching for Common Ground*, Advances in Mathematics Education,
DOI 10.1007/978-94-007-7473-5_8,

Introduction

Problem solving was put on mathematics education agenda under the tremendous influence of the book "How to solve it?" written by a prominent mathematician George Pólya in 1945 (e.g., Schoenfeld 1992). Keen interest in problem solving emerged in the mathematics education community since there and has been sustained until today for a clear reason: mathematical problem solving (including problem posing, conjecturing and proving) is *the* central activity in mathematics as a living science, and thus it has been hoped that it would also become *the* central activity in mathematics education. For a long time, the idea of understanding how mathematicians treat and solve problems, and then implementing this understanding in instruction design, was pivotal in mathematics education research and practice (e.g., Pólya 1945/1957; NCTM 1980; Schoenfeld 1985).

A somewhat unexpected result of the extensive development of the idea, "Let's teach our students to treat problem solving as mathematicians do," was that problem solving became both an over- and underrepresented topic in mathematics education. It is overrepresented because literally thousands of studies have been devoted to its different aspects, and a great deal of knowledge on how problem solving occurs and what can be learned through problem solving has been accumulated. It is underrepresented because neither research nor practice resulted in a clear identity for problem solving in mathematics education, and because many fundamental issues related to the role of problem solving in mathematics education, as compared to its clear role in mathematics, are still unresolved (e.g., Mamona-Downs and Downs 2005).

Some of these issues are discussed in the four contributions to this chapter. Gerald Goldin points out that besides the importance of teaching students how to solve certain classes of problems, problem solving is a means of achieving some more general purposes pertaining to mathematics learning. He argues that in order to achieve these more general purposes, it is important to unravel tacit processes of learning during problem solving that lie behind the problem solving of experts and the mathematically gifted. Israel Weinzweig develops, through an elaborated example, the claim that certain sequences of mathematical questions, even at the elementary school level, can provide students with problem-solving experiences similar to those of research mathematicians, and that such experiences are beneficial for promoting students' conceptual understanding. Shlomo Vinner discusses the role of schemata and creativity in mathematical problem solving, and argues that the notions "problem solving in mathematics" and "problem solving in exam-oriented mathematics instruction" are incompatible. Roza Leikin presents a synopsis of a large-scope study, a part of which is aimed at identifying unique cognitive traits of intellectually gifted students who excel in solving insight-based mathematical problems and have the potential to become research mathematicians in the future. The chapter concludes with a reflective summary, in which the points made by the four contributors are considered as parts of a longer-term debate on the relationships between problem solving in mathematics and in mathematics education, a conversation that has developed over the years according to a certain spiral pattern.

Mathematical Learning Through Problem Solving: Toward what Purposes?

Gerald A. Goldin

During the 1980s it became fashionable to highlight problem solving as the loftiest goal of mathematics education. In the United States, the National Council of Teachers of Mathematics published the widely-circulated document, *An Agenda for Action: Recommendations for School Mathematics of the 1980s* (NCTM 1980). The first recommendation could hardly have been worded more strongly or succinctly: "The National Council of Teachers of Mathematics recommends that problem solving be the focus of school mathematics in the 1980s"—not "one focus", but *the* focus. And problem solving was selected as the theme for the NCTM's (1980) Yearbook (Krulik 1980).

At the time, researchers in cognitive science, mathematics education, and science education were devoting much effort to studying problem solving—characterizing and studying the processes of effective and less effective problem solvers, exploring ways of defining and measuring problem-solving outcomes to include processes as well as products, identifying the influences of problem tasks on problem solving, including the role of problem structure and the transfer of learning between structurally related problem domains, and modeling problem-solving processes (Goldin and McClintock 1984; Jeeves and Greer 1983; Larkin et al. 1980; Lester and Garofalo 1982; Newell and Simon 1972; Schoenfeld 1985; Silver 1985). However, the *purposes* underlying the NCTM's recommendation remained unclear—*why* should problem solving be *the* focus, rather than skills, concepts, abstraction, modeling, pattern generation and recognition, problem posing, or other important mathematical activity? Sometimes hidden assumptions as to the purpose of teaching mathematical problem solving influenced the problem solving research, affecting the questions asked and the ways in which they were answered.

A major, long-standing dichotomy I would like to highlight here has to do with interpreting the focus on problem solving in mathematics education as (1) having the goal of *teaching students to solve classes of mathematical problems*, as distinct from (2) being the means of *accomplishing some other, possibly more general purposes* pertaining to mathematics learning. Thus Schroeder and Lester (1989) distinguish teaching *for* problem solving from teaching *about* or *through* problem solving.

The first kind of goal was expressed clearly in the mid-1980s. For example Heller and Hungate (1985), citing Greeno (1980), write:

> ... there is a strong need for more effective instruction, particularly to prepare students to solve problems with understanding.
> Such instruction should be specifically tailored to prepare students to solve the kinds of problems they will encounter. For the design of such instruction, an understanding of the *knowledge* [emphasis in original] required for solving problems is of paramount importance.

Greeno (1980) summarized this point well:

> To teach students how to solve a class of problems, first analyze the knowledge that they need in order to solve that class of problems, and then carry out the instruction that will result in their acquisition of the required knowledge. (p. 13)
> ...both descriptive and prescriptive efforts have contributed powerful methods for identifying that knowledge. In order to design effective instruction in mathematical problem solving, then, we should continue to apply those methods in different problem domains to identify the specific knowledge that would need to be taught in each of those domains. (Heller and Hungate 1985, p. 98)

Thus one characterizes domains of problem tasks according to relevant mathematical criteria, identifies the knowledge needed through task analyses and cognitive analyses, and teaches problem solving primarily by providing that knowledge. Of course, the more broad the characterization of the problem domains, the more powerful and general (but, possibly, the more vaguely-defined and less efficient) the identified knowledge is likely to be.

When the teaching of problem solving is oriented toward standardized mathematics assessments, there is a strong—indeed, nearly irresistible—impetus toward characterizing the problem domains narrowly, so as to make possible knowledge representations that are as efficient as possible. The flow chart depicted in Shlomo Vinner's presentation (below) is illustrative of such knowledge representations. Then "what is learned" when problem solving is the focus, and when the knowledge needed for each class of problems is taught, reduces logically to fluency in routine procedures, representations, and strategies. Indeed, when the term "mathematical achievement" is taken to be synonymous with standardized test performance, we are limiting ourselves *a priori* to characterizing the potential for learning through task analysis.

The second kind of goal involves specifying some *other* learning purposes which mathematical problem solving can help students to attain (more so, perhaps, than some alternative kinds of mathematical activity). Among such purposes might be, for example: students' acquisition and development of sophisticated systems of mathematical representations, including methods for constructing new representations (Goldin and Kaput 1996; Goldin 1998); students' development of generalizable heuristic processes and strategies, including ways of inventing new strategies (Pólya 1962/1965); students' increasing sophistication of intuition in mathematical contexts, of visual imagery (Eisenberg and Dreyfus 1991), and spatial and kinesthetic encoding; students' mathematical concept formation (see the problem solving example presented by Israel Weinzweig below), including the development of links between and among different representations; students' creative problem generation and mathematical exploration (Mason et al. 2009); and students' development of powerful patterns of mathematical affect.

The latter is of interest not only as a possible outcome of certain kinds of problem-solving activity, but also in relation to our understanding of mathematical ability and giftedness. Powerful affect includes patterns of in-the-moment emotional feelings, such as curiosity evoked by novelty; meta-affective excitement, determination, and intrigue evoked by frustration and impasse; the joy of mathematical discovery, and the anticipatory joy of working toward an insight that may result in an "aha" experience (Goldin 2000; DeBellis and Goldin 2006). It also includes beliefs and self-concepts that empower the student toward the further learning of mathematics and more sophisticated problem-solving challenges, and accompanying social-contextual interactions (Goldin 2002; Goldin et al. 2011).

Of course, one might seek to characterize the knowledge embodied in such more generalized purposes as *pertinent* to solving problems in various mathematical domains—and, indeed, it is. The NCTM sought, perhaps, to suggest this by elaborating their recommendation as follows: "This recommendation should not be interpreted to mean that the mathematics to be taught is solely a function of the particular mathematics needed at a given time to solve a given problem. Structural unity and the interrelationships of the whole should not be sacrificed" (NCTM 1980). But the kinds of knowledge discussed here are not *necessary* for *any* particular problem domain—one does not need sophisticated strategies, profound visual capabilities, or curiosity-driven persistence to carry out a well-learned, practiced problem-solving process, even when the process is a complex one (again, see the discussion by Shlomo Vinner below). The task analysis or cognitive analysis will not generally incorporate the knowledge associated with more general purposes. Such knowledge *develops* in ways that often defy prior specification, in the context of solving problems that involve *substantial impasse*—where solution strategies are not known or trivially accessible to the solver.

Then the ways in which we set out to study expert mathematical problem solvers—mathematicians, scientists, or exceptionally talented students—depend naturally on which type of purposes we have in mind, possibly less-than-explicitly, for problem solving in mathematics education. The goal of teaching the knowledge needed to solve classes of mathematics problems suggests research toward characterizing and representing various experts' previously-learned, highly sophisticated techniques, and how they make use of them—an "expert systems" approach. But if one sees experiences in mathematical problem solving as the means toward reaching other educational goals, including the realization or attainment of exceptional mathematical ability, what becomes interesting are the (often tacit) *processes of learning* during problem solving that lie *behind* the expertise—cognitive and affective, meta-cognitive and meta-affective, personal and social processes. It then becomes a challenge for mathematics educators to bring such processes to light, and to make use of them in our educational practice. I think this perspective is implicit in Roza Leikin's point (below) that mathematical giftedness is multidimensional, and that excellence in school mathematics can be unrelated to some of these dimensions.

The tension between the disparate purposes for problem solving in mathematics education has existed for many decades, and is not likely to disappear soon. In this short contribution, I have sought to put some of those issues "on the table" for more explicit discussion.

Concept Development Through Problem Solving

A. Israel Weinzweig

Introduction

For the most part, when mathematicians do mathematics, they first pose questions and then try to answer them. In the process, they create new objects and begin to define them. In contrast, teachers of mathematics tend to answer questions that not only have not been posed, and that moreover, most students would never think of asking. Consequently, the objects they introduce have little relevance to their students and yield little understanding. I contend that when appropriate problems are posed and students are given the opportunity to work on them collaboratively, they encounter the need for certain objects and concepts that can then be introduced. Since it is they who have felt the need for the concept and a use for it, they develop an ownership of the concept, and a much deeper understanding. I believe that this provides a much better approach to the introduction of many of the concepts of elementary arithmetic. I illustrate this approach with an example that I have used successfully.

The 252 Problem

Students are presented with the number 252. They are asked to find two numbers whose product is 252 such that:

1. Their sum is an even number;
2. When the smaller number is divided into the larger one, the quotient is an odd number.

Students[1] immediately pull out their calculators and divide 252 by 2, 3, ..., and each time the quotient is a whole number, they add the quotient and the divisor to check whether the sum is an even number. If it is, they then divide the quotient by the divisor to check whether they get an odd whole number. They quickly arrive at the numbers, 2 and 126 whose sum is 128, an even number, and dividing 2 into 126 yields the quotient 63, an odd number. I then ask if there are other solutions. They soon come up with 6 and 42, whose sum is 48, an even number, and the quotient is 7, an odd number. Are there any other solutions? They soon decide there are none. How do they know? They have exhausted all the possibilities! I refer to this as proof by exhaustion.

I then change the problem slightly. We now want the sum of the two numbers to be an odd number and the quotient to be an even number. They immediately get to

[1]Teachers, prospective teachers and middle school students.

work on their calculators and come up with the numbers 3 and 84 whose sum, 87, is odd and the quotient of 84 divided by 3 is 28, which is an even number. I again ask if there are other solutions. They are convinced there are none. I then ask about 1 and 252. Their sum is 253, an odd number, and the quotient is 252, an even number. They are surprised. They never thought of dividing by 1 on their calculators.

I change the problem again. By this time the context has been well established so the language can be looser. We now want the sum to be even and the quotient to be even. Again they take their calculators and, using proof by exhaustion, assert that it cannot be done.

Can they find a number other than 252 for which there will be a solution to the even sum, even quotient problem? They try random numbers. Eventually someone comes up with one, 504. They then explore 504, for its appropriateness in all the four problems:

1. Even sum, odd quotient;
2. Odd sum, even quotient;
3. Even sum, even quotient;
4. Odd sum, odd quotient.

So far, they discovered that the number 252 that I provided yields (two) solutions to the first two; they obtained a numbers, by trial and error, 504 that yields (two) solutions to the second and third.

Can they find a number for which there is a solution to the last problem? After some experimentation, they conclude that it is not possible to find such a number. I point out to them that there is a difference between saying that *they* are unable to find such a number, and saying that there is *no such a number*!

At this point I introduce some terminology to facilitate the discussion. Instead of saying that 2 divides evenly into 504 (or 24, or 252) we say that 2 is a factor of that number. When we divide the number by a factor, the quotient is the complimentary factor of the factor in that number. Thus, 2 is a factor of 504 with complimentary factor 252 so that $504 = 2 \times 252$. But 2 is also a factor of 252 with complimentary factor 126, so that $504 = 2 \times 2 \times 126$. Now 2 is also a factor of 126 with complimentary factor 63, so that $504 = 2 \times 2 \times 2 \times 63$. Note also that 3 is a factor of 63 with complimentary factor 21. Hence, $504 = 2 \times 2 \times 2 \times 3 \times 21$. Moreover, 3 is again a factor of 21 with complimentary factor 7, so that $504 = 2 \times 2 \times 2 \times 3 \times 3 \times 7$. Now, 2 has exactly two factors, 1 and 2. Similarly, 3 and 7 each have exactly two factors. These are prime numbers. Each of the above representations of 504 as a product of factors is a factorization of 504. Since all the factors in the last representation are prime factors, this is a prime factorization of 504.

I next introduce *multiplicity* of factors. Thus 2 in the prime factorization of 504 has multiplicity of 3, 3 has multiplicity of 2 and 7 has multiplicity of 1 and write the prime factorization of 504 as $504 = 2^3 \times 3^2 \times 7^1$.

Observe that I have not explicitly *defined* multiplicity, but students grasp this concept quite easily and accept the notation. I find that students have far less difficulty with the concept of multiplicity than with the concept of exponents. It is much more natural to them and the fact that $7^1 = 7$ requires no great explanation.

I then ask, "What is the multiplicity of 7 in 6?" The usual reaction is to assert that 7 is not a factor of 6 so that 7 has no multiplicity in 6, at which point it usually dawns on them that the multiplicity of 7 in 6 is 0, so that the prime factorization of 6 could be represented as $6 = 2^1 \times 3^1 \times 7^0$.

At this point I call attention to the fact that the multiplicity of each prime factor of 504 is the sum of the multiplicities of these factors in 6 and 84. Moreover, when 504 is divided by 6, in the prime factorization of the quotient—the complementary factor 84—the multiplicity of each prime factor of 84 is the multiplicity of each prime factor of 504 reduced by the multiplicity of that prime factor in 6. A clear advantage of focusing on multiplicity rather than exponents is that they grasp the central features from a few examples. I refer to them as generalizable examples.

I then raise the question "What is an even number?" Students can usually tell whether a given number is even or odd, but have difficulty explaining why. After some discussion, they recognize that if 2 is a factor of a number, then that number is even. If 2 is not a factor of a number, then dividing that number by 2 will yield a remainder. Since the remainder is less than the divisor, it will be 1 or 0. In the latter case, the number is even and in the former case the number is odd. Hence, an odd number is the sum of 2 times the quotient plus 1, an even number and 1. Thus, 127 is an odd number.

When is the sum of two numbers even or odd? If both numbers are even, say 6 and 84, then they can be written as $6 + 84 = 2 \times 3 + 2 \times 42 = 2 \times (3 + 42)$, an even number. For any other even numbers, the numbers in the parenthesis can be changed appropriately and the result is still an even number! Then the students are asked to explore in a similar way the structure of the sum of two odd numbers and the sum of an odd number and an even number.

Then we are ready to return to the question "Can you find a number for which there is a solution to the odd sum, odd quotient problem?" In the introduced terms, the students realize that such a number would have to have an odd factor and the complementary factor would have to be even, or vice versa.

As we have already observed, for one number to divide a second number, the multiplicity of every prime factor of the second one must be equal to or greater than the multiplicity of that prime factor in the first number. In an even number, 2 has multiplicity of at least 1, and in an odd number, multiplicity 0. Hence an even number cannot divide an odd number, but an odd number can divide an even number. However, in this case, 2 will have multiplicity of at least 1 in the quotient, so the quotient will never be odd! In view of this, there is no solution to the fourth question of finding a number with an odd factor, such that the complimentary factor is even and the factor divides the complimentary factor with an odd quotient.

I call attention to the fact that this is not a question of being unable to find such a number, but that no such number exists! Then the students come back to the initial problems, with 252 and 504, and realize why there were two solutions to the first two problems with the number 252 and two solutions to the second and the third problems with 504, but none to the first.

In particular, for 252 we could never get an even sum and an even quotient, whereas for 504 we can never get an even sum and an odd quotient. The fact that the multiplicity of 3 is 2 gives rise to the two solutions. They can also observe that 7 plays a very passive role. We could drop the 7, and the situation would be unchanged; 36 would still have two solutions to each of the first two problems but none to the third, 72 would still have two solutions to the second and third problem but none to the first. However, if we increased the multiplicity of 2 in 72 to an even multiplicity, say 4, then 144 would have two solutions to each of the three problems, $144 = (2^2) \times (2^2 \times 3^2)$ and $144 = (2^2 \times 3) \times (2^2 \times 3)$ for two solutions to the even sum and odd quotient problem, and $144 = (2) \times (2^3 \times 3^2)$ and $144 = (2 \times 3) \times (2^3 \times 3)$, two solutions to the even sum and even quotient problem. Finally, $144 = (1) \times (2^4 \times 3^2)$ and $144 = (3) \times (2^4 \times 3)$—two solutions to the odd sum and even quotient.

It is the multiplicity of 3 that determines how many solutions there will be to each of the problems. If the multiplicity is less than 2, there will be only one solution. If the multiplicity of 3 is 2 or 3, there will be two solutions. If the multiplicity is 4 or 5, there will be three solutions. Students discover these patterns quite quickly!

Summary

The original problem was quite innocuous, yet it opened up a number of interesting additional problems. In the course of working on these problems, the moment is made appropriate to meaningfully suggest a new concept which then does not drop artificially from the sky but its introduction serves a purpose. Note that in the lesson plan described above it is not the children that explicitly express the need for a new concept (as might have been the case in the process of a mathematician at work), it is the teacher who takes advantage of the situation to create an appropriate moment to introduce a concept which serves a propose, and is then perceived as useful and meaningful by the students. In this way a number of concepts are introduced: factor of a number, factorization of a number, prime numbers, prime factorizations of a number, multiplicity of a factor in a number. The students had to develop a more precise definition of an even (natural) number and an odd number. They formulated generic proofs (Mason and Pimm 1984) for the fact that the sum of two odd numbers or two even numbers is even and only the sum of an odd number and an even number is an odd number. They had to formulate a generic proof that there was no number that would produce an odd sum and an odd quotient.

In summary, the students were engaged in doing mathematics rather than only studying mathematics. This provides an example of an important purpose of problem solving: how to "ride" on the shoulders of problems in order to meaningfully and purposefully introduce concepts (even perhaps procedures, notation and other ideas). Consequently, they learn much more and internalize what they have learned. It is not so much problem solving per se, but choosing appropriate problems for solution.

The Irrelevance of Research Mathematicians' Problem Solving to School Mathematics

Shlomo Vinner

Prologue

Because of space limits I have to be short. Therefore, I have to simplify. Simplification and oversimplification are quite close and both are considered sometimes by some people as superficiality. I am aware of this and I am ready to face the charges.

How Do They Really Solve It?

In some of my courses for in-service mathematics teachers I raise the embarrassing question: what is mathematics? I distinguish between mathematics as a product and mathematics as an activity. Mathematics as an activity (for research mathematicians) is inventing or discovering mathematical structures and exploring them. Whether it is inventing or discovering depends in which church or synagogue you practice your philosophical beliefs. Some of these structures were invented or discovered centuries ago and others are quite new. Exploring mathematical structures is the research mathematician's job. The outcomes of this exploration are theorems about the structures under consideration. Thus, mathematics as a product is a collection of mathematical theories like Number Theory, Group Theory, Game Theory, and many other theories, even if the notion of theory is not part of the theory name.

What part of this collection can be presented to the students in school mathematics? The classical curriculum in the majority of countries is restricted to elementary algebra, calculus, geometry and trigonometry. From this the curriculum can focus on computations, equations, identities of various kinds, function investigation and, of course, all the related theorems.

From cognitive psychology point of view mathematics is a problem-solving activity. However, problem solving in school mathematics is a totally different activity than the research mathematician's activity. Because of the currently prevailing educational policy, passing the mathematical exams has become the top priority of teachers and students. The main goal of teaching is, therefore, to prepare the students to pass the exams. This is done by providing the students with a tool box, by means of which they are supposed to be able to solve the mathematical problems presented to them in homework assignments and on tests. Such problems should be routine problems or routine problems under disguise. Why? Because it is unfair to present on an examination a problem for which there does not exist a solution strategy in the student's tool box.

Therefore, a student who is asked to solve a routine problem is supposed to have the following:

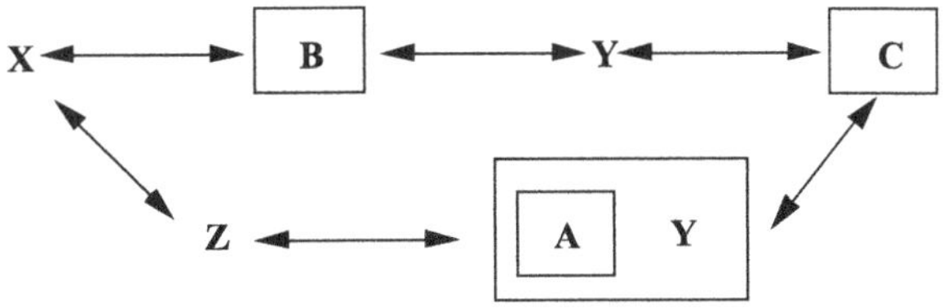

Fig. 8.1 A model describing a thought process occurring when a student solves a routine mathematical problem

1. A pool of solution procedures. Call it A.
2. Mental schemes by means of which the type of a given mathematical problem and its particular structure can be determined. Call them B.
3. Mental schemes by means of which a solution procedure can be assigned to a given mathematical problem, whose type and structure were previously determined. Call them C.

When a mathematical problem X is posed to the student, the following process is supposed to occur (see Fig. 8.1):

1. B, the identifying schemes, are activated by X. An analysis is carried out which determines the type and structure of X. Denote that type and structure by Y.
2. C, the solution selector schemes, are activated by Y. They select a solution procedure from A. Call it Z.
3. Z is applied to X and produces a solution for X.

I am using two way arrows in order to indicate that the involved thought processes go back and forth from one stage to another in case the problem solvers are stuck before reaching a solution.

As a matter of fact, if you compare the above model to pages xvi–xvii of Pólya's "How to solve it?" (Pólya 1945/1957), you can claim that the above model is a flowchart representation of Pólya's recommendations of how to solve word problems. Pólya describes meaningful thought processes that should be involved in a successful solution. If you examine these thought processes, you may come to the conclusion that we do not face here a big intellectual challenge. Real creativity is not involved and, in fact, cannot be prescribed. Creativity can be described by the research mathematician after it occurred. Some mathematicians do that in their memoirs (e.g., Hadamard 1945/1996). Their texts are extremely interesting for people who are interested in the psychology of mathematical invention. They are not relevant to the decisive majority of students in mathematics classes in high schools and even not in Bachelor mathematics classes at universities. Reflecting on my years as a mathematics student, I recall how I wondered at all kinds of mathematical inventions when I faced them in class or in books. I could not figure out how the Chinese remainder theorem got into the proof of Gödel's first incompleteness theorem. I had no idea how Taylor's formula was discovered and proved, and so on and so forth. When I taught calculus in high school, even when proving relatively simple theorems, like the theorem about the derivative of a product of functions, where adding and subtracting a certain expression does the job—my students asked me: *How, on earth, could I figure it out by myself*? My answer was: *Relax, you are not supposed to. Just enjoy the beauty of the simple device that I presented to you. It was presented to me as well when I was a high school student. I also did not figure it out myself.*

Thus, if we return to my model, about which I claim that it represents Pólya's recommendations of how to solve mathematical problems, I believe that we should be pleased if our students' thought processes progress accordingly.

However, there are too many cases in which students fail to follow this path. They choose quite often the solving tool accidently relying on superficial similarities and uncontrolled associations. Psychologists nowadays speak about two modes of thinking which they call *System 1 and System 2* (e.g., Stanovich 1999). System 1 is characterized by the following adjectives: *associative, tacit, implicit, inflexible, relatively fast, holistic and automatic*. System 2 is characterized by: *analytical, explicit, rational, controlled and relatively slow*. Thus, notions that were used by mathematics educators can be related now to System 1 or System 2 and therefore this terminology is richer than the previously suggested notions. Fischbein (1987) spoke about *intuition* and it can be considered as System 1. Skemp (1976) spoke about two systems which he called *delta-one and delta-two* which can be considered as *intuitive and reflective*, or using the new terminology, *System 1 and System 2*, respectively. I myself (Vinner 1997b) have used the notions *pseudo-analytical and pseudo-conceptual* which can be considered as System 1.

In mathematical contexts, the required thinking mode is that of System 2. Unfortunately, in many cases, a correct answer can be obtained even by System 1 thought processes. Daniel Kahneman (2011), a 2002 Nobel prize laureate, claimed that System 1 mode of thinking serves us successfully in almost all our common everyday situations. Therefore, the chance to change our thinking habits is quite small. But Kahneman is not an educator. Educators, especially mathematics educators, are supposed to try to improve thinking and behavior. The domain of mathematical problem solving is quite relevant to that.

Good problem solvers are those who are not locked on the first solution procedure, which occurs to their mind, in case it is an inappropriate one. They are capable of identifying inappropriate solution procedures and they are capable of considering additional solution procedures, which are seemingly relevant and sometimes even not seemingly relevant, at first sight, to the problem under consideration. This requires flexibility of thought and a rich pool of associations. Some of us believe that we can train our students to become flexible and to improve their problem solving skills. However, teaching non-routine problem solving is, in my opinion, an oxymoron. Solving non-routine problems is the research mathematician's job. He or she has to solve problems the solution strategies for which do not yet exist. It is true that there are some recommended ways to do so, for instance, see Pólya's various books. But the advice given there does not guarantee a solution.

Thus, we can enjoy the beauty and originality of the way some mathematical discoveries occurred. We can read with great interest books by Hadamard (1945/1996), Poincaré (1952), etc. However, this has a very little relevance to the activity of problem solving in school mathematics. It is irrelevant to the majority of teachers and students.

"What is mathematics education for?" asks Underwood Dudley, an emeritus Mathematics Professor, in his 2010 essay (Dudley 2010). His answer is: "*Mathematics education is for, and has always been for: to teach reasoning, usually, through the medium of silly problems*." An example of such a silly problem is:

> Give 100 loaves to five men so that the shares are in arithmetic progression and the sum of the two smallest is 1/7 of the three greatest. (Rhind Papyrus, Egyptian textbook of mathematics, 1650 BC).

Of course, Dudley's expression "silly problems" is not politically correct. A common dilemma in the academic life, as well as in politics and everyday life is choosing between truth and political correctness. I prefer truth to political correctness, at least in the academic life, where we are supposed to be committed to truth.

A quote from Pólya's lecture on teaching mathematics in primary schools (Pólya late 1960s) is quite coherent with Dudley claims. Contrary to Dudley, Pólya's enthusiasm was not spoiled by sarcasm.

> Mathematics in the primary schools has a good and narrow aim and that is pretty clear in the primary schools.... However, we have a higher aim. We wish to develop all the resources of the growing child. And the part that mathematics plays is mostly about thinking. Mathematics is a good school of thinking. But what is thinking? The thinking that you can learn in mathematics is, for instance, to handle abstractions. Mathematics is about numbers. Numbers are an abstraction. When we solve a practical problem, then from this practical problem we must first make an abstract problem.... But I think there is one point which is even more important. Mathematics, you see, is not a spectator sport. To understand mathematics means to be able to do mathematics. And what does it mean doing mathematics? In the first place it means to be able to solve mathematical problems (retrieved from http://cmc-math.org/members/infinity/polya.html).

According to Pólya in this text, mathematics education is mainly for improving mathematical problem-solving ability. He does not claim like Dudley that its role is to teach reasoning in general. Pólya is more modest in his claim. This makes his claim more acceptable. Indeed, if we challenge Dudley's claim, there is no experimental evidence that good mathematical problem solvers are also good in solving problems in other domains.

Epilogue

I assume that the reader has realized by now that I am extending the relevance question which appeared in the title of my contribution. At this stage I am challenging the relevance of mathematics to the ultimate goals of education of young people in our school systems. I started doing that 15 years ago (Vinner 1997a) by referring my readers to Confrey (1995). It was claimed there that *in the vast majority of countries around the world mathematics acts as a draconian filter to the pursuit of further technical and quantitative studies*. This is an unpleasant claim and the majority of people who are involved in mathematics education prefer to ignore it. There are several reasons for that. The majority of us believe that there is much more in mathematics education than preparing students to pass crucial examinations. There are also economic reasons for that. If we stop using mathematics as *a draconian filter to the pursuit of further technical and quantitative studies* many people will lose their jobs—teachers, mathematics supervisors, textbook authors, publishers, etc.

Niss (2011) has raised the question: Why do we do research on the teaching and learning of mathematics? His answer is: "*We do research on the teaching and learning of mathematics because there are far too many students of mathematics... who get much less out of their mathematical education than would be desirable for them and for society*" (p. 1293). I absolutely agree. However, I would like to add to the research agenda also the question "what does mathematics education contribute to the students and to the society?"

Can All High Achievers in School Mathematics Become Professional Mathematicians?

Roza Leikin

Rationale

Mathematical problem solving has been the focal point of mathematics educators, mathematicians, and educational researchers who are seeking a better understanding of the mechanisms of mathematical reasoning and of the development of mathematics understanding and deeper analyzing mathematical proficiency. For instance, high level problem-solving expertise (e.g. success in solving Olympiad problems) often serves as an indicator of mathematical giftedness (MG).

High achievements in school mathematics usually reflect students' problem-solving proficiency in the topics that students have studied in school. As such, high achievements can be perceived as an indication of MG. Mathematics teachers usually evaluate students' mathematical abilities based on their scores in mathematics tests. However, MG is a complex construct which is ill-defined. Different researchers use different criteria for its evaluation. Some researchers associate MG with extraordinary cognitive abilities, connect MG with general giftedness (i.e., IQ scores) and evaluate MG with SAT-M (Lubinski and Benbow 2006). Often MG is connected with mathematical creativity by distinguishing between 8 levels of creativity (Sriraman 2005).

While the notion of MG (or mathematical talent as realized giftedness) is quite clear with respect to research mathematicians, it is rather vague with respect to school students. This uncertainty reflects the distinction between *absolute* and *relative* creativity (Leikin 2009). *Absolute creativity* is associated with "great historical works" (in the words of Vygotsky 1930/1984), with discoveries at a global level (e.g., as seen in discoveries of Fermat, Hilbert, Riemann). *Relative creativity* refers to discoveries of a specific person in a specific reference group. This type of creativity refers to the human imagination as it creates something new independently of the scope of the idea (Vygotsky 1930/1984).

The relationship between MG as associated with relative creativity and MG as associated with absolute creativity is not well established. While asking the question "Can all high achievers in school mathematics become professional mathematicians?" I have learned that a frequent response, even by experts in education of the gifted, is that "this question has a simple 'yes' answer". I am skeptical about this point of view. While the answer to the question is not trivial, its importance is associated with understanding of ways in which future mathematicians should be taught in school. Better understanding of the nature of MG—both as associated with absolute and relative creativity—can inform mathematics educators about the ways in which school mathematics should be taught to students who can become research mathematicians. This understanding can lead to a special instructional design and kinds of mathematical curriculum that can be suitable for these students including the choice of mathematical problems for MG students. Some insight about teaching MG students can be learned from Kolomgorov's mathematical schools in Russia (Vogeli 1997). However, in past two decades, characteristics of MG students are overlooked in the mathematics education research.

The Study

Inspired by the observations presented above, a research team[2] from the Faculty of Education and the RANGE (Research and Advancement of Giftedness and Excellence) Center at the University of Haifa carries out MULTIDIMENSIONAL INVESTIGATION OF MATHEMATICAL GIFTEDNESS. This study is aimed at providing neuro-cognitive explanations for cognitive-behavioral characteristics of giftedness. It combines a *neuropsychological investigation* (Ph.D. dissertation in progress by Ilana Waisman) associated with solving mathematical problems with the study of *cognitive abilities and processes* (Ph.D. dissertation in progress by Nurit Paz-Baruch), such as: memory, attention, IQ, executive functions, awareness (in different fields), linguistic ability, and *creative ability* (Ph.D. dissertation in progress by Miri Lev).

A sample of 200 students was chosen from a population of 1200 10th–12th grade students (16–18 years old). The sampling procedure was directed at investigation of the effect of EM (Excellence in Mathematics) and G (general giftedness) factors as defined bellow. Students for G groups were mainly chosen from classes for gifted students (IQ > 130). All research population was examined with Raven's Advanced Progressive Matrix Test (RPMT). Students were sampled as excelling in mathematics (EM) if they studied mathematics at high level with scores higher than 92. Additionally, excellence in mathematics was examined with the SAT-M test (Scholastic

[2]The team includes Roza Leikin with responsibility for the mathematical content of the study and research on creativity and giftedness; Mark Leikin who is responsible for cognitive and neuro-cognitive research dimensions; Shelly Shaul who is an ERP-research specialist at the Faculty of Education. The team of researchers collaborates in supervision of a group of Ph.D. students in the design of a multidimensional research puzzle: Ilana Waisman, Nurit Paz and Miri Lev.

Aptitude Test in Mathematics). Students who were chosen for the research sample, were subdivided into four experimental groups by the combination of EM and G factors: *G-EM group*: students who are identified as generally gifted and excelling in mathematics; *G-NEM group*: students who are identified as generally gifted but do not excel in mathematics; *NG-EM group*: students excelling in mathematics who are not identified as generally gifted; *NG-NEM group*: students who were neither identified as generally gifted nor excelling in mathematics.

Findings and Hypotheses

Creativity Dimension

The part of the study that explores the relationship between mathematical creativity and mathematical ability (Leikin and Lev 2013) accepts the distinction between relative and absolute creativity in order to address personal creativity as a characteristic that can be developed in schoolchildren. By employing Multiple Solution Tasks (MSTs) we demonstrate that effects of EM and G factors are task dependent and are a function of mathematical insight embedded in the mathematical task. Precisely, the G factor has main effect on creativity associated with solving problems whose solutions are largely based on insight, while the EM factor has an effect on creativity associated with solving problems in which variety of the solutions is based on strategies which are based on the school curriculum. Problems whose solutions combine both pre-learned multiple strategies and insight-based multiple strategies allow distinguishing between the four groups of participants with main effect of EM and G factor on students' flexibility. The task dependence of the effect of EM and G factors as reflected in students' fluency and flexibility when solving MSTs raises the hypothesis that EM and G traits are interrelated but different in nature.

Cognitive Dimension

A similar hypothesis was raised based on the investigation of memory mechanisms, speed of information processing, attention and other cognitive traits. For example, the results of cognitive investigation related to memory capacity in different groups of participants demonstrate differences in memory mechanisms related to G and EM factors: The study reveals that the G factor is related to a high level of short term memory (STM) for both phonological loop and phonological central executive mechanisms. It was also found that the EM factor is associated with a high level of visual-spatial memory (VSM), in particular with the visual central executive mechanism. We also found an interaction between G and EM factors regarding WM (working memory). The central executive mechanism appeared to be related to both G and EM factors (Leikin et al. 2013).

Test	Examples		
 Stage in ERP recording	Presentation of givens / of a situation	A question	Verification: True or false
Arithmetic fact retrieval in simple multiplication problems	2×5		20
Geometry tasks that requires transition from a geometrical object to a symbolic representation of its property	The triangle is a right triangle 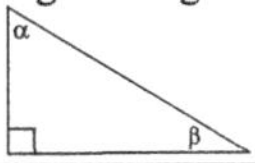	The connection between the angles is:	$\alpha + \beta = 90^{\circ}$
Insight-based mathematical problems	There are 8 corners in a polygon. One corner is cut.	The number of corners in the new polygon is:	7

Fig. 8.2 Examples of the items in different tests in the ERP experiment

Neuro-cognitive Dimension

Research on brain activity is performed using ERP—Event-Related Potentials—methodology (Leikin et al. 2012; Waisman et al. 2012; Shaul et al. 2012). The study includes different types of short mathematical problems ranging from simple multiplication, problems that require transition from a geometrical object to a symbolic representation of its property, transition from visual representation of an algebraic object to its symbolic representation, and insight-based problems (see Fig. 8.2).

The study differs from all previous ERP-based studies on mathematical processing in the level (higher than in previous studies) of mathematical problems included in the tests and by the way of sampling the target population. Similarly to creativity and cognition-related parts of the study, we found that effects of the EM and G factors are task-dependent. For example, application of a simple multiplication task revealed the effect of the EM factor only. In contrast, for geometry and algebra tasks the differences in brain activity appeared only in students from the EM group with higher brain activity in NG than in their G counterparts. For insight-based tasks, we found that gifted participants who excel in mathematics are more accurate and faster when solving the problems. The G factor has a significant effect on accuracy but not on the reaction time, whereas the EM factor has a significant effect both on accuracy and reaction time. The electrophysiological data have revealed the differences in mean amplitude of brain activity and the time course at different stages of solving the tasks (introducing a situation, question presentation,

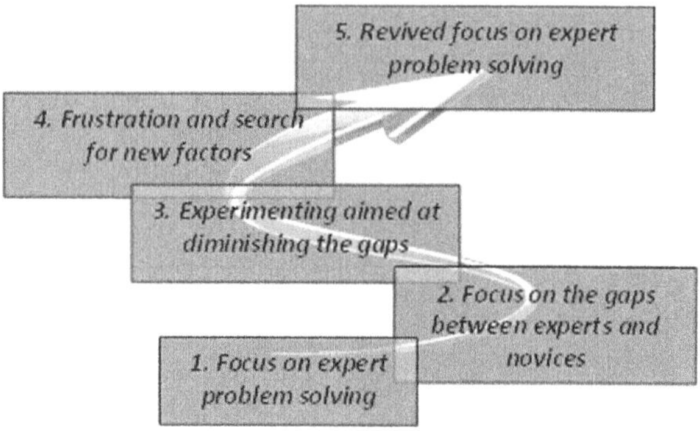

Fig. 8.3 A developmental cycle of research on problem solving

answer verification; see Fig. 8.2) for the G and EM factors. This finding supports again the hypothesis about differences in the nature of G and EM traits.

In sum, multidimensional investigation of mathematical giftedness demonstrates that group differences (effect of the EM and G factors) in all three study dimensions are task dependent. Thus we hypothesize that the G and EM factors are different in nature. We are happy to report that mathematical excellence can be achieved not only in G students, while not all G students are excelling in mathematics. School mathematics rarely introduce insight-based problems even to high level students, which—according to our study—are essential for realization of mathematical potential in G-EM students who have to be challenged by the mathematics presented to them. Finally, based on our observations, we argue that excellence in school mathematics is a necessary but not sufficient condition for MG and that students from the G-EM group have high potential to become professional mathematicians in the future.

Reflective Summary

Boris Koichu

The four contributions to this chapter address different but interrelated aspects of problem solving in mathematics and mathematics education. The differences are apparent. The interrelation is highlighted in this section by considering the main points made by the contributors as integrative parts of a spiral developmental pattern, in which problem-solving models and the associated attempts to use them in mathematics education seem to have evolved over the last decades. One cycle of a spiral is schematically presented in Fig. 8.3.

Simply put, each cycle of the spiral begins with the study of how mathematicians or mathematically advanced individuals solve problems (phase 1 in Fig. 8.3). For instance, Pólya's (1945/1957) four-phase model of problem solving emphasizing the crucial role of heuristics and Schoenfeld's (1985) model proposing problem-solving attributes such as mathematical resources, cognition, control, affect and practices, were produced in this way. The same holds for some current multiple-dimensional problem-solving models attempting to capture the interplay between problem-solving attributes and cycles (e.g., Carlson and Bloom 2005).

In the second phase of a cycle, the gap between expert or gifted problem solvers and novice or regular problem solvers is expressed in terms of a particular model. The model or its aspects then serves as a source for formulating objectives for mathematics education research and practice (phase 3 in Fig. 8.3). For instance, since we know that mathematicians and mathematically gifted individuals use an extended pool of problem-solving heuristics, let's try to teach our regular students how to do so (e.g., Schoenfeld 1979; 1983; Koichu et al. 2006; 2007). Or, since we know that the mathematically gifted are attentive to the elegance of their solutions, let's try to teach the rest of the students to appreciate aesthetics in problem solving (e.g., Dreyfus and Eisenberg 1986). Or, since mathematicians invent new concepts only when they need them in problem solving, let's try to introduce mathematical concepts to our students in this way (e.g., Harel 2013), and so on.

As a rule, these objectives are achieved in a relatively small number of experimental settings, but regular mathematics classrooms change slowly and much less than is hoped by the proponents and followers of a particular model (e.g., Hembree 1992; Mamona-Downs and Downs 2005). Such observations lead to a certain level of frustration and, simultaneously, to the search for new factors that have not yet been taken into consideration (phase 4 of Fig. 8.3). For instance, the focus of research attention in the 1990s shifted from problem solving *per se* to exploration of social and socio-mathematical norms related to problem solving in mathematics classrooms (e.g., Yackel and Cobb 1996). This research venue appeared to be immensely useful, in particular, for deepening our understanding of differences between problem solving in a classroom and in a mathematician's office; but it also essentially changed the research agenda. As a result, the approach "let's teach our students to treat problem solving as mathematicians do" fell out of the mainstream for a while and the approach "let's study what our students actually do in a mathematics classroom" was put forward. After a while, interest in mathematical problem solving as a central theme of mathematics education revived (e.g., Schoenfeld 2007). Now a new developmental cycle begins, as suggested by our spiral cycle, with consideration of a new, more sophisticated, model of expert problem solving (e.g., Carlson and Bloom 2005).

One unequivocally positive result of this process is related to the gradual refinement of problem-solving models, making them applicable in a greater number of contexts. Another positive result is the growing understanding of the complexity of the use of problem solving in mathematics education. However, as has been mentioned above, the continuing gap between the promise of the use of problem solving and the realization of that promise often becomes a source of frustration. It is also worth mentioning that the real developmental process, which of course is much more sporadic and complicated than was presented in the above paragraphs—for example, the roles of technology and political influences were not even mentioned—includes changes in research methods and fluctuations in the foci of research attention. As a result, the very role of problem solving in mathematics education is periodically questioned.

Based on the above observations, I suggest that problem solving has entered a new developmental cycle. The cycle can (tentatively) be characterized by the fact that it seeks to provide new answers to some old questions; in particular, by employing and bridging methods, models and theories that have been produced both within and outside mathematics education. Another characteristic of the current cycle is that it is based on a huge body of accumulated experience, both positive and negative, and thus, its phases tend to essentially overlap: simultaneous attention is brought to issues that previously have been considered only separately.

Looking again at the contributions to this chapter, one can see that they represent positions that can be associated with different phases of this new cycle. The work of Roza Leikin and her colleagues promises to shed new light on cognitive and brain processing involved in solving insight-based mathematical problems, and thus, may potentially lead to producing a problem-solving model sufficiently complex to overcome some of the limitations of past models (phase 1) and pinpoint the differences between mathematically gifted and regular problem solvers (phase 2). This work is nicely aligned with one aspect of Gerald Goldin's contribution, namely, the call to study the tacit cognitive and affective processes that lie behind expertise and giftedness in mathematics.

Another aspect of Goldin's contribution—making use of knowledge about how mathematicians *learn* mathematics through problem solving—aligns with Israel Weinzweig's contribution. This contribution can be associated with the third phase of a cycle, namely, with attempts to involve students in act-as-a-mathematician problem-solving activities. Weinzweig concluded his presentation by saying that choosing a sequence of problems to solve rather than problem solving *per se* was what enabled his students to *do* (as opposite to study) mathematics; this point puts forward the importance of problem-based task design aimed at enhancing conceptual understanding rather than acquiring techniques needed to prepare for various test and exams (cf. Isoda and Katagiri 2012, for an elaborated discussion of this point in the context of the Japanese approach). Finally, Shlomo Vinner's point about the irrelevance of problem solving by mathematicians to mathematics education, or at least to the exam-oriented part of mathematics education, can be associated with both the second and the fourth phases of the cycle: Vinner discusses the gap between problem-solving practices of mathematicians and of school students (phase 2) and expresses some frustration regarding the role ascribed to problem solving in current school mathematics (phase 4). It is appropriate to note here that the role of problem solving in teaching is hotly debated nowadays also from the cognitive architecture perspective (e.g., see Kirschner et al. 2006, and Hmelo-Silver et al. 2007, for pros and cons of teaching through worked-out examples vs. teaching through problem solving).

In summary, the next several years will probably be indicative of the future of problem solving in mathematics education. The enduring questions, "How do mathematicians and the mathematically gifted solve problem and learn through problem solving?" and "How can knowledge about problem solving by mathematicians and the gifted be used in mathematics education for all?" require our further attention. Given the proposed spiral evolution of problem solving within mathematics education, it seems imperative that mathematicians, mathematics educators and cognitive scientists re-join forces in order to address these (and such) questions in a truly interdisciplinary effort, which has always being advocated by Ted Eisenberg (e.g., Eisenberg 1975; Eisenberg and Fried 2009).

References

Carlson, M., & Bloom, I. (2005). The cyclic nature of problem solving: an emergent multidimensional problem-solving framework. *Educational Studies in Mathematics*, *58*, 45–75.

Confrey, J. (1995). Student voice in examining "splitting" as an approach to ratio, proportions and fractions. In L. Meira & D. Carraher (Eds.), *Proceedings of the 19th international conference for the psychology of mathematics education* (Vol. 1, pp. 3–29). Recife: Universidade Federal de Pernambuco.

DeBellis, V. A., & Goldin, G. A. (2006). Affect and meta-affect in mathematical problem solving: a representational perspective. *Educational Studies in Mathematics*, *63*, 131–147.

Dreyfus, T., & Eisenberg, T. (1986). On the aesthetics of mathematical thoughts. *For the Learning of Mathematics*, *6*(1), 2–10.

Dudley, U. (2010). What is mathematics education for? *Notices of the American Mathematical Society*, *57*, 608–613.

Eisenberg, T. A. (1975). Behaviorism: the bane of school mathematics. *International Journal of Mathematical Education in Science and Technology*, *6*, 163–171.

Eisenberg, T., & Dreyfus, T. (1991). On the reluctance to visualize in mathematics. In W. Zimmermann & S. Cunningham (Eds.), *MAA notes series: Vol. 19. Visualization in teaching and learning mathematics* (pp. 25–37). Washington: Math. Assoc. of America.

Eisenberg, T., & Fried, M. N. (2009). Dialogue on mathematics education: two points of view on the state of the art. *ZDM. The International Journal on Mathematics Education*, *41*, 143–150.

Fischbein, E. (1987). *Intuition in science and mathematics—an educational approach*. Dordrecht: Reidel.

Goldin, G. A. (1998). Representational systems, learning, and problem solving in mathematics. *The Journal of Mathematical Behavior*, *17*, 137–165.

Goldin, G. A. (2000). Affective pathways and representation in mathematical problem solving. *Mathematical Thinking and Learning*, *2*, 209–219.

Goldin, G. A. (2002). Affect, meta-affect, and mathematical belief structures. In G. C. Leder, E. Pehkonen, & G. Törner (Eds.), *Beliefs: a hidden variable in mathematics education?* (pp. 59–72). Dordrecht: Kluwer Academic.

Goldin, G. A., & Kaput, J. J. (1996). A joint perspective on the idea of representation in learning and doing mathematics. In L. Steffe, P. Nesher, P. Cobb, G. A. Goldin, & B. Greer (Eds.), *Theories of mathematical learning* (pp. 397–430). Hillsdale: Erlbaum.

Goldin, G. A. & McClintock, C. E. (Eds.) (1984). *Task variables in mathematical problem solving*. Philadelphia: Franklin Institute Press.

Goldin, G. A., Epstein, Y. M., Schorr, R. Y., & Warner, L. B. (2011). Beliefs and engagement structures: behind the affective dimension of mathematical learning. *ZDM. The International Journal on Mathematics Education*, *43*, 547–560.

Greeno, J. G. (1980). Trends in the theory of knowledge for problem solving. In D. T. Tuma & F. Reif (Eds.), *Problem solving and education: issues in teaching and research*, (pp. 9–23). Hillsdale: Erlbaum (cited in Heller and Hungate 1985).

Hadamard, J. (1945/1996). *The mathematician's mind: the psychology of invention in the mathematical field*. Princeton: Princeton University Press.

Harel, G. (2013). Intellectual need. In K. R. Leatham (Ed.), *Vital directions for mathematics education research* (pp. 119–151). New York: Springer.

Heller, J., & Hungate, H. (1985). Implications for mathematics instruction of research on scientific problem solving. In E. A. Silver (Ed.), *Teaching and learning mathematical problem solving: multiple research perspectives* (pp. 83–112). Hillsdale: Erlbaum.

Hembree, R. (1992). Experiments and relational studies in problem solving: a meta-analysis. *Journal for Research in Mathematics Education*, *23*, 242–273.
Hmelo-Silver, E., Duncan, R., & Chinn, C. (2007). Scaffolding and achievement in problem-based and inquiry learning: a response to Kirschner, Sweller, and Clark. *Educational Psychologist*, *42*, 99–107.
Isoda, M., & Katagiri, S. (2012). *Mathematical thinking: how to develop it in the classroom*. Singapore: World Scientific.
Jeeves, M. A., & Greer, B. (1983). *Analysis of structural learning*. London: Academic Press.
Kahneman, D. (2011). Cognitive limitations and the psychology of science. A public lecture given on December 29, 2011, at the center for the study of rationality at the Hebrew University of Jerusalem, on the occasion of its 20-th anniversary.
Kirschner, P., Sweller, J., & Clark, R. (2006). Why minimal guidance during instruction does not work: an analysis of the failure of constructivist, discovery, problem-based, experiential, and inquiry-based teaching. *Educational Psychologist*, *41*, 75–86.
Koichu, B., Berman, A., & Moore, M. (2006). Patterns of middle school students' heuristic behaviors in solving seemingly familiar problems. In J. Novotna, H. Moraova, M. Kratka, & N. Stehlikova (Eds.), *Proceedings of the 30th international conference for the psychology of mathematics education* (Vol. 3, pp. 457–464). Prague: Charles University.
Koichu, B., Berman, A., & Moore, M. (2007). The effect of promoting heuristic literacy on the mathematic aptitude of middle-school students. *International Journal of Mathematical Education in Science and Technology*, *38*, 1–17.
Krulik, S. (Ed.) (1980). *Problem solving in school mathematics (1980 NCTM yearbook)*, Reston: NCTM.
Larkin, J., McDermott, J., Simon, D. P., & Simon, H. A. (1980). Expert and novice performance in solving physics problems. *Science*, *208*(4450), 1335–1342.
Leikin, R. (2009). Exploring mathematical creativity using multiple solution tasks. In R. Leikin, A. Berman, & B. Koichu (Eds.), *Creativity in mathematics and the education of gifted students* (pp. 129–145). Rotterdam: Sense Publishers.
Leikin, R., & Lev, M. (2013). Mathematical creativity in generally gifted and mathematically excelling adolescents: what makes the difference? *ZDM. The International Journal on Mathematics Education*, *45*, 183–197.
Leikin, R., Waisman, I., Shaul, S., & Leikin, M. (2012). An ERP study with gifted and excelling male adolescents: solving short insight-based problems. In T. Y. Tso (Ed.), *Proceedings of the 36th international conference for the psychology of mathematics education*, Taiwan, Taipei (Vol. 3, pp. 83–90).
Leikin, M., Paz-Baruch, N., & Leikin, R. (2013). Memory abilities in generally gifted and excelling-in-mathematics adolescents. *Intelligence*, *41*, 566–578.
Lester, F. K. Jr., & Garofalo, J. (Eds.) (1982). *Mathematical problem solving: issues in research*. Philadelphia: Franklin Institute Press.
Lubinski, D., & Benbow, C. P. (2006). Study of mathematically precocious youth after 35 years: uncovering antecedents for the development of math-science expertise. *Perspectives on Psychological Science*, *1*, 316–345.
Mamona-Downs, J., & Downs, M. (2005). The identity of problem solving. *The Journal of Mathematical Behavior*, *24*, 385–401.
Mason, J., & Pimm, D. (1984). Generic examples: seeing the general in the particular. *Educational Studies in Mathematics*, *15*, 277–289.
Mason, J., Burton, L., & Stacey, K. (2009). *Thinking mathematically* (2nd ed.). London: Pearson Education.
NCTM (1980). *An agenda for action: recommendations for school mathematics of the 1980s*. Reston: National Council of Teachers of Mathematics. http://www.nctm.org/standards/content.aspx?id=17278. Accessed 20 June 2012.
Newell, A., & Simon, H. A. (1972). *Human problem solving*. Englewood Cliffs: Prentice Hall.
Niss, M. (2011). Reflection on the state of and the trends in research on mathematics teaching and learning—from here to utopia. In F. K. Lester (Ed.), *Second handbook of research on mathe-*

matics teaching and learning (Vol. 2, pp. 1293–1312). Reston: National Council of Teachers of Mathematics.

Poincaré, H. (1952). *Science and hypothesis*. New York: Dover.

Pólya, G. (1945/1957). *How to solve it*. Princeton: Princeton University Press.

Pólya, G. (1962/1965). *Mathematical discovery: on understanding, learning, and teaching problem solving* (Vols. I and II). New York: Wiley.

Pólya, G. (late 1960s). A lecture on teaching mathematics in the primary schools. http://cmc-math.org/members/infinity/polya.html. Assessed 27 Dec 2012.

Schoenfeld, A. (1979). Explicit heuristic training as a variable in problem-solving performance. *Journal for Research in Mathematics Education, 10*, 173–187.

Schoenfeld, A. (1983). *Problem solving in the mathematics curriculum: a report, recommendations, and an annotated bibliography*. Washington: Math. Assoc. of America.

Schoenfeld, A. (1985). *Mathematical problem solving*. New York: Academic Press.

Schoenfeld, A. (1992). Learning to think mathematically: problem solving, metacognition, and sense-making in mathematics. In D. Grouws (Ed.), *Handbook for research on mathematics teaching and learning* (pp. 334–370). New York: Macmillan Co.

Schoenfeld, A. (2007). Problem solving in the United States, 1970–2008: research and theory, practice and politics. *ZDM. The International Journal on Mathematics Education, 39*, 537–551.

Schroeder, T. L., Lester, F. K. Jr. (1989). Developing understanding in mathematics via problem solving. In P. R. Trafton (Ed.), *New directions for elementary school mathematics. Yearbook of the national council of teachers of mathematics* (pp. 31–42). Reston: National Council of Teachers of Mathematics.

Shaul, S., Leikin, M., Waisman, I., & Leikin, R. (2012). Visual processing in algebra and geometry in mathematically excelling students: an ERP study. In *The electronic proceedings of the 12th international Congress on mathematics education (Topic study group-16: visualization in mathematics education)*, Coex, Seoul, Korea (pp. 1460–1469).

Silver, E. A. (Ed.) (1985). *Teaching and learning mathematical problem solving: multiple research perspectives*. Hillsdale: Erlbaum.

Skemp, R. (1976). Relational understanding and instrumental understanding. *Mathematics Teaching, 77*, 20–26.

Sriraman, B. (2005). Are giftedness and creativity synonyms in mathematics? An analysis of constructs within the professional and school realms. *The Journal of Secondary Gifted Education, 17*, 20–36.

Stanovich, K. E. (1999). *Who is rational?* Mahwah: Erlbaum.

Vinner, S. (1997a). From intuition to inhibition—mathematics, education and other endangered species. In E. Pehkonen (Ed.), *Proceedings of the 21st international conference for the psychology of mathematics education*, Lahti, Finland (Vol. 1, pp. 63–79).

Vinner, S. (1997b). The pseudo-conceptual and the pseudo-analytical thought processes in mathematics learning. *Educational Studies in Mathematics, 34*, 97–129.

Vogeli, B. R. (1997). *Special secondary schools for the mathematically and scientifically talented. An international panorama*. New York: Columbia University Press.

Vygotsky, L. S. (1930/1984). Imagination and creativity in adolescent. In D. B. Elkonin (Ed.), *Child psychology: Vol. 4. The collected works of L.S. Vygotsky* (pp. 199–219). Moscow: Pedagogika (in Russian).

Waisman, I., Shaul, S., Leikin, M., & Leikin, R. (2012). General ability vs. expertise in mathematics: an ERP study with male adolescents who answer geometry questions. In *The electronic proceedings of the 12th international Congress on mathematics education (Topic study group-3: activities and programs for gifted students)*, Coex, Seoul, Korea (pp. 3107–3116).

Yackel, E., & Cobb, P. (1996). Sociomathematical norms, argumentation, and autonomy in mathematics. *Journal for Research in Mathematics Education, 27*(4), 458–477.

Mathematical Literacy: What Is It and How Is It Determined?

Chapter 9
"Mathematical Literacy": An Inadequate Metaphor

E. Paul Goldenberg

Abstract I'm reminded of Brian Harvey's 1983 paper titled "Stop Saying 'Computer Literacy'!" His lament was partly that the analogy to *literacy*-literacy is, at best, thin. I've recently been adopted onto a project that keeps talking about numeracy—another adaptation of the L-word. Though I keep referring to my focus as 'mathematics'—which will guide me also in *this* talk—I've become curious about what people mean when they use these literacy-like terms. Googling didn't help except to connect the varied and vague usage with the Real World. Whatever that is. I've struggled with the Real World for years. The real world of children or adults? They're different. What about the real worlds of the barely-subsisting subsistence farmer, the fairly wealthy city-dweller, and the blue collar laborer? And is that what really catches peoples' interest? What about the very *real* world of the mind? To take seriously the idea of serving people well and to avoid limiting or pre-judging their eventual paths, we might focus on the latter.

Many educational terms share a common problem: When *you* or I use the terms, we *do* know what we're talking about. At least sort of. But whoever we're talking with might well have a different understanding, because the terms have no universally shared definition. Without trying to declare what mathematical literacy *should* mean—that would be yet another usage, unshared except among us—I'll punt. I'll take our question "what is mathematical literacy?" to mean "what is a mathematics education that is 'useful' to people?" and will focus not just on the topics of mathematics but on the thinking, the real world of the mind.

Keywords Mathematics education · Mathematical literacy · Numeracy · Mathematical habits of mind

Literacy, as used in the sense of "literacy rate" is the ability to decode and produce culturally useful written forms and sufficient vocabulary to derive meaning from them.

E.P. Goldenberg (✉)
Education Development Center, Inc., 43 Foundry Ave., Waltham, MA 02453, USA
e-mail: PGoldenberg@edc.org

M.N. Fried, T. Dreyfus (eds.), *Mathematics & Mathematics Education: Searching for Common Ground*, Advances in Mathematics Education,
DOI 10.1007/978-94-007-7473-5_9,

The Organisation for Economic Co-operation and Development (OECD) defines *mathematical* literacy as "an individual's capacity to identify and understand the role that mathematics plays in the world, to make well-founded judgements and to use and engage with mathematics in ways that meet the needs of that individual's life as a constructive, concerned and reflective citizen." This goes beyond culturally useful written forms and sufficient vocabulary, but leaves us to invent our own meanings for "well founded judgements"[1] and "use and engage with mathematics." Moreover, because individuals vary, so does the meaning of "meet the needs of that individual's life." And, does "meet the needs of that individual's life" mean life *as it is*, or life *as it might be* if education promoted or even permitted change? Too often—in developing countries and in underserved communities in the United States—abysmal innumeracy is met by "mathematical literacy" expectations that are higher but still abysmal, *better* meeting the needs of lives people already have, but not opening up genuinely new life options. We have a definition, but no clear meaning.

Numeracy, another spawn of the L-word, also has no agreed-upon meaning (Donoghue 2002). Focusing on number, the name claims less breadth than "mathematical literacy." US policymakers and educators tout *basic* numeracy as a survival skill: defense against getting the wrong change, being misled by statistics, buying lottery tickets (mathematician Michelle Manes calls that a tax on people who don't understand probability) and predatory lending practices. In a developing country, numeracy may be a step toward economic development. Marc Tucker writes "Over the long term, basic skills only give you the right to compete against the Third World for Third World wages."[2] Al Cuoco often quips that the real value of mathematics is not as a defense against getting the wrong change, but to let you choose a job that pays well enough that you don't *care* whether the change is correct.

The social goals of promoting economic development and protecting people from a world of exploitative commerce are truly important—perhaps along with promoting health and peace and protecting the planet, these are the most important goals we have. If we think long range—not common in policymaking because politicians need results fast—they set a high bar for learning but, even so, not for any *particular* area of learning, mathematics or otherwise. Moreover, at least in the US, my best guess is that they helped erode intellectual discipline in mathematics education, partly through an increased focus on applications[3] (not within mathematics) and partly through minimalism: the common belief that only a few people need or

[1]Though this calls for logical grounding, the field's response has been mixed. Even the National Council of Teachers of Mathematics recommended (in NCTM 1989) a *reduction* in the emphasis on proof. Though that statement may have been just a poor choice of wording—the same document clearly called for greater emphasis on reason than on rote—many in mathematics education hailed that reduction in emphasis.

[2]Quoted in National Research Council (1989).

[3]A focus on the pragmatic utility of mathematical results works against the development of mathematical sensibility. If a mathematical result's value is its utility, one hardly needs to understand *why* it works or go through the effort to *prove* it, as long as an authority has approved the result. Proof becomes "academic."

can learn high level mathematics. *Mathematics* is seen as exclusive; the rest of us need the less high level and more egalitarian *mathematical literacy*, a "mathematics for all." As a metaphor, literacy also suggests a first step, a remediation, a threshold essential for all, not a long-range goal.

When I was asked to contrast mathematics education with "mathematical literacy," I was reminded of Brian Harvey's 1983 paper titled "Stop Saying 'Computer Literacy'!" His lament was partly that the analogy to literacy was thin. So was the content. Judging by the curricula, computer literacy meant little more than knowing new jargon like RAM and floppy disk.

The literacy metaphor is seductive. A "computer literacy" of technical terms and symbols lets one communicate clearly, ask for help, and read the relevant literature to learn more. It *is* culturally useful. But it is not computer science. Mathematics absolutely *requires* precise communication so much so that it has elaborate and extensive formal languages and a vast technical vocabulary. But its *essence* is in its logic, manner of thinking, investigative style, and unique approach to establishing truth and causality: proof. No natural extension of the meaning of "literacy" captures this essence of mathematics.

Why does this matter? Ideas drive instruction. If we start with an image of a subject that does not capture its essence, we lose a lot for whatever else we might gain, and we should be aware of both sides of the coin as we decide what direction to take.

The problem is this: *any* "mathematics for all" must serve two kinds of students without our knowing, in advance, which are which. It must serve students who will never use even the rudiments of what we teach beyond about fifth grade. After all, how many people actually *need* to know how to add $\frac{3}{7} + \frac{2}{5}$, let alone the quadratic formula or the cosine law? At the very least, these students' time and effort must not be sacrificed: time spent on mathematics is time taken from a new language, psychology, learning to cook or maintain a home or car, or studying music, personal health and nutrition, agriculture.... That time must be justified. At the same time, a "mathematics for all" must not sacrifice the students who will someday choose mathematics-intensive fields, or would have if they had not already been turned off or defeated by drivel passed off as mathematics. Since we don't reliably know which students are which, we cannot build separate tracks, at least not early.[4] To craft a sensible "mathematics for all," standards and curriculum designers must base decisions not on which particular facts, skills, or concepts to teach—beyond the most basic, virtually none are "for all"—but on how to organize the facts that only *some* will need in a way that teaches the kind of thinking that *all* will need: habits of mind that are the foundation for sound reasoning, creative and logical inventiveness, and effective problem solving. Bluntly, "mathematics *for all*" is a sham if it isn't *mathematics* for all. Mathematical literacy can certainly be done right, but I think that the

[4]Culture can exert strong local influences over an individual's interests and efforts. Early tracking can lock in those effects long after experiences grow, tastes and interests mature, and an individual has developed the ego to break with conventions, expectations, and stereotypes. By creating distinct math/non-math tracks too early, we virtually guarantee that the non-math tracked students never make it back into the running.

metaphor, "literacy" in the name, makes it easier not to notice if the mathematics gets thin.

Developing mathematical ways of thinking alone, without learning some key results of such thinking, is not sufficient; nor, frankly, is it possible. To paraphrase Seymour Papert, we can't learn to think without learning to think about something. The assertion that "no particular fact is essential" no more justifies the conclusion that facts don't count, than the statement "there are many right answers" justifies the conclusion that there are no wrong ones. It is not enough to know good ways of thinking—or habits of mind—and how to look up facts when one needs them.

This is obvious in science, where not having extensive knowledge *in one's head* means that one cannot take advantage of serendipity. In medicine, for example, in the pursuit of an answer to one problem one often trips unexpectedly over the solution to a seemingly unrelated problem. One would not recognize the solution without knowing about the problem or the territory within which that problem falls. Likewise, one has no perspective on a historical fact without having a great deal of knowledge into which that fact fits, or against which that fact stands out. Knowledge much broader than one's immediate task is probably always an essential ingredient in creative, productive work.

This is equally true in mathematics, even elementary arithmetic. For example, it requires very particular prior knowledge, not just addition skill, to become interested in the pattern of sums 1, $1+3$, $1+3+5$, $1+3+5+7$, and so on. After all, any set of addition problems must have *some* set of answers, and the answers 1, 4, 9, 16 don't feel special unless they are *already* "good friends" popping up in an unexpected setting. Only prior knowledge gives that sense of surprise—very different processes producing the same results—that leads one, like the serendipitous discovery of some cure, to suspect that there might be some connection worth further attention. Here, literacy may be the *perfect* metaphor: "knowing the literature," not just "basic literacy." In mathematics that might mean being intimate with frequently encountered sets or sequences of numbers, common structures, useful proof strategies, and so on, maintaining the perspective that these are neither the *essence* of the learning, nor can they be treated as the boring first chapter delaying students' encounter with the real subject.

If mathematical habits of mind become a goal, students need *regular* experience inventing mathematical ideas and methods themselves. Not all mathematics need be learned through discovery—there's no time for that anyway—but discovering/inventing is part of *doing* mathematics, "engaging" with it, as OECD puts it, but with fidelity. Perhaps more importantly, it teaches people how to solve problems they have not already been taught how to solve, new problems, as life keeps throwing at us. Learning to be a competent problem solver allows a student to adapt to change, and to remain a valuable and valued resource. That actually does fit with OECD's "literacy" in a deep way.

I Googled to see what else is said about mathematical literacy. The results mostly connect mathematical literacy with the Real World.[5] Whatever *that* is. *Which* real

[5]"The real world is overrated." —Al Cuoco.

world? The real world of children or adults? They're different. What about the real worlds of the barely-subsisting subsistence farmer, the wealthy city-dweller, and the blue-collar laborer? And is the Real World really what catches peoples' interest? It seems hard to believe that what fuels kids' (or adults') fascination with dinosaurs, black holes, *Harry Potter*, puzzle books, and video games is what educators have in mind when they call for Real World applications. In fact, puzzle books are so popular with the general public that even supermarkets sell them. Real World utility is not what drives those sales.

Or maybe it *is*. Maybe we're thinking about the "Real World" in entirely the wrong way! The fact that people *are* fascinated with these things suggests a real world "utility" that the daily-life, application-based, utilitarian movements in mathematics education aren't properly acknowledging. The real world utility lies not in the context to which the thinking is applied, but in the nature of the thinking. Puzzles *are* useful because they exercise the mind.

June Mark, Al Cuoco, and I have long talked about the very *real* world of the mind. *That* real world includes puzzles and mathematics along with *Harry Potter* and black holes—a world of curiosity and imagination, with intense focus and concentration and (except in the pure fantasy worlds) logic. *That* real world is where we need to focus if we are to take seriously being *useful* in people's lives without limiting or pre-judging those lives. If we were kittens, *fun* would be stalking and pouncing, because the pleasure centers in our brain reward us for honing skills that help us survive including keeping our claws sharp. But we're not kittens. Our survival depends on keeping our brains sharp and figuring out the complicated world around us. So, fun for a child is solving puzzles—not just the invented ones we buy, but natural ones like how to get stuff off a high shelf, figuring out the buttons on the TV remote, learning to stay upright on a bicycle, sorting out the world. In mathematics education, *that* real world application—satisfying curiosity and honing the mind—is served by focusing not only on topics or applications but on real problems and the *thinking* that mathematicians (and other "mathematically literate" people) do.

Two decades ago, Al, June, and I took that on when we developed and later published the beta-test version of what we called "mathematical habits of mind" (Cuoco et al. 1996). The divide between the mathematics and math education communities worried us, but what motivated our habits-of-mind work was our observation that the disconnect between the discipline of mathematics and school mathematics was *not* new. School mathematics tended to be a set of mathematical results, often fragmented and stylized, sometimes (presumably) to make them more learnable, sometimes to reflect common (often obsolete) applications.[6] This set included

[6]Standard elementary school content is a legacy of methods optimized for large accurate computations by hand. The algorithms *are* particularly safe and easy for adding long columns of numbers, or performing large multiplications. But *other* methods, even other "basic facts," would better serve today's needs for algebra readiness and mental calculation. Consider the carry-method for, say, 39×65, that begins "five times nine is forty-five, write down the five and carry the four..." That algorithm camouflages the fact that four products are being found and added. If, instead, stu-

what mathematicians knew, but did not reflect the ways mathematicians thought or worked. Our concern was not, of course, that K-12 students weren't studying K-theory or working on open questions, but that what they *were* doing, in domains that were accessible to them, did not resemble what mathematicians do in domains accessible to *them*. What students were studying was *content* from mathematics, not mathematics itself, or its nature. Worse yet, what they were studying was not what develops the real world of the mind.[7] The mathematical habits of mind we articulated (Cuoco et al. 1996, 2010) feed far more than mathematics. They support logical thinking in every realm; they are the skills that allow people to make "well founded judgments" and to meet the needs of an individual's life not only as it is, but *as it might become*.

How Did We Get from Facts to Habits of Mind?

Any set of facts or ideas can be arranged in more than one way to form a coherent curriculum. To find a "best" arrangement, one must consider the goals of the course ("mastering the content," for example, is rarely a complete description of the goals), and such other factors as the students' backgrounds and goals, and the teacher's inclinations and style. What makes an arrangement *coherent* is that it has a "story line," a message about mathematics that is emphasized along with the explicit content.

What Is a Mathematical "Story Line"?

There *are* mathematics texts, especially at the K-12 level, that seem to lack much of a story line at all, or whose "story" is little more than that mathematics consists of facts, skills, and procedures.

---

dents learned "30 × 60 is 1800, write that down; 9 × 60 is 540, write that down," and so on, they would get a better first approximation, see the four products (and thus a better model of the algebraic steps which have no "carry") and would have a generally more accessible method for mental computation.

[7]Developing the real world of the mind might be easier in an alternative curriculum that slices knowledge a different way, replacing traditional subjects like mathematics with "courses" such as Communication (comprised of and needed in mathematics, poetry, politics, law, managing our health…), Reasoning Under Constraint (mathematics, personal budgeting, law, ecology, business management…), Troubleshooting and Problem Solving (mathematics, science, diagnosing a car, computer, or person), and so on. Mathematics can help teach these, but so can other subjects. Replacing current disciplines with such "courses," however, is impractical. On the other hand, if each *current* discipline, with its own unique contexts and facts, were internally organized by the elements of reasoning that make it a discipline, the inevitable areas of overlap and commonality would surely become a mutually supportive theme making transfer among disciplines more natural.

Another common organization, at least in US curricula, has favored *applications* as a story line. The implied message about mathematics is that its value lies in its utility in service of other aims. While the purpose of this article is not to discuss applications, the relationship to the mathematical literacy definition and the popularity of this curricular school of thought warrants attention. Personally, I worry about these issues: (1) Is there solid evidence that applications reliably raise students' interest or are the only (or "best") way to do so? Students for whom the "real life" approach is assumed to be most needed may well be more motivated by good puzzles than by more "real life." (2) Applications tend, at best, to be only pseudo-real, the truly real ones being far too hard, tedious, and (often) boring. (3) "Real" for adults is not guaranteed to be "real" for kids. And where have we learned that kids, even of college age, are notoriously pragmatic in their approach to life? (4) As was said earlier, if mathematics is measured by its utility, it hardly needs proof.

A misunderstanding of "mathematics for all" may be partly responsible for the widespread, uncritical acceptance of the applications-first stance—the assumption being that while not everyone will be a mathematician, you can't survive without mathematics. The first of these assumptions is irrelevant—not everyone will be a historian, either. The second is untrue and also unbelievable—every kid knows many adults who claim to know no mathematics, yet they all seem to survive.

Other story lines are also used to organize mathematics curricula. One traditional view is ladder-like: mathematics builds one step at a time from "basic" building blocks to "more advanced" concepts. In this model, real mathematical thinking is neglected for years, and many (most?) students, never get to see that thinking is part of mathematics.

Historical development or problem solving could also be used as organizers. The same facts and procedures can be organized to support any of these approaches but, just as in literature, what we learn is often determined more by the story line than by the details. All great stories contain the same basic elements—love, power, fear, greed, hate, bravery, self-sacrifice—but the *story* that one remembers is how these elements are put together.

A Proposed Alternative Story Line About Mathematical Thinking

Another view of mathematics is that its story is not about the facts themselves, but about how mathematicians *find* the facts.

Such a story cannot be told *without* facts, of course. To weave a good story whose elements interrelate in interesting ways one must develop the setting and characters well. And a mathematical story cannot be appreciated without students developing some skills that enable them to process the facts with facility, just as one's sense of a piece of literature is marred if one cannot read it fluently.

So facts and skills must still be included, but the selection and organization tell a new story: mathematics is (in part) a way of thinking, a set of "habits of mind."

Reasoning at the Core: Habits of Mind

When Al, June, and I were responding to issues in *mathematics* education, we chose habits of mind as an organizing principle because, as I'll try to illustrate later, that provides fidelity to mathematics and has application beyond mathematics. By "habits of mind," we mean ways of thinking that one acquires so well, makes so natural, and incorporates so fully into one's repertoire, that they become, well, mental habits: not only *can* one draw upon them easily, one is *likely* to do so. We believe that putting habits of mind at the focus in curriculum development is valuable not only for K-12 mathematics learning, but even at the undergraduate level, at least prior to highly specialized courses. The "new" idea here—if indeed it is new at all—is not that mathematics (or some other disciplined study) could be good for one's thinking. That notion goes back at least a couple of thousand years, and probably more. What we propose is not an act of faith that taking mathematics seriously gives one the mathematics directly and (also) improves one's thinking, but almost the reverse: if, among the various principles one needs for organizing mathematics (or other) curricula, one gives top priority to certain ways of thinking, one gets the thinking skills directly *and also improves one's mathematics.*

Some of the habits of mind that we articulated—reasoning by continuity, looking for extreme cases and passing to the limit, seeking invariants, delaying evaluation to seek structure in calculations, and so on—seem distinctively mathematical. Others—tinkering, visualizing, performing thought experiments, generalizing from examples, conjecturing, seeking and describing pattern—are more generic. But even the mathematical ones have common "generic" forms, which should really be no surprise at all. Mathematics is not an alien thought-form, but an extended and sharply honed specialization of normal, human, widespread, *effective* ways of thinking.

And that's the key to why we care at all. While any particular mathematical fact or method will be useful only to some people, virtually *all* people will need the ways of thinking, even the polished and extended versions of them, that mathematicians use.

To reiterate—how many times now?—mathematical habits of mind can't be taught content-free: *content remains important.* Moreover, to avoid limiting future choices that neither our students nor we can predict now, we *must* provide that content. But even as we sometimes apply the content to problems in varied contexts (including mathematical ones) the *organization* of that content, the "story line" conveyed by the content, can not be about daily-life, as if we could predict what students' daily lives, or even citizen-science concerns, would require: it must be about seeing the world mathematically through the kinds of thinking that will allow for flexible, creative problem solving and that will equally well serve a physicist, physician, composer, auto-mechanic, lawyer,[8] investigative reporter, and mathematician.

[8]Certain logic puzzles that appear in recreational mathematics websites and magazines and books and that show up in math classes on rainy Thursday afternoons to kill time (but only for students who have finished all of their "real" work) are used as basic exercises in law, so important that an entire section of the Law School Admission Test (LSAT) is devoted to these "Logic Games".

It's hard to justify a *compulsory* mathematics education *for all* any other way. Eagerness for more people to fall in love with a truly beautiful subject gives mathematics no edge over art, music, or literature. And it's hard to defend a compulsory mathematics of facts and how-to's, even very important ones, on purely utilitarian grounds. That would hardly give mathematics an edge over drawing, home repair, psychology, or understanding one's own body and health well enough to manage one's medical care.

We often sell mathematics with claims like "mathematics is all around you." False. The fact that doors are rectangle-shaped is no more mathematical than the fact that trees are tree-shaped. Things whose properties could be described using the Pythagorean Theorem will continue to work whether or not I know that theorem. *Things* are all around us: the mathematics is what our mind does with those things. Mathematics is in our *minds*.

Illustrating the Claim that Habits of Mind Serve More than Mathematics

Following are five habits of mind that are central to mathematics and have applications or analogues outside of the discipline.

The Inclination to Build Systematic Explanations and Proof Proof, in developmentally appropriate ways, can be part of all levels and all subdisciplines, not just high-school geometry. The *form* is not what counts, and some aspects of its nature will change with the grades, but the act of proving and the structure of proof are essential to mathematics.

Mathematical proof is unique because of the strict criteria by which mathematics judges its reasoning, but the underlying habit of mind—showing how one idea follows from others—is a discipline central to good story-writing, science, legal argument, and, in general, clear thinking. *All* students need this basic idea. Of course, we don't want students to confuse stating sources and reasoning in an essay with stating the givens and theorems in a mathematical proof, and so to be mathematically literate, students need *more* than the basic idea. But the idea that one can chain thoughts coherently in any discipline should be emphasized as we state and analyze proofs in mathematics.

The Inclination to Translate Between Visually and Verbally Presented Information In geometry, we must often make visual sense of verbal descriptions (e.g., "Let point M be the intersection of two medians of triangle ABC inscribed in circle k…"), and translate in the other direction as well. Describing the behavior of a function requires moving between ideas about number, order, visual space, and words. Such translation skills are invaluable outside of mathematics, too: in giving clear directions to a traveler (or interpreting directions), painting a verbal picture of a beautiful landscape, diagramming a corporate structure or theoretical framework, and so on. The mathematical and not-so-mathematical situations differ, of course, but more in detail than in principle.

The Inclination to Tinker Take a moment to think up some uncommon examples of cylinders. The usual *image* is so tied to objects roughly the size and shape of a soup can that people must often work hard to think of extreme cases like a coin or a stiff strand of spaghetti. Why bother? Mostly to become conscious of the *independence* of defining attributes: in this case, recognizing that base diameter and height can vary independently.

Solving a problem, or gaining deeper understanding of it, may also be aided by looking for its independent attributes, changing them, and seeing what results. A problem posed on the Euclidean plane may be re-examined on a sphere or cylinder or torus, or with a taxicab metric. After students have proved that the interior angle sum for a triangle is 180°, let them consider a big triangle drawn, say, on the parking lot; and, still larger, on the northern hemisphere entirely paved over! Figuring out what part of the proof failed, even without considering all the complexities of spherical geometry, helps students notice the assumptions they start with. A problem posed with integers may be re-examined with a superset or subset. Which numbers, for example, are "prime" (defined as having exactly two divisors, itself and the unit) in the set that includes only $2n$ (or $\frac{1}{2}n$) and 1, where $n \in \mathbb{Z}^+$? What if that set included only $4n$ and 1?

When students tinker, they come to recognize the independent attributes of a problem situation; in mathematics, such tinkering can lead to extensions or generalizations of theory. When curriculum tinkers, it provides the counterpoint that allows important ideas to stand out sharply. When medical researchers, administrators, teachers, and engineers tinker, they discover new ways of looking at problems and solving them.

Interpreting Diagrams Popular communication makes extensive use of visual representations of essentially *non*-visual information, like health statistics or diagrams of corporate structure. So does mathematics. But to use visualization well, one must respect its power, recognize its limitations, and know its forms and applications. Diagrams like the one shown here are often given as "visual proof" of the algebraic relationship below it. But this is not, and *should* not be, a proof at all to one who lacks the background and perspective to know (1) what information is overspecified and should be ignored (the absolute and relative sizes of a and b are shown, but irrelevant), and (2) what is underspecified and must be assumed (the angles must be taken as right angles even if, in a rough sketch, they happen not to be), and (3) though the algebraic identity is provably true for a much larger domain, the diagram's "validity" requires a and b to be real and positive. Many curricula use diagrams like this, but fail to help students learn how to produce or translate such diagrams, or understand their affordances and limitations.

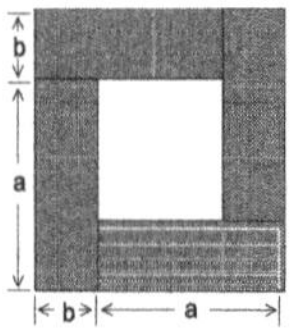

$(a+b)^2-4ab=(a-b)^2$

Reasoning at the Core: Mathematical Practices

Recently, a focus on habits of mind entered American education in a serious way. The Common Core State[9] Standards in Mathematics (CCSSM: NGA and CCSSO 2010) are quite different from the prior individual state standards and even the NCTM *Standards*. For one thing, the CCSSM has two sections, one focusing on mathematical content and the other on mathematical practice. The Content section is new in focus and coherence, but familiar in form—a list of things to know and understand, organized more or less by topic. The eight Standards for Mathematical Practice, however, do *not* just hone or rearrange prior standards. The ideas they contain are not new; the habits of mind that Cuoco, et al., had spelled out in 1996 and 2010 are quite evident. But the idea of treating these as *standards*—that is, *mandating* an emphasis on mathematical practice—is very new and very important. Nearly all states have now adopted CCSSM, requiring attention not just to the facts and procedures that are the results of mathematical work, but to the ways that mathematically proficient individuals *do* that work. The Standards for Mathematical Practice—mathematical habits of mind and action—cut across *all* content topics, pervading K-12 mathematics curriculum and pedagogy in age-appropriate ways.

I will briefly illustrate three of these eight standards, giving its title, a very brief excerpt from the original description, and then my own interpretive examples and commentary.

Look for and Make Use of Structure

> Mathematically proficient students... can see complicated things, such as some algebraic expressions, as single objects or as being composed of several objects.—CCSSM

Taking structure sense (Linchevski and Livneh 1999) seriously is important not just because it appears to be effective pedagogy (e.g., Hoch and Dreyfus 2009), but because it increases fidelity to our subject; it reflects how mathematicians *see* the objects we play with.

The CCSSM description of this practice—and much of the literature on structure sense—makes it seem mostly beyond the level of abstraction appropriate for elementary school. But even young children can learn to see and *reason about* the structure of expressions and relations. One key habit of mind is *deferring evaluation* for certain kinds of tasks. For example, when we present second graders with

[9]In the United States, all control and authority over education must remain at State level; our Constitution does not allow the Federal government to establish central standards or control. The Balkanization of US, education eventually led States to organize against the chaos and establish a common set of standards for education.

$7 + 5 \bigcirc 7 + 4$ and ask them to use $<$, $=$, or $>$ to compare the two expressions, they typically perform the calculations first and then compare; often enough, that's what they're explicitly *told* to do. Random arithmetic practice often has that character—there is nothing *but* the calculation to notice—but the point of this exercise is for students to see *structure*,  $+ 5 \bigcirc$  $+ 4$ or $\clubsuit + 5 \bigcirc \clubsuit + 4$, rather than focus on arithmetic. As long as we don't muddy the idea by talking about "symbols standing for numbers," which is distracting and irrelevant, second graders readily see that, whatever might be in my hand,  $+ 5$ is more than  $+ 4$. The *structure* is "something plus 5 compared to the same thing plus four." An inclination to defer evaluation—to put off calculation until one sees the overall structure—helps older students realize they don't need common denominators for $1\frac{3}{4} - \frac{1}{3} + 3 + \frac{1}{4} - \frac{2}{3}$. Students who do not immediately plunge into left-to-right evaluation can find structure that makes this a trivial mental computation. When students begin to solve algebraic equations, a related idea helps them see $3(5x - 4) + 2 = 20$ as "something plus 2 equals 20" and conclude—using common sense, not mechanically applied "rules"—that $3(5x - 4) = 18$.

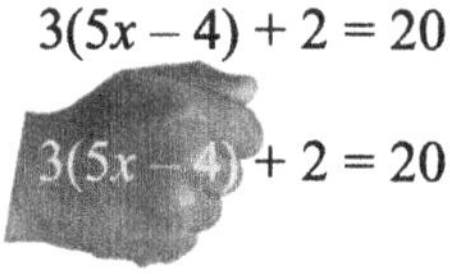

In fact, from such reasoning, they can *derive* the rules for algebraic manipulation, rules that may otherwise seem totally arbitrary.

Look for and Express Regularity in Repeated Reasoning

> Mathematically proficient students notice if calculations are repeated, and look both for general methods and for shortcuts. Upper elementary students might notice when dividing 25 by 11 that they are repeating the same calculations over and over again, and conclude they have a repeating decimal.... Noticing the regularity in the way terms cancel when expanding $(x - 1)(x + 1)$, $(x - 1)(x^2 + x + 1)$, and $(x - 1)(x^3 + x^2 + x + 1)$ might lead students to the general formula for the sum of a geometric series.—CCSSM

Again, though the examples use advanced arithmetic or algebra, this habit of mind can be developed much earlier. For example, children who know most of their "multiplication facts" and know how to use those to square small multiples of 10, like 30×30, can enjoy the following activity in which the teacher does not talk at all, but "enacts" a pattern and invites the children to join in. The teacher sketches a segment of number line, draws a pair of arrows from some number and, where the arrows meet, writes the result of multiplying that number by itself.

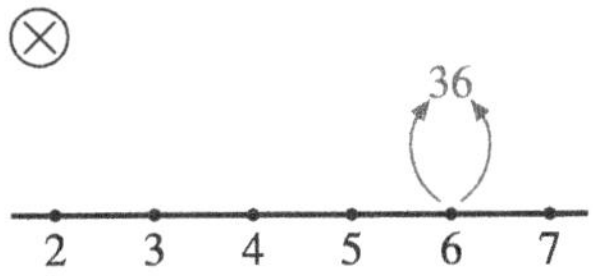

Still without speaking, the teacher draws another pair of arrows, from the nearest neighbors of that number and writes their product, perhaps touching the number and the × sign to clarify what she is doing. (The silence is more than a dramatic gimmick. The lack of teacher talk does rivet visual attention on the action, but more importantly, it avoids camouflaging or competing with the pattern children hear in the numbers they call out.)

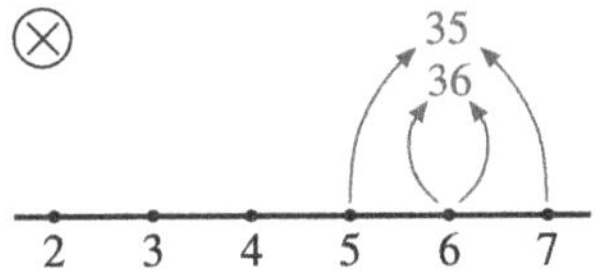

Then the teacher chooses another number, anywhere on the line, draws a pair of arrows from it and writes its square. At this point, most children see what this notation means, so when the teacher draws arrows from that new number's neighbors, she can turn to them and offer the marker, or silently beckon them to call out what number to write. The teacher continues, choosing a new "center" number and draw-

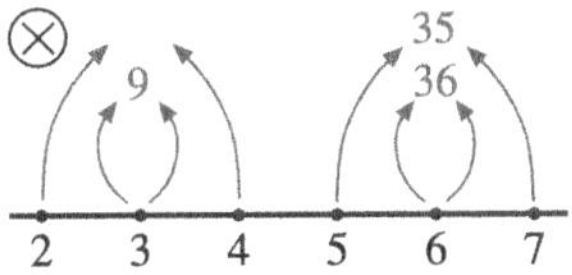

ing its two arrows, but because children see the structure of the activity (though not yet any numeric pattern), the teacher no longer writes any numbers spontaneously but *always* seeks a called-out number from the class. After continuing in this way for several more numbers, the display might look like this.

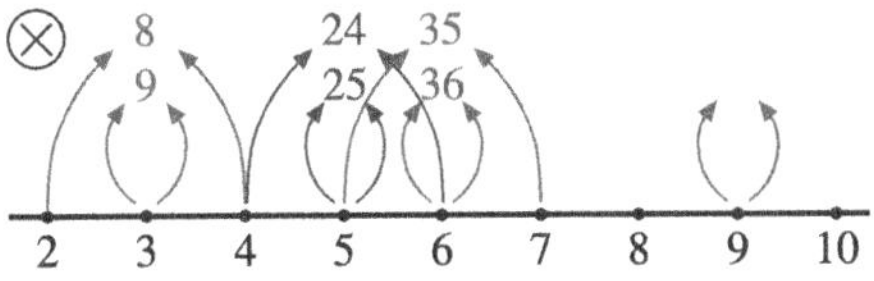

A few steps later, it might look like this.

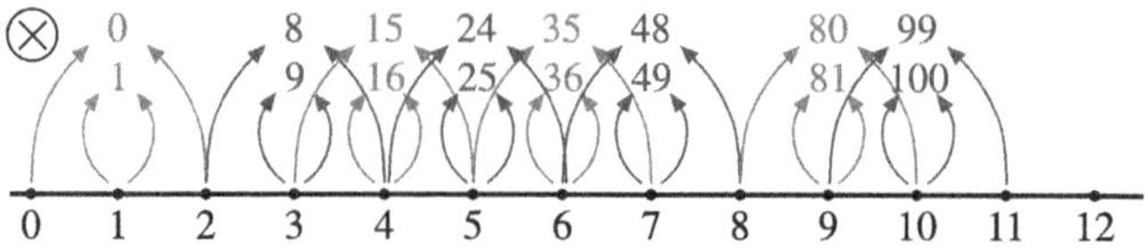

Everything so far has involved children practice facts they already know. Perhaps, for some, the pattern already helped them with facts they were unsure of, like using 8×8 to help them remember 7×9. But now they do something new.

Still without asking the children anything *about* the pattern they've been using, or commenting herself, the teacher draws a new number line segment like this and again beckons children to call out the number.

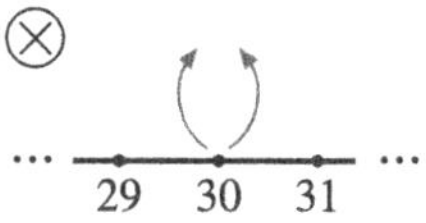

As before, the teacher draws arrows from the nearest neighbors. But this time, if they call out 899, they are *using* a pattern that they abstracted from repeated reasoning.

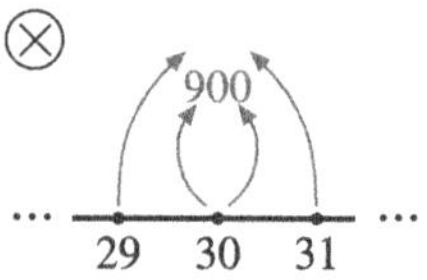

Students have not written the pattern in the language of algebra, or even expressed the pattern orally in their own native language, but they have expressed, in *action*, that they have seen the regularity and can use it.[10]

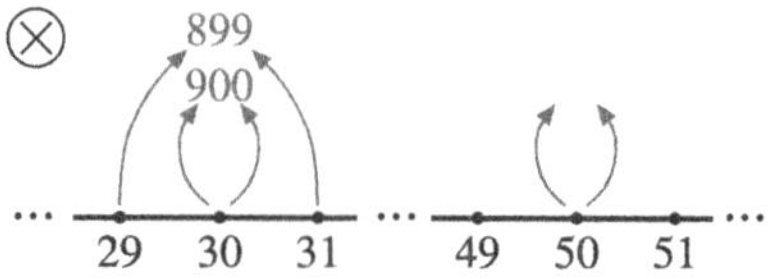

[10] See http://thinkmath.edc.org/index.php/Difference_of_squares for one approach to developing the language of algebra "naturally" with elementary school students as they attempt to describe the pattern they've seen and applied.

Make Sense of Problems and Persevere in Solving Them

The title is almost a cliché. We've all talked about sense-making for forever, and we're always admonishing students to "keep trying," so what's new here? The descriptive text that follows the title does include those old chestnuts, but also includes two key ideas that are at the very core of problem solving but are often overlooked.

> Mathematically proficient students start by… looking for entry points to [a problem's] solution…. They monitor… their progress and change course if necessary.—CCSSM

Many (most?) students see mathematics as a collection of rules to know and follow. Genuine problems—whether in mathematics or in life outside of school—are not so cut and dried; they appear without asking what chapter we've just studied. So, too, the pseudo-real problems of good high-stakes tests require students to *think*, to go beyond the rules. Even standard word problems require one to figure out where to start and what to do next. *There is no "formula" for that*, and that is one reason why students find them hard. The same is true of finding proofs. Figuring out what might be a good place to start is key. And, while students do need the patience and stamina to persist in a potentially profitable direction—what the standard calls perseverance—equally important is the ability (and inclination) to "monitor… their progress and change course, if necessary."

Well-constructed, non-arbitrary, logical puzzles place those particular skills front and center. In crossword puzzles, one must often attempt several "across" and "down" clues before finding one that can be filled in with confidence. Each step helps with others that were too hard at first. Sudoku and KenKen® also typically require you to search before finding something you can do. At least in this one sense, (certain) puzzles model mathematical investigation, whether or not their content is mathematical. Suitably designed, puzzles can also be vehicles for mathematical *ideas*—the content, not just the practice of mathematics—in a way that crossword and Sudoku puzzles are not. They must also not be more work than they justify in curricular worth. The same goes for word problems: the work of decoding the verbal camouflage should not distract from the mathematical ideas. Too often, the extra verbiage is the only hard part, while the actual mathematical content turns out to be trivial. Good puzzles must also not feel like arbitrary "tricks" to the student, more about psyching out the puzzle's author than about logic. A satisfying puzzle is one that you don't know how to solve at first, but can figure out logically, without a "trick."

Puzzles (always modified by "suitable") have some advantages over "standard problems" especially for students who are not yet committed mathematicians. When students don't know how to start a problem, they often feel either that *they* are inadequate, or that their teacher failed to teach them enough. But puzzles are *supposed* to be puzzling! They give permission not to know how to start before starting. Giving oneself a little extra time can, with nurturing, become a habit of mind. Puzzles build stamina and confidence for problem solving. They are genuine problems to solve—true to real life—not exercises in following a rule or template. They exercise mathematical habits such as experimenting, juggling multiple constraints and, depending on the nature of the puzzle, others (e.g., visualizing) as well. Even that elusive "Real World" highly values the ability to solve puzzles—for example those Logic Games of the Law School Admissions Test—which have been part of the "literature" of society, part of our culture, as far back as historians have managed to plumb.

Again, mathematics is not *just* everyday common sense, but a sharply honed extension of natural ways people think. Moreover, people crave the thinking, and even the honing of it, enough to make up stuff that gets them to think. When people don't have lots of prerequisite mathematical machinery or complicated things to think about, they invent opportunities in the form of puzzles. Puzzles—at least some of them—are "pure mathematical thinking" without the usual content trappings.

This long defense of puzzles is not to suggest that puzzles replace other important parts of a mathematical education, but that they be taken seriously! They don't replace mathematics problems (except dumb ones); rather, they *are* a kind of mathematics problem, a kind that is respected in the mathematics community but often considered mere entertainment in schools. Nor is the goal to include objects specially crafted to be called "puzzles." Genuine mathematical research—"open questions" to students if not to mathematicians—give learners much of the same valuable experience seeking an entry point that puzzles give. Time is tight, but puzzles or research problems that are well selected and crafted and supported by a teacher whose understanding of mathematics is deep enough can pay for the time they cost by focusing on mathematical practice, embedding the facts as a byproduct, and providing the needed practice by being more engaging and therefore commanding more attention. If we are not harassed by tests that value miscellany over thought, research experiences and puzzles are excellent vehicles for content, and provide structure and strategies that optimize students' abilities to learn the content.

Redefining Literacy May Be too Hard. Redefining "The Basics" Might Be Possible

Reasoning is Basic

Arithmetic facts are basic. But if we start with a more ambitious view of the mathematics we want students to learn—for example, if we favor mental algorithms over paper ones, and if we keep algebraic goals in mind—we might well favor a different set of basic facts, even to serve purely computational goals. Most children can become quite good mental calculators if they build a strong set of ideas[11] about 10: pairs of numbers that make 10, adding or subtracting 10 to/from anything, multiplying by 10, subtracting or adding 10 and then adjusting (in order, e.g., to add/subtract 8 or 12 to anything mentally), and extending all of these to multiples of 10 and then to powers of 10. Starting humbly with "how many fingers *don't* you see?" to learn pairs that make 10 and learning those playfully to full mastery,[12] nearly *all* seven year olds fairly quickly make the transition to derive pairs to 20, and when they succeed and are good at it, they almost trivially extend it to pairs to 30 and feel brilliant because the numbers are "so big." Pairs to 100, first starting with multiples of 10 (seventy, thirty), come easily as simple extensions of the same pair to 10 (seven, three). Then children learn to adjust when they hear seventy-one. This is not just a way to practice "basic facts," but a use of *structure* in mathematics, a way to build a generalizable sense of mathematical properties of arithmetic. Another basic for the early grades is doubling and halving—*all* numbers, not just "basic facts"—entirely mentally, not with a paper-and-pencil rule. As with the "everything-about-ten" idea described above, this *does* practice "basic facts" and give a solid and extendible foundation for mental arithmetic, but again it does more. It uses the "natural" (pre-multiplication) ideas students have, later to be formalized as the distributive property, and builds a strong base for applying that property in a formal way in later grades. It supports memory, but it is more importantly about a mathematical structure, and not *just* memory. Such games—pure drill, but no kill—get children comfortable handling the tens and ones independently, adjusting, using approximation to support *exact* computations and, by grade four, handling most two-digit (and some three-digit) addition, subtraction, and multiplication naturally in their heads in a way that foreshadows the algebra that we want them later to find "natural" (recall footnote 6 above).

Literacy is basic, too. Even "literacy" in mathematics—the symbols, lingo, and common knowledge of the discipline—is basic. But *mathematics* is more than *any* interpretation of literacy. Mathematical content certainly *is* a service to other disciplines, but the essence, the real gift, of mathematics is the high refinement of our natural ways of reasoning, its logic and the precision of its expression of that logic. To educate students who "make well-founded judgements and use and engage with mathematics in ways that meet the needs of that individual's life as a constructive, concerned and reflective citizen," we need to keep that essence of mathematics—which is to say, *mathematics*—at the center of curricula we develop and of the preparation of mathematics teachers at all levels.

[11] For more about how this approach is used in at least one curriculum, see http://thinkmath.edc.org/index.php/Addition_and_subtraction#All_about_10.

[12] "My fingers are tired, so I'll just *tell* you how many I'd hold up, and you say how many you don't see. OK? Seven. (three) Eight. (two) Two. (eight)" and so on. Playfully and at a lively pace.

References

Cuoco, A., Goldenberg, P., & Mark, J. (1996). Habits of mind: an organizing principle for mathematics curriculum. *The Journal of Mathematical Behavior*, *15*(4), 375–402.

Cuoco, A., Goldenberg, P., & Mark, J. (2010). Organizing a curriculum around mathematical habits of mind. *The Mathematics Teacher*, *103*(9), 682–688.

Donoghue, J. (2002). Numeracy and mathematics. *Irish Mathematical Society Bulletin*, *48*, 47–55.

Harvey, B. (1983). Stop saying "computer literacy"! http://www.cs.berkeley.edu/~bh/stop.html.

Hoch, M., & Dreyfus, T. (2009). Developing Katy's algebraic structure sense. In *Proceedings of CERME 6*. 28 Jan–1 Feb, Lyon, France. www.inrp.fr/editions/cerme6.

Linchevski, L., & Livneh, D. (1999). Structure sense: the relationship between algebraic and numerical contexts. *Educational Studies in Mathematics*, *40*(2), 173–196.

National Council of Teachers of Mathematics (1989). *Curriculum and evaluation standards for school mathematics*. Reston: NCTM.

National Governors Association Center for Best Practices, & Council of Chief State School Officers. (2010). Common Core State Standards: mathematics. http://www.corestandards.org/assets/CCSSI_MathStandards.pdf.

National Research Council (1989). *Everybody counts*. Washington: Nat. Academy Press.

Chapter 10
Reflections on Mathematical Literacy

What's New, Why Should We Care, and What Can We Do About It?

Anna Sfard

Abstract Today's widespread concern with mathematical literacy indicates the educators' commitment to the old principle, formulated more than three decades ago by Hans Freudenthal: "The child should be able to use in society what [he or she] has learned at school." As obvious and uncontestable as this maxim seems to be, the question of how the term *mathematical literacy* should be interpreted and implemented has been an object of debates and disagreements. One position on this issue has been presented by Paul Goldenberg in the preceding chapter. With Paul's critical statement serving as a point of departure, the authors of this chapter tackle such questions as *What do we have in mind while talking about mathematical literacy? Should we give in to the utilitarian approach to school mathematics that seems to transpire from this latter term, whatever its interpretation? And even if we agree that the students' ability to broadly apply mathematics should be fostered, will we ever be able to overcome the inherent situatedness of learning? How?* Coming from a diverse groups of writers—two researchers in mathematics education, a psychologist, a mathematician, and a policy maker—the five attempts at answering these questions may sounds too diverse to end up in agreement. Still, they certainly make for an interesting, important, and ultimately useful conversation.

Keywords Literacy · Mathematical literacy · (Formal) mathematical competency · Transfer · Situatedness of learning · Mathematical discourse

With contributions by
Abraham Arcavi, Weizmann Institute of Science, Rechovot, Israel
Ron Livné, Mathematics Institute, The Hebrew University of Jerusalem, Jerusalem, Israel
Iddo Gal, Ph.D., Department of Human Services, The University of Haifa, Haifa, Israel
Hannah Perl, Israel Ministry of Education, Jerusalem, Israel

A. Sfard (✉)
Faculty of Education, The University of Haifa, Haifa 31905, Israel
e-mail: sfard@netvision.net.il

M.N. Fried, T. Dreyfus (eds.), *Mathematics & Mathematics Education: Searching for Common Ground*, Advances in Mathematics Education,
DOI 10.1007/978-94-007-7473-5_10,

Introduction

These days, the word *literacy* is in everybody's mouth. Either alone or preceded by a noun or adjective—*computer, technological, digital* or *mathematical*—it features prominently in educational discourse and appears with increasing frequency in written and electronic media. And yet, as is often the case with buzzwords, this "next big thing" of the educational parlance often seems to be bringing more confusion than enlightenment. As observed by Paul Goldenberg in the preceding chapter, the term *mathematical literacy* is capable of baffling even the almighty Google. Clarifying the meaning of the expression is one of the goals of the present chapter, in which two researchers in mathematics education (Arcavi and Sfard), a mathematician (Livné), an applied psychologist interested in adult numeracy skills (Gal), and a policy maker (Perl) respond to Goldenberg's challenging position on the issue of mathematical literacy.

To initiate this conversation and help ourselves with getting a better sense of the nature and significance of the topic, let me outline the phenomena that can be seen as having given rise to the present prominence, indeed, to the very appearance of the expression *mathematical literacy*. Our wariness of educational fads aside, let us try to understand the weighty problems that this term may be signaling.

As far as I can tell, the story of mathematical literacy begins in the last quarter of the previous century, and this is true even if the term *mathematical literacy* itself did not appear until later. Its seeds can be found, among others, in the seminal work of Hans Freudenthal, guided by the belief that "[t]he child should be able to use in society what it (!) has learned at school" (Freudenthal 1978, p. 44). As straightforward as this principle might have appeared to its followers, research findings soon began showing that it might be rather difficult to implement. In the late 1980s researchers specializing in cross-cultural and cross-situational studies—with Jean Lave (1988) being probably the most outspoken among them—came up with the disquieting declaration: Most people, even those who had been quite successful in their school mathematical learning, tend not to use mathematics in day-to-day situations in which such use could be of help. True, the mathematics-eschewing problem solvers may still be able to manage, but this, as well as the question of how they are able to do this, is a different story. What is relevant to our present topic is the fact that mathematical knowledge stored in the successful high-school graduates' minds does not seem to make it to those places for which it was meant. Although this phenomenon was not altogether unknown before Jean Lave and her socioculturally-minded colleagues published their findings—indeed, psychologists had been talking about it ever since the word "transfer" was coined by Thorndike and Woodworth (1901)—it was the work of the sociocultural researchers that made the decisive impact and highlighted the need to rethink schooling. These latter scholars' new rendering of the old story offered a novel vision of mechanisms of learning and brought home to us the significance and possible ramifications of the phenomenon that, in its revised version, is now known as *situatedness of learning*.

The fact that what is being learned is tightly dependent on specificities of the situation in which it is learned may have far reaching implications for mathematics and its teaching. Mathematics has always been an obligatory part of school curricula. The traditional argument in support of this practice has been that mathematics, because of its abstractness and the special reasoning techniques, is a universal means for describing the world around us and thus constitutes a necessary ingredient of everybody's problem-solving toolbox. This belief goes hand in hand with the conviction that equipping students with the tool suffices to ensure that they would use this tool whenever necessary. In the light, however, of what the socioculturalists claimed to have shown in their research, this belief loses ground.

To put it differently, it seems that there are two types of mathematical competencies, neither of which develops spontaneously in the mere presence of the other. As it turns out, only one of these competencies, which I will call *formal mathematical*, or simply *mathematical*, is usually fostered in schools. This competency expresses itself in the familiarity with the tool and in knowing *how* to use it. The other competency involves the capability for deciding *when* to use mathematical concepts and techniques. To the best of my understanding, it is this latter competency that people have in mind while speaking about *mathematical literacy*.

The fact that those mathematical abilities that we need throughout our lives are not the same as the skills and understandings developing in schools has been officially acknowledged in the beginning of this century, when the international comparative study PISA was first launched. According to its founders, PISA tests the participants' mathematical literacy, and thus complements the older comparative study, TIMSS, which measures mathematical competency that is being developed in school.

This brief historical account explains the present attention to the issue of mathematical literacy and the current calls for a change in school curricula. Aware that our students' ability *to summon mathematics when it is most needed* will not develop by itself, policy makers require that those who teach mathematics stop acting according to the principle "take care of the 'how', and the 'when' will take care of itself". Attempts to turn the development of applicable mathematical skills into an explicit goal of school learning are a global trend these days.

And yet, be the reasons for these new initiatives as good as they appear, not everybody would readily jump in; at least not without some critical reflection. This is best instantiated by inspired and inspiring ideas presented in Paul Goldenberg's chapter. As Paul tries to make clear, almost any aspect of this new global trend toward mathematical literacy can—and should—be questioned. To begin with, what do we have in mind while talking about mathematical literacy? In particular, what can be done, if anything, to bar the unwanted entailments of the metaphor of literacy, against which Goldenberg warns us already in the title of his chapter? However the term "mathematical literacy" is interpreted, it seems to have to do with the ability to treat mathematics as a means to some practical ends rather than as a goal in itself. It is this usefulness of mathematics, therefore, that becomes the main justification for our as yet unshaken belief in the slogan "mathematics for all." But should we really give in to such a utilitarian approach, renouncing the time-honored tradition of teaching mathematics for what it is: one of the most astounding achievements of the human intellect? And besides, even if we do agree that the students' ability to broadly apply mathematics should be fostered, will we ever be able to overcome the inherent situatedness of learning? How? These and similar questions are more than enough to start us on an interesting, important, and ultimately useful conversation.

Reflecting upon the Adequacy/Inadequacy of the Mathematical Literacy Metaphor—Some Thoughts Inspired Paul Goldenberg's Paper

Abraham Arcavi

I enjoyed very much reading Paul Goldenberg's thought provoking paper. In my view, it challenges the pervasive preoccupation (or perhaps a fad?) with mathematical literacy initiated and invigorated by the OECD through the introduction of the PISA exams, which are deeply influencing almost all educational systems (including those who do not participate in the exams).

Whereas advocates of mathematical literacy deal with refinements of answers to the question "what is it?" and they advance reasons for "why is it important?", Paul Goldenberg (according to my reading of his paper) poses some challenges to the idea. He seems to be posing the question "is it?" (i.e. a worthwhile idea). Paul suggests that the teaching of mathematics can (or should) be organized around other organizing principles than mathematical literacy (commonly envisioned as related to "applications").

I found myself in agreement with almost all of Paul's articulate and eloquent exposition, and in particular with the following:

- Stressing not only the "all" but also the "mathematics" in the phrase "mathematics for all".
- Sound background knowledge is a necessary pre-requisite for creativity, for serendipity and for a strong sense of surprise in mathematics learning.
- "Knowing the literature" may be a better characterization/metaphor of what to pursue than "basic literacy".
- There is a multiplicity of distinct "real worlds", and thus the so called applications may not necessarily engage students.
- There is a pressing need for an organizing principle, a "story line", around which to pursue the teaching of mathematics—applications is not the only one, and possibly not the most appropriate.
- The organizing principle for the teaching of mathematics proposed is "habits of mind" which have to be anchored in mathematical content.
- "Habits of mind" include those which are distinctively mathematical (e.g. reasoning by continuity, seeking invariants, looking for extreme cases and passing to the limit, observing regularities and seeking structure) and those which are more generic (e.g. building systematic explanations, translating between different forms of representation of information, visualizing, performing thought experiments, conjecturing, and my favorite—the inclination to tinker).

In my reaction, I would like to concentrate on a selection of the many thoughts Paul's talk provoked in me. Firstly, it is worth discerning certain points of convergence between opposing views regarding mathematical literacy. Secondly, I would like to propose my own views on some of these convergence points.

First convergence point. Both the arguments of the advocates of mathematical literacy and Paul's criticisms of it are a reincarnation of the discussion around a quite old question: "why math?" Whereas the answers to this question offered by both perspectives may differ substantially, they seem to tacitly agree that mathematics should remain a school subject. Let us remind ourselves that this agreement should not be taken for granted, given, for example, the suggestion to subsume mathematics within other subjects in elementary school (Benezet 1935), or the dramatic calls such as "Let's eliminate math from schools" (Schank 1987), or here in Israel, Professor Yair Karo's appeal of some years ago to end the mathematical studies of a large percentage of the general population at the end of 9th grade.

Second convergence point. Whereas there are substantial differences in how to achieve it, a main goal of education seems to be shared by both perspectives, as stated outright by Paul: "The social goals of promoting economic development and protecting people from a world of exploitative commerce are truly important—perhaps along with promoting health and peace and protecting the planet, these are the most important goals we have." It seems that mathematics education can and should play an important role towards the achievement of these goals.

Third convergence point. Whereas "habits of mind" seem to be in stark opposition to "applications" as the organizing principle for teaching mathematics, both are presented from an 'utilitarian' point of view—namely best serving the needs of future citizens living within a rapidly changing world.

I fully endorse the first convergence point by stating that, in my view, mathematics should not be questioned as a school subject throughout K-12. However, I would like to take issue with the other two.

Demanding that one of the main goals of mathematics education be the advancement of social equity, protecting people from exploitation, promoting health and peace and saving the planet is laudable; however, it may be a bit unrealistic. There is only so much one can load on the shoulders of just one school subject, regardless of how noble these goals may be.

Regarding the next convergence point, are the utilitarian goals the ultimate/only reason for learning mathematics? Whether we take applications or habits of mind as the organizing principles, would the only purpose for that be to furnish the future citizen with a survival kit, and helping the planet? I would like to remind to us of old fashioned, maybe even romantic goals for teaching and learning mathematics, such as producing enjoyment, intellectual pleasure, fulfillment of a sense of aesthetics, and also just gaining plain old non-applicable knowledge and appreciation of an important component of our cultural heritage.

Finally, I would like to propose a possible meeting point of the different perspectives. From Paul's talk, one may be tempted to conclude that the habits of mind are exclusively related to abstract and formal mathematics, and thus they seem to be at odds with applications of mathematics. It does not have to be so. It is possible to find interesting and authentic applications upon which the habits of mind advocated by Paul not only can be nurtured and developed but they are needed as meaningful tools for those applications. Good curricula have shown that this is possible and perhaps these will support intellectual enjoyment too.

Mathematical Literacy: For Whom?

Ron Livné

In her introduction to this chapter, Anna Sfard presented Paul Goldenberg's "position statement" as our starting point. Having read Paul's chapter, I agree with her. The article describes an interesting suggestion in Mathematics Education—developing "Habits of Minds"; it criticizes another—mathematical literacy. My argument will be that these suggestions were adopted, by the CCSSM[1] and by the PISA,[2] because they fit their respective policies. My conclusion will be that neither should be taken as an over-arching principle for everybody, but rather, as ingredients which can contribute to an efficient curriculum if used correctly and with the right dosage. Moreover, the right dosage is neither the same in all circumstances nor for everyone. Throughout, I will discuss only the High School level.

The views of PISA were shaped by the European educational systems: children are sorted into different strands with rigid curricula; their possibilities after school are largely dependent on the strand in which they find themselves and their results in a national exam (matriculation, baccalaureate) at the end. On the other hand, in the American system the curriculum is less dictated: it mainly has to satisfy standards. The high school certificate is local and less regulated, the sorting is less directed. The SAT[3] exam is only for college; the AP[4] Calculus tests are for 6–7 % of the age group.[5]

The High School students in Europe must buy a package deal: the strand that fits their goals best among those available to them. In the USA, the students make their own package with their goal as guide. The minority (6–7 % of the age group) who will go into math-rich further studies often take the AP tests. Very few people are

---

[1]Common Core State Standards Initiative: see http://www.corestandards.org/, retrieved August 2, 2012.

[2]Program for International Student Assessment: see http://www.oecd.org/pisa/, retrieved August 2, 2012.

[3]Scholastic Aptitude test: see http://sat.collegeboard.org/home, retrieved August 2, 2012.

[4]Advanced Placement: see http://apcentral.collegeboard.com/apc/Controller.jpf, retrieved August 2, 2012.

[5]I was unable to get this and similar statistics from an official source, and had to compute them by dividing the number of people taking the exam (given by the administrating authority) in a given year by a official estimate of the age group (the Bureau of Census of the country or the CIA tables; the country entry in Wikipedia is a good place to start). These figures should be viewed as approximations.

worried about pedagogical considerations there. For the much larger group who go to college, algebra is considered a key for future opportunities, an opinion popular in the last two decades, adapted by CCSSM and by Achieve,[6] for example. Indeed, algebra is useful for a larger segment of College students than is the material in the AP tests (which presupposes algebra). But as it is commonly taught, algebra is a procedural affair. In the lower levels, Math requires students to perform algorithms (one thinks of a Montessori task in Kindergarten, in which filling a jug of water and from it to a cup is broken to 28 steps in strict order; perhaps extreme, perhaps fun). The higher levels require finding and delineating the solution more independently. The "Habits of Mind" approach goes beyond such needs by emphasizing the way one reaches solutions. It aims to teach structured research, with the researching mathematician as paradigm. This point of view is helpful in algebra and onwards; it has something to offer even earlier, in pre-algebra.

On the other hand, in European systems the mathematical literacy approach aims at the lower echelons: people who want the national certificate but who will not study further or will not need formal mathematics later. The package deal of the national certificate enables these systems to make such students study mathematics beyond their future economic needs. This ability could be used for civics. The media is full of mathematical data; number sense, order of magnitude, change of scale, ability to digest visually displayed numerical data and to understand its implications; rudimentary statistics; curves that show evolution in time. All these do not require a chain of logical operations and decisions, but rather the connection with reality and the translation between forms of representation. These require more maturity than is available in pre-high school students. Connections with the real world, as reflected for example in the media, might be a very real factor in making such an approach appealing and motivating to students who are motivated to study but are not math oriented. This approach also mixes very well with other subjects in the curriculum. It is possible to use connections to reality to motivate—admittedly to a lesser degree—geometry and even rudimentary trigonometry; but geometry, even in Europe, is taught mainly as a mean to develop the mind. In analogy with this, for the same population, The "Habits of Mind" approach that Goldenberg described and helped develop is an appealing suggestion to bring in some amount of algebraic-type, multi-step reasoning.

While these approaches have developed in specific contexts, global factors are bringing them into contact. The OECD[7] is tripling the size of its population, as more than a billion Chinese, more than a billion Indians, and a few hundred millions from other Asian nations are joining the more-than-a-billion of those already there. Technology, and behind it Science and Math, are very easy to import or export (the voguish term is "to offshore"). Chinese manufacturing power seems almost unlimited; Indian programmers and engineers, speaking good English, go abroad or

[6]For Achieve's agenda of "college and career readiness" see http://www.achieve.org/.

[7]Organisation for Economic Co-operation and Development: see http://www.oecd.org/, retrieved August 2, 2012.

even work from India. Even if the demographical influences fell to zero today, the effect will still be felt till 2030 or a little later.

The influence on the less mathematically oriented is even more significant (and came earlier). A receptionist for a Chicago hotel can be physically in Delhi. Production lines, anything that can be automated/outsourced, become less profitable because of China (products) and India (services), while growing unemployment is an ever-present worry in the west. Jobs in large number—and still increasing—are globalizing. The official opinion of the OECD is that the quality of high school education is a major asset in the global competition.

For the higher level math students there seems to be an agreement as to what should be done: the economical value of a first degree in Science, Programming, and Engineering comes down to India-China prices; but the value of an appropriate second degree, certain interdisciplinary subjects, or a combination with a Business degree may actually increase. This has a major influence on the high end of High-School Math Education (AP tests in the USA; A-level tests in the UK: 6–7 % of the age group): the separation between the mathematically gifted and the less so gets steeper. The 6–7 % may contract to an even smaller proportion of technological leaders.

There is less agreement on what needs to be done with the less mathematically sophisticated children, and this is where the debate over mathematical literacy takes place. The difference of opinions is made more acute by the fact that the international exams are pushing for internationally uniform norms. For the USA the PISA is particularly problematic: it mainly values Mathematical Literacy. The math content is too low; in particular, the level required in algebra is minimal, and dexterity in handling mathematical expressions is not tested at all because—so seem to think the PISA administrators—Mathematics is not valuable in itself, only as a part of a "real" context. This may be true if by Mathematics one means just technique: the Habits of Mind approach can make a better case for including more math and algebra in the capacities required in the PISA if they can be combined into it. (With this, one should not forget that even in the mathematical literacy aspect the PISA is not adequate, though I have seen elsewhere good PISA-type questions.)

The principles we have seen in Goldenberg's chapter are useful to handle algebra intelligently. And the suggestion to connect algebraic and visual is very sensible too. But why not combine a story and some math? Here's an example. At a price of $100 per ticket, 500 will be sold; for every $ less per ticket 10 more tickets will be sold. When will the revenue be maximal? The apex of a parabola gives a very useful model. This idea can be developed; the behavior of a parabola around its apex (say the apex is a maximum) shows that the benefit (revenue) is optimal in a middle ground compromise between two policies (increasing ticket price or selling more tickets); near the optimum a small change in policy in either direction is tolerable; but exaggerate in either direction and the revenue falls steeper and steeper. This completely schematic model applies to many questions of choice between competing policies; it probably applies to the two approaches in math education we have been discussing, helping those who need more algebra and those who need more real life connections. Needless to say, quantification is a different matter...

Mathematical Literacy: Internal and External Perspectives

Iddo Gal

A discussion about the meaning of mathematical literacy and its place in mathematics education is a timely one. The current attention to this topic is fueled in part by a force that many mathematics educators and mathematicians may see as external to mathematics education itself, i.e., the emphasis on mathematical literacy and on numeracy in OECD's two key assessment programs, PISA (students) and PIAAC (adults). My comments pertain to two related aspects: the conceptual lens from which we hold a discussion on the nature of mathematical literacy, and the educational approach to developing mathematical literacy.

Conceptual Issues The debate about the meaning and place of mathematical literacy is naturally held in large part by mathematicians, mathematics educators, and academics and trainers interested in mathematics education, i.e., insiders to the domain. However, to understand mathematical literacy and why it is essential to address it as part of mathematics education, it is important to bring in an external viewpoint. Instead of asking, "what is the goal of mathematics education?" or, "what mathematics is most important to teach?" I believe a broader question should be asked first:

- What are the skills or competencies needed by and expected of citizens from all walks of life for effective functioning in the information age [at work, at home, in civic life, etc.]?

Many stakeholders have been pondering for years the nature of desired skills that adults should possess, and hence that educational systems should develop or worry about (e.g., Secretary's Commission on Achieving Necessary Skills (SCANS) 1991). Recent years, though, have seen a focus on a broader notion of 'competencies'. In its groundbreaking work, OECD's project DeSeCo (Definition and Selection of Competencies) has defined competency as: "[The] interest, attitude, and ability of individuals to access, manage, integrate, and evaluate information, construct new knowledge, and communicate with others in order to function effectively in the information age." (Rychen and Salganic 2003, p. 8). Given this definition, which has been adopted by many countries (and by OECD), a second external question then emerges:

- Are educational systems developing the competencies needed by adults?

I argue that a discussion about the nature and importance of mathematical literacy should be held with the above external questions in mind. However, an external perspective presents a challenge to mathematics educators. They are asked neither whether they teach important mathematics nor whether their students can do what mathematicians do (a popular instructional goal among many mathematics educators), but a much broader and different question—whether the outcomes of their hard labor are sufficiently synchronized with the (future) needs of the people they educate (and their employers and communities), and aligned with the nature of the tasks facing adults in modern societies and information-rich economies. Of course, different people may have different futures, and the future itself is changing as we speak due to social and technological changes, hence the answers are not simple. Still, these questions bring a very different focus, on educational outcomes and human capital, and on competencies, which are posited as building blocks of human capital.

Mathematics educators are now called upon by external stakeholders to demonstrate that virtually all of their students are coming out with a broad range of needed competencies. So far, cumulative results both from PISA and from large-scale surveys of adult skills (PIAAC Numeracy Expert Group 2009; OECD 2013) suggest that too many people, both adults and students, are not engaging real-life mathematical tasks very well. Thus, whatever is being done in mathematics classes is not working as well as it should. Why? Is it because we are not doing well in teaching students to think like mathematicians and should try harder in this regard? I believe that we should look for different answers. Given space limits, let me emphasize just two: we need to better understand the nature of the target competencies themselves, and we have to be aware of the cumulative research findings regarding the complexity of skill transfer and of the many factors affecting people's ability to cope with new kinds of mathematical or functional tasks in different life contexts (Lovett and Greenhouse 2000; Burke and Hutchins 2007).

From Conceptual Issues to Educational Actions Schools have historically stressed, explicitly or implicitly, students' ability to handle abstract, college-related topics such as advanced algebra and calculus. Yet in most countries I know fewer than 50 % of all students who graduate from high-school enter college, and many of these never study mathematics any further in college. Why does school mathematics focus so much on abstract, "academic" aspects and give relatively little attention to functional skills and to skill transfer? One very likely explanation is that the mathematics education community overemphasizes internal views of the goals of mathematics education and does not sufficiently balance them against external views.

From an external view, we need to make sure that school graduates are able to act in a numerate way. Numeracy has been defined (PIAAC Numeracy Expert Group 2009) as "the ability to access, use, interpret, and communicate mathematical information and ideas, in order to engage in and manage the mathematical demands of a range of situations in adult life". Due to space limits, let me highlight just three of the terms in this description of adult numeracy:

- "Interpret" reminds us of the role of adults as critical consumers of quantitative information, as interpreters of a wide range of quantitative messages. Such information is often presented via different types of texts. Yet, texts are often seen by mathematics teachers as a distraction, as external to the world of pure mathematics, and shunned from the classroom.
- "Engage" is not about having engaging instruction, as some may think, but about preparing learners to effectively engage a very wide range of real-life situations

that present mathematical demands. To engage such tasks successfully, one needs not only a range of cognitive skills and knowledge bases, but also positive or supporting dispositions, i.e., beliefs, attitudes, and a critical stance, coupled with productive habits of mind. We want people to feel comfortable about being able to approach and cope with a range of tasks, including tasks that involve ambiguity, that call for decision making and solving problems embedded in real contexts, and the like.

- "Manage" refers to the fact that adults do not normally "solve" problems as in a math class. Most numeracy situations do not have "solutions" that can be classified as right or wrong. Rather, adults manage situations, and can decide on one of several courses of action, based on their assessment of personal goals and situational demands, severity of consequences, and personal and situational resources.

To develop a transferable competency such as numeracy or mathematical literacy, and to increase the chance our graduates can autonomously engage a wide range of real-life mathematical or statistical tasks and situations, we need to rethink the mix of tasks used in instruction, and the associated teaching sequences and assessment methods. Among other directions, we need to increase the dosage of tasks involving ill-structured problems, that contain text-based messages conveying various quantitative and statistical arguments or requiring critical interpretation of texts, that present statistical information of the kinds normally encountered in the media or in workplace and civic action contexts, and that demand the kinds of coping behaviors that adults are called upon to demonstrate in real life.

Developing Mathematical Literacy as Fostering Habits of Communication

Anna Sfard

One cannot but admire Paul Goldenberg's inspired and inspiring vision of what school mathematics should be all about. One gets swept by his passionate argument. In my presentation, I will nevertheless begin by taking a critical look at Paul's ideas. I will continue with an endorsement—and an extension—of his call for cultivating mathematical habits of mind.

Here is the point of disagreement: I wish to take exception with Paul's vehement rejection of the argument of utility as the one to use while justifying the principle of "mathematics for all". Well, it is not easy for me to say this. Like Paul, I have been brought up to love mathematics for what it is. Like him, I have been born into the modernist world ruled by logical positivism, where mathematics was treated as a queen even when it acted as a servant. Still, I am acutely aware of the fact that times change and that these days, the modernist romanticism is often at odds with the postmodernist pragmatism.

Many arguments can be brought to show that in our postmodern communication-driven world, where the need for the kind of mathematical knowledge we had once pursued for its own sake no longer goes without saying and where activities such as thinking and getting to know are seen as closer to surfing than to diving, mathematical competency as described by Paul, although desirable for all the reasons he counts, seems a goal more distant than ever. The very fact that the interest in mathematics is constantly decreasing[8] and that teaching mathematics requires justification signals that young people may be lacking the motivation needed for the success of learning. Paraphrasing Emmanuel Levinas' famous claim on morality one can say that asking "Why do we need to learn mathematics?" is, in a sense, the end of mathematics; or at least the end of mathematics as we know it. Above all, however, the young people's universe changes too quickly and is way too noisy these days to allow for what Paul is seeking: the "intense focus and concentration and... logic". Indeed, in the world that never stops, the long lonely diving into intricacies of mathematics cannot be expected to be the young person's first preference. Although diving means going deeper, it also means remaining unseen and staying in one place; and when the world itself is moving under one's feet, staying put and unnoticed may, in fact, mean going backwards. The young person's preoccupation with communicating with others leaves little room for substantial episodes of in-depth, intense, well focused self-communication. All the more so, when her inner voice is silenced by music flowing from the iPod directly into her ears.

This much for my critique of Paul's ideas. It is now time to admit that there is a lot in what he says that I applaud. Thus, in spite of the criticism above, I do sympathize with Paul's suggestion to use habits of mind as a leitmotif around which to organize mathematics curriculum. Like him, I am convinced that mathematical ways of thinking, more than the specific mathematical universes explored in school, should be seen as, potentially, the student's most valuable reward for the twelve years of grappling with numbers, function, and geometric figures. In fact, I am willing to take Paul's proposal a step farther and suggest that we use the words "habits of communication" rather than "habits of mind". For me, this new term is simply an extension of Paul's idea. Since mathematics can be seen as a discourse—a special type of communication[9]—the term *habits of communication* includes all that is meant by *habits of mind*, and more: it also contains those habits that regulate communicating with other people.

---

[8]This is evidenced by numerous publications on the drop in enrollment to mathematics-related university subjects (e.g. Garfunkel and Young 1998; Gilbert 2006; OECD 2006) and by the frequent calls for research projects that examine ways to reverse this trend (see e.g. TISME initiative in UK, http://tisme-scienceandmaths.org/). The decline in young people's interest in mathematics and science is generally considered these days as one of the most serious educational problems, to be studied by educational researchers and dealt with by educators and policy makers.

[9]This basic tenet, that builds on ideas of the philosopher Ludwig Wittgenstein and the psychologist Lev Vygotsky, defines the approach to the study of thinking and learning known as *communicational*, or *commognitive* (this last term, the portmanteau of the words *communication* and *cognition*, signals that the two ingredients are members of the same category). For justification and elaboration of the idea, as well as for a survey of its implications for the theory and practice of teaching and learning mathematics see Sfard (2008).

I can see a number of reasons why extending Paul's idea of mathematical habit in this way may be a good thing to do. First, the new term brings to the fore the interpersonal dimension of mathematics, one that practically disappears when we restrict the conversation to habits of *mind*. The word *communication*, when reinserted, reminds us that mathematics originates in a conversation between mathematically-minded thinkers, concerned about the quality of their exchange at least as much as about what this exchange is all about; and that what Paul calls habits of mind—the habits of systematic explanation, of proving, of translating between modalities, of tinkering and of interpreting diagrams—developed over history as the necessary characteristics of maximally effective interpersonal communication. More specifically, one of the major driving forces for the formation of these habits was the mathematicians' (impossible) dream about infallible discourse, which would also have an unlimited capacity for generalization. And indeed, Paul's five "inclinations" are what immunizes interpersonal exchanges against ambiguity and provides means for solving problems in a way that leads to unshakable consensus.

And there is more. My second argument is that the importance of the communicational habits one develops when motivated by the wish to prevent ambiguity and ensure consensus goes well beyond the mathematics itself. I am prepared to go so far as to claim that if these habits were regulating all human conversations, from those that take place between married couples to those between politicians, our world would be a happier place to live. Third, presenting mathematics as the art of interpersonal communication is, potentially, a more effective educational strategy than focusing exclusively on intra-personal communication. The interpersonal approach fits the preferences of today's young people's and it is easier to implement. After all, shaping the ways students talk to each other is, for obvious reasons, a more straightforward job that trying to mould their thinking directly. Finally, there are grounds to believe that framing the task of learning mathematics as perfecting one's ability to communicate with others creates a better chance for overcoming the inherent situatedness of learning. Challenging students to find solutions that would convince even the worst skeptic will likely help them develop the life-long habit of paying attention to the way they talk (and thus think!). This kind of attention, being focused on the person's own actions, may bring about habits that are less context-dependent and more universal than those developed when the learner is almost exclusively preoccupied with mathematical objects. For all these reasons, and more, I suggest that we teach mathematics as the art of communicating.

Mathematical Literacy: What Does It Mean, to Whom, and Do We Really Need to Teach It?

Hannah Perl

In his chapter, Paul Goldenberg's touched upon many aspects of teaching and learning mathematics which I have struggled with, within myself and together with many of the people I have worked with, in the last seven years, as the chief superintendent of mathematics in the Israeli Ministry of Education. Although Goldenberg's habits of mind approach appeals to my perception of teaching mathematics, mathematical literacy has some merits, which cannot be dismissed and it is necessary to reconcile the two conflicting approaches in some manner. My comments are based on my experience of implementing mathematical literacy curricula in schools.

Since Israel is a member of the OECD, it is expected of our students to participate in international surveys and assessments in which PISA plays an important role. We are not only expected to participate, we have to do reasonably well and if possible, extremely well. Therefore it is reasonable that mathematical literacy should be fostered throughout the mathematics curriculum. The definition of mathematical literacy, for us, at the Ministry of Education has to be, at least in its core, the PISA definition[10] as reflected in the PISA framework and questions. This definition includes the contents, the processes, and the contexts as defined in the framework. We also have to take into consideration the mathematical tools of the 21st century, with which the students will be assessed. We can extend the definition of mathematical literacy in the curriculum as we see fit, but the PISA definition has to be its foundation.

The mathematics chief superintendent and the curriculum developers are expected to incorporate this particular "mathematics literacy" into the curriculum and into teaching beginning in middle school (maybe even earlier), and continuing into high school targeting students 15–16 years old. Resources are made available to make this process possible. These resources include development of new materials, teacher professional development and special classroom instructors. As a result, we want our students to be able to demonstrate their ability to solve a wide range of problems embedded in "real-life" contexts.

The ability of students to identify and apply mathematics when it is needed does not develop by itself even with mathematically oriented students and has to be taught explicitly to both mathematically strong students and those who are not mathematically inclined. Therefore mathematical literacy should be taught at all levels and cannot be ignored. Solving a problem that has, as its ingredients, a story, some visual representations and algebra can be interesting and mathematically challenging. With careful effort the concept of mathematical literacy can be extended so that the PISA definition will be embedded in it. This encompassing definition will be suitable for higher-level students as well.

What kind of mathematics should we teach students who are not mathematically oriented? One answer is "mathematical literacy" as defined by PISA: skills and

[10]The PISA 2012 Definition of Mathematical Literacy: "Mathematical literacy is an individual's capacity to formulate, employ, and interpret mathematics in a variety of contexts. It includes reasoning mathematically and using mathematical concepts, procedures, facts, and tools to describe, explain, and predict phenomena. It assists individuals to recognize the role that mathematics plays in the world and to make the well-founded judgments and decisions needed by constructive, engaged and reflective citizens." (PISA 2012 Mathematics Framework, Draft, November 30, 2010 http://www.oecd.org/pisa/pisaproducts/46961598.pdf).

understanding which students will be able to use throughout their life. Maybe it is not enough, but it is a must. For example, to understand medical laboratory test results one needs to understand probability concepts and percentages. Maybe such realistic problems that may seem "relevant" to their life may motivate students to learn mathematics.

Mathematically oriented students should be challenged with more complex problems that integrate a broader range of topics, advanced mathematical texts and the use of higher mathematical concepts and competencies. The term "realistic" should be broadened.

It is important to understand that language literacy is intertwined with mathematical literacy. In order to solve problems that are situated in contexts, students must be able to read and decipher texts that are frugal and dense (every word counts). This adds to the level of difficulty of the problem and we have found that this is the most common complaint of our teachers who claim that reading and understanding the problem is the main issue. Identifying and applying mathematics come second and third.

Reflective Summary: Where to from Here?

The position papers presented by the authors of this chapter, when taken together, substantiate Paul Goldenberg's claim about the ambiguity of the term "mathematical literacy". The six interpretations, no two of which appear quite the same, seem to be drawing on three basic views, differing from each other in their vision of the relation between *mathematical competency* and *mathematical literacy*.

The first, "not-much-new" view, represented by Paul Goldenberg and Abraham Arcavi, practically equates the two competencies. According to these authors, the claim that somebody is mathematically literate is not much different from saying that this person knows mathematics (or mathematical literature—see Abraham Arcavi's contribution), except that the word "literacy" stresses the aspect of utility. According to these authors, the rhetoric of literacy, about which they are not overly enthusiastic, conveys the message that the main value of mathematical knowledge lies in its "real-life" usefulness. It also implies that the learning of mathematics should be organized around its applications.

The other two approaches make a clear distinction between formal mathematical competency and mathematical literacy. According to the *minimalist view*, the requirement of mathematical literacy is somewhat less demanding than the call for formal mathematical competency. As stated by Ron Livné, mathematical literacy is meant for "less mathematically sophisticated children" or for "the lower echelons: people who want the national certificate but who will not study further or will not need formal mathematics later". This view is echoed by Hannah Perl who, after admitting that "mathematical literacy should be taught at all levels", adds that it is the only realistic option for students who "are not mathematically oriented".

The proponents of the last, *maximalist view*, with Iddo Gal and myself (Anna Sfard) among them, tend to claim the opposite: mathematics literacy is more demanding than formal mathematical competency. While having the formal mathematical knowledge as its subset, mathematical literacy includes knowing when to turn to mathematical discourse and what parts of this discourse to use. This said, mathematical literacy is not "formal mathematics plus" that is, does not result from a simple addition of the "when" (applications) to the "how" (the formal mathematical competency). The *how* of literate mathematical discourse—the repertoire of its routine ways of doing things—is not quite the same as that of formal mathematical discourse. For one thing, it is richer and more varied, flexibly adaptable to specificities of the situation in which one is required to implement the routines.

Clearly, these three conceptualizations of mathematical literacy are mutually incompatible, and it is therefore not surprising that the authors occasionally sound as contradicting one another. Still, it is clear that they agree on one basic issue: mathematics, in one form or another, must be an obligatory part of everybody's school experience. Indeed, all the contributors seem to care for mathematics and to be genuinely concerned about its future and the future of those who are going to need it. Even the proposal to significantly reduce the compulsory curriculum, put forth by some of the authors, grew out of this very authentic concern. And there is yet another consensual view: the contributors are unified in their belief in the need for a change. In one way or another, each of them claims that much additional thought must be given to the questions of "*What* mathematics should be taught?", "*To whom*?", and "*How*?" In other words, the authors propose unanimously that we rethink school mathematics.

This, however, is where the consensus ends. The debate does not bring agreed answers to the questions everybody was asking. But responding in unison was not among the aims of this conversation. Rather, the idea was to brainstorm, think together, and end up with an assortment of possibilities that can trigger and fertilize a future discussion. The proposed reforms in school mathematics vary from minor to radical, and from affecting only some aspects of curricula to requiring a total overhaul. The proposals are ordered below according the degree of change implied by their response to the *what*-, *to-whom*-, and *how*-questions.

Minor Reform—Change the "How" This approach, which goes hand in hand with the "nothing new" view of mathematical literacy, can be described as guided by the slogan "Keep the *what* and *to whom* of the current mathematics curricula, while revising the *how*". Professed in this volume by Paul Goldenberg and Abraham Arcavi, who discuss the idea of organizing the teaching of mathematics around the habits of mind rather than around applications, this recommendation aims at continuing and refining the principle of "mathematics for all".

Moderate Reform—Change the "To Whom" This approach, reflective of the minimalist view of mathematical literacy, can be described as "Mathematical literacy for those who can't (or don't want to) manage formal mathematics". The foundational assumption of this proposal is that mathematical literacy, as a reduced, less demanding version of the regular school mathematics, has a chance to succeed where the 'normal' mathematics fails. This approach is reflected in the practice of grouping the less successful mathematics students in separate classes, supposed to learn according to special curricula dedicated to mathematical literacy.[11]

Radical Reform—Change All This approach, which calls for "Mathematics literacy for all, formal mathematics for some" can be seen as, in a sense, the opposite of the former proposal. It goes hand in hand with the maximalist view, according to which mathematics literacy does not lower the bar, but on the contrary, raises the learning and teaching standards to new heights. To be successful in dealing with everyday tasks requiring mathematics one should know all that has always been taught, but must know it in a new, more complex way. Indeed, the radical reform, as outlined here, is quite different from the one that merely stresses the applications of formal mathematics. It calls for a reflexive, cyclic process in which formal mathematical discourse is being derived from other discourses and then, after considerable elaborations, returns to these other discourses in order to refine, to enrich and to extend them. Deep, far-reaching changes are thus required by the proponents of this kind of reform in all aspects of school mathematics: in what to teach, to whom, and how.

Clearly, this latter radical stance clashes frontally with the "minimalist" idea of mathematical literacy as the "second best" for underachieving students. The "radicals" are likely to claim that the minimalist view results from a faulty interpretation of certain research findings. True, many studies have shown that even unschooled people can be impressively skillful in numerical or geometric tasks that constitute a routine part of their daily activities. And yet, this fact does not imply that mathematical literacy should or can function as a safety belt for students with a history of drowning in numbers, functions and geometrical figures. Indeed, however one looks at the impressive findings about "everyday mathematics", the message seems to be just the opposite: this mathematics is neither easily learnable—it requires a great deal of practice; nor truly useful—the skills developed through repetitive performance of a small number of everyday tasks are highly situated, that is, they depend on the specificities of these tasks and of the situations in which they are routinely executed, and they do not transfer to anywhere else.

Unfortunately, none of the three proposals came with a clear statement on how to overcome the situatedness of learning that has always been the major obstacle to our efforts to promote students' mathematical literacy. How to foster students' ability and willingness to "speak mathematics" whenever this kind of talk may be of help remains the question of questions, one that has not yet been properly dealt

[11]Such tracking has been practiced in a number of countries, with England (the special program is known as *Functional Mathematics*; see QCA 2005) and South Africa (see Venkatakrishnan and Graven 2006) among them. A similar tendency exists in Israel, where there is a plan to allow students with the history of low achievement to "compromise" on mathematical literacy (see, for instance, the newspaper publication in Hebrew, http://www.nrg.co.il/online/1/ART2/333/661.html, retrieved on 21.07.12).

with in this conversation. This topic is weighty and complex enough to merit another meeting, and probably much more than one.

References

Benezet, L. P. (1935). The teaching of arithmetic: the story of an experiment. *The Journal of the National Education Association, 24*(8), 241–244. http://www.ithaca.edu/compass/storyI-III.htm. Accessed 21 Mar 2013.
Burke, L. A., & Hutchins, H. M. (2007). Training transfer: an integrative literature review. *Human Resource Development Review, 6*(3), 263–296. doi:10.1177/1534484307303035.
Freudenthal, H. (1978). *Weeding and sowing: preface to a science of mathematics education*. Dordrecht: Reidel.
Garfunkel, S. A., & Young, G. S. (1998). The sky is falling. *Notices of the American Mathematical Society 45*(2), 256–257.
Gilbert, J. (Ed.) (2006). *Science education in schools*. TLRP.
Lave, J. (1988). *Cognition in practice*. New York: Cambridge University Press.
Lovett, M. C., & Greenhouse, J. B. (2000). Applying cognitive theory to statistics instruction. *American Statistician, 54*(3), 196–206.
OECD (2006). Evolution of student interest in science and technology studies. http://www.oecd.org/dataoecd/16/30/36645825.pdf.
OECD (2013). *OECD skills outlook 2013: first results from the survey of adult skills*. Brussels: OECD Publishing. doi:10.1787/9789264204256-en
PIAAC Numeracy Expert Group (2009). *PIAAC numeracy: a conceptual framework*. OECD Education Working Papers, No. 35, OECD Publishing. www.oecd-ilibrary.org/education/piaac-numeracy-a-conceptual-framework_220337421165. Accessed 21 Mar 2013.
QCA (2005). Functional skills: essential for life, learning and work. http://www2.ofqual.gov.uk/files/qca-07-3170_functional_skills.pdf. Accessed 21 Mar 2013.
Rychen, D. S. & Salganic, L. H. (Eds.) (2003). *Key competencies for a successful life and a well-functioning society*. Gottingen: Hogrefe & Huber.
Secretary's Commission on Achieving Necessary Skills (SCANS) (1991). *What work requires of schools: a SCANS report for America 2000*. Washington: US Dept. of Labor.
Schank, R. (1987). Let's eliminate math from schools. *Whole Earth Review*, 58–62.
Sfard, A. (2008). *Thinking as communicating: human development, the growth of discourses, and mathematizing*. Cambridge: Cambridge University Press.
Thorndike, L. E., & Woodworth, R. S. (1901). The influence of improvement of mental function upon the efficiency of other functions (I). *Psychological Review, 8*, 247–261.
Venkatakrishnan, H., & Graven, M. (2006). Mathematical literacy in South Africa and functional mathematics in England: a consideration of overlaps and contrasts. *Pythagoras, 64*, 14–28.

Visualization in Mathematics and Mathematics Education

Chapter 11
Fifty Years of Thinking About Visualization and Visualizing in Mathematics Education: A Historical Overview

McKenzie (Ken) A. Clements

Abstract This chapter surveys meanings given to the term "visualization" in mathematics, mathematics education, and psychology, and considers the evidence for the oft-heard assertion that mathematics learners tend to prefer to think algorithmically rather than visually. The analysis reveals that students who do very well on pencil-and-paper "visualization" tests often prefer *not* to use visual methods when attempting to solve mathematical problems; and those who do not do well on standard visualization tests often describe themselves as "visual thinkers", and prefer to use visual methods when attempting to solve mathematics problems. The influence of various mathematics educators, and especially Alan Bishop—who thought of visualization in terms of a person's use of visual images when posing and solving mathematics problems—of Norma Presmeg, and of a group of mainly Israeli mathematics educators who developed the construct "concept image", is also examined. Views of some mathematicians are also taken into account. In the early 1990s, Zimmermann and Cunningham (Visualization in teaching and learning mathematics, 1991) wrote of how David Hilbert had spoken of two tendencies in mathematics—one that sought to crystallize logical relations, and the other to develop intuitive understanding, especially through "visual imagination" (p. 2). In addressing that theme, Ted Eisenberg and Tommy Dreyfus (Visualization in teaching and learning mathematics, pp. 25–37, 1991) spoke of mathematics students' preference for "algorithmic over visual thinking" (p. 25). The paper draws special attention to the work of two lesser-known mathematics education researchers, Nongnuch Wattanawaha and Stephanus Suwarsono. It was Suwarsono who devised and applied a method whereby learner preferences for visual or verbal thinking, as well as the "visualities" of the mathematics tasks themselves, could be measured and calibrated on the same scale, using item response theory.

Keywords Visual image · Visualization · Mathematics learning · Problem solving · Item response theory · Verbal-visual

M.A. Clements (✉)
Department of Mathematics, Illinois State University, Campus Box 4520, Normal, IL 61790-4520, USA
e-mail: macleme@ilstu.edu

M.N. Fried, T. Dreyfus (eds.), *Mathematics & Mathematics Education: Searching for Common Ground*, Advances in Mathematics Education,
DOI 10.1007/978-94-007-7473-5_11,

More than two centuries ago, Johann Friedrich Herbart, a German philosopher, psychologist and educator, emphasized the need for teachers to try to become aware of links and preferences in individual learners' minds. Herbart urged teachers to design lessons that took into account what learners were likely to be thinking about when they attempted to solve problems (Ellerton and Clements 2005). During the twentieth century there gradually developed, among mathematicians (e.g., Hadamard 1945; Hilbert 1999; Zimmermann and Cunningham 1991), psychologists (e.g., Kosslyn 1980; McGee 1979; Lohman 1979; Paivio 1971; Piaget and Inhelder 1971; Pylyshyn 1973; Richardson 1977; Shepard and Metzler 1971; Thomas 1989; Thurstone 1938), and mathematics educators (e.g., Bishop 1980; Clements 1982; Eisenberg and Dreyfus 1991; Skemp 1972; Suwarsono 1982), an interest in the extent to which people use visual imagery when tackling mathematical tasks.

Toward the end of the 1970s, Gagné and White (1978) argued that each individual's working memory with respect to a given topic or problem comprises a uniquely related set of five components: verbal knowledge (e.g., the definition of an equilateral triangle), skills (e.g., how to complete the square in algebra); imagery (e.g., a visual image of an isosceles triangle), attitudes (e.g., belief about a personal inability to cope with geometrical proof); and episodes (memories of pertinent personal events). Gagné and White referred to a learner's unique configuration in working memory, with respect to a topic, as that learner's *cognitive structure* with respect to the topic. In mathematics education, a similar construct—that of a *concept image*—would be accorded center stage (Vinner and Hershkowitz 1980).

Tall and Vinner (1981) maintained that imagery, and especially visual imagery, could be a key ingredient of a learner's concept image. They used the term concept image "to describe the total cognitive structure that is associated with the concept, which includes all the mental pictures and associated properties and processes" (p. 152). The contention that idiosyncratic and unique links in cognitive structure determined an individual learner's concept image with respect to some mathematical concept or principle, and that therefore educators needed to take account of and seek to influence students' concept images, would be emphasized by numerous researchers during the 1980s (e.g., Dreyfus and Eisenberg 1982, 1983; Tall and Vinner 1981; Vinner and Hershkowitz 1980; Vinner and Dreyfus 1989).

This paper summarizes some of the research carried out since 1970 which has sought to identify how students think when attempting to solve mathematical problems. The preparation of the paper has been a moving experience for me, in two ways. First, after re-reading the writings of Ted Eisenberg, Tommy Dreyfus, and others within a very hardworking and highly-achieving group of Israeli mathematics education scholars, I was reminded that their work was—I choose these adjectives carefully—courageous, creative and powerful. Some of their publications—like, for example, Eisenberg and Dreyfus's (1991) "On the Reluctance to Visualize," Dreyfus and Eisenberg's (1986) "On the Aesthetics of Mathematical Thought," and Eisenberg's (2008) "Flaws and Idiosyncrasies in Mathematicians…"—have had a significant impact on my own thinking. The second influence on me, personally, was to reignite my interest in the role of visual thinking in mathematics teaching and learning. I trust that this paper will not only attest to my re-emerging interest but, in so doing, pay tribute to the pioneering and persistent work in the field by Miriam Amit, Alan Bishop. Tommy Dreyfus, Ted Eisenberg, Michael Fried, Rina Hershkowitz, Norma Presmeg, Richard Skemp, Schlomo Vinner, and David Tall.

"Visualization" Has Different Meanings Among Different Groups of Scholars

The term "visualization" means different things to different groups of scholars. I shall refer, especially, to meanings developed: (a) by psychologists who used factor analytic techniques; (b) by mathematics educators; and (c) by mathematicians. Although researchers within and between these groups have offered different definitions of visualization, the differences within the groups have been less than those between the groups.

Factor Analysts' Perspectives on Visualization

I first became acquainted with the psychological literature on visualization during 1976 and 1977 when Nongnuch Wattanawaha and I investigated gender differences in spatial ability, and how these might have an impact in mathematics education. We became aware that factor analysis (in psychology) tended to reveal two major factors—"spatial orientation" and "visualization"—as well as numerous other minor factors (for further details, see, Lohman 1996).

The following statement by Burin et al. (2000) is representative of the kind of language to be expected of psychologists working from a factor-analytic perspective:

> Two general kinds of solution strategies for *Vz* tasks are described in the literature. One is an *analytic* or *feature comparison* approach, in which the examinee seeks to verify the identity of key features of the probes to match them with the target stimulus. A variant of this analytic strategy is verbal labeling of the features. The other is a *holistic* or *spatial manipulation* strategy, which involves mental movements of the probes, such as rotation, translation, synthesis. (p. 277)

Burin et al. (2000) argued that these two "*Vz*" strategies involved "analytic" and "holistic" thinking (p. 278), and that persons employing analytic strategies tended to solve visualization tasks more quickly than those using holistic strategies.

We also discovered that there was a verbalizer-visualizer hypothesis (Richardson 1977), which asserted that some people preferred to think in verbal/analytic ways, and others in visual ways. This hypothesis seemed to be supported by Vadim Krutetskii's (1976) research, which indicated that although some gifted mathematics students preferred to use visual approaches to mathematical problem solving, others did not, and that, indeed, some tried to avoid using visual approaches. Krutetskii classified mathematics students into three categories:

1. Geometric thinkers (with a preference for visual-pictorial thinking);
2. Analytic thinkers (with a preference for verbal-logical thinking); and
3. Harmonic thinkers (who preferred to use a mixture of visual-pictorial and verbal-logical methods.

Mathematics Educators' Perspectives on Visualization

Following Bishop (1973, 1980), Presmeg (1986a, 1986b, 2006) maintained that when a person creates a spatial arrangement there is a visual image in the person's mind, guiding this creation. For Presmeg, visualization includes processes of constructing and transforming both visual mental imagery and the inscriptions of a spatial nature. This interpretation of visualization takes on a constructivist position whereby a person establishes links between the unique, but temporary, arrangements of that person's mental structures, which allows and precipitates unique constructions of meaning.

Presmeg (1985) distinguished between five kinds of imagery:

- *Concrete-pictorial imagery* occurs when imagery resembles real-life objects or situations;
- *Pattern imagery* is employed when pure relationships are imagined in some form, and the image so formed is devoid of concrete details;
- *Memory images of formulas* are used when abstract information is manifested in concrete images;
- *Kinesthetic imagery* is associated with muscular activity (e.g., gestures by which a shape or pattern is traced out);
- *Dynamic imagery* occurs when images are mentally transformed and manipulated.

According to Presmeg (2006), in the 1980s quantitative research which measured performance, visuality, and other such creations, began to be regarded as a relic of a positivist, behaviorist era. Cognitive researchers, and those working from semiotic perspectives, sought ways to investigate what transpired in the minds of students through introspective and retrospective analyses of interview data, observation data, or data from responses to questionnaires. Since it was impossible to "photograph" what went on in the brain, triangulation, in which data from various vantage points were analyzed and compared, became important (Owens 2005; Owens and Outhred 2006).

Mathematicians' Perspectives on Visualization

In discussing the question "what is mathematics visualization?" Zimmermann and Cunningham (1991), two mathematicians, wrote:

> We take the term *visualization* to describe the process of producing or using geometrical or graphical representations of mathematical concepts, principles or problems, whether hand drawn or computer generated. (p. 1)

Zimmermann and Cunningham acknowledged that this view of visualization differed from common usage and from the view in psychology, where the meaning was "closer to its fundamental meaning 'to form a mental image'" (p. 3).

Of many statements made by scholars about the role of visualization in their own theorizing, problem posing and problem solving, perhaps the most famous is that by Albert Einstein:

> The words or the language, as they are written or spoken, do not seem to play any role in my mechanism of thought. The psychical entities which seem to serve as elements in thought are certain signs and more or less clear images which can be "voluntarily" reproduced and combined.... The above mentioned elements are, in my case, of visual and some of muscular type. Conventional words or other signs have to be sought for laboriously only in a secondary stage.... In a stage when words intervene at all, they are, in my case purely auditive, but they interfere only in a secondary stage as already mentioned. (Quoted in Hadamard 1945, p. 142)

Probably, agreement will never be reached, among mathematicians, mathematics educators, psychologists, and linguists, on the meaning of the term "visualization." That makes it important to be able to determine what someone means when she or he is using the term.

For me, visualization is something which someone does in one's mind—it is a personal process that assumes that the person involved is developing or using a mental image. From that perspective, one of the most appealing descriptions of visualization, for me, could be associated with the so-called DIPT classification framework for spatial tasks developed by the Thai mathematics educator, Nongnuch Wattanawaha.

Wattanawaha's DIPT Classification Framework

Wattanawaha (1977) pointed to four fundamental general characteristics of spatial tasks:

1. The **D**imension of thinking required by the task;
2. The degree of **I**nternalization required;
3. The manner in which the task required an answer to be **P**resented; and
4. The **T**hought Process required, and in particular whether the sequence of mental operations needed was given to, or had to be worked out by, the person doing the task.

Having decided that these four characteristics were of primary importance for the classification of a person's response to a task, Wattanawaha identified logical hierarchies within each characteristic. Thus the Dimension characteristic was taken to have three values, depending on whether the task required 1-, 2-, or 3-dimensional

Table 11.1 Wattanawaha's (1977) DIPT classification framework for responses to spatial tasks

Symbol	Name	Value labels, and corresponding definitions
D	Dimension (3 values) 1, 2, 3	1. The response used 1-D thought (but not 2- or 3-D thought) 2. The response used 1-D thought, (but not 3-D thought) 3. The response used 3-D thought
I	Internalization (3 values) 0, 1, 2	0. The task was done at the perceptual level. There was no attempt to evoke a visual image, or the only visual image was a "duplicate" of a given stimulus, or an image corresponding to a simple transformation of the stimulus, or parts of it 1. A visual image was evoked, but in order to do the task, thinking only needed to about aspects of the image—that is to say, it remained fixed in the mind 2. Not only was a visual image evoked, but in working on the task that image was operated upon ("transformed") in the mind
P	Presentation (3 values) 0, 1, 2	0. The form of the expected response to the task did not require a mental image to be described, identified, or drawn on paper 1. The expected response had to be chosen from a number of different images (presented in pictorial form) or actions. The pictorial images or actions corresponded to the final visual image associated with the task 2. The expected response required the final visual image to be drawn on paper, or otherwise described using words or hand or other movements
T	Thought process (2 values) 0, 1	0. The task specified the mental operations that had to be carried out 1. The task did not specify the mental operations that had to be carried out, but enough information was given to enable an appropriate sequence of operations to be determined

thought. In a similar way, the number of values allowed for the Internalization, Presentation, and Thought Process characteristics were 3, 3, and 2 respectively (see Table 11.1, taken from Wattanawaha and Clements 1982).

Wattanawaha (1977) analyzed responses to 72 pencil-and-paper spatio-mathematical tasks from a representative sample of 2346 students in grades 7 through 9 (1201 males, 1145 females). Although females did not significantly outperform males on any of the 72 tasks on the *Monash Spatial Thinking* test, males significantly outperformed females on 25 of the tasks. Of those 25 tasks, 21 had an "Internalization" value of 2. Males especially tended to do better on tasks with $D = 3$ and $I = 2$ (Wattanawaha and Clements 1982).

In a separate study, Clements and Wattanawaha (1978) observed and, where necessary, questioned, 328 interviewees (from grades 7, 8 and 9) as they worked through 16 spatio-mathematical tasks suitable for lower-secondary school students. They concluded that "the range of strategies employed... was amazing" (p. 434). With one of the tasks, for example, a student, on being handed a model of a cube and shown what an edge was, was asked to state how many edges the cube had altogether. About 30 percent of the students did not answer correctly, and those who

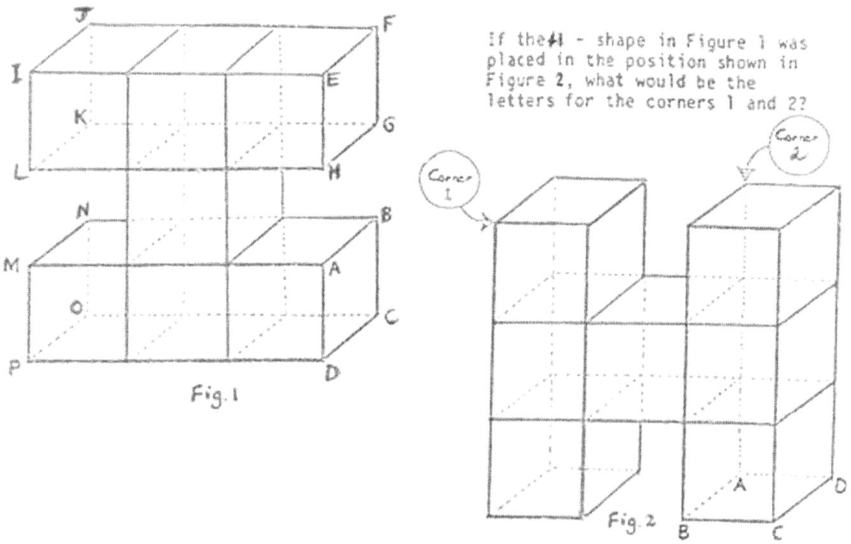

Fig. 11.1 The H-task from Wattanawaha's (1977) Monash Spatial Thinking Test (If the H-shape in **Fig. 1** was placed in the position shown in **Fig. 2**, what would be the letters for the Corners 1 and 2?)

did answer correctly used a wide range of methods to get answers. Some responses were very "visual," others were not, and for some it was difficult to decide one way or the other (e.g., how could one classify the response: "There are six faces, each has four edges. Six multiplied by 4 is twenty-four which, which when divided by 2 is 12"?).

Clements and Wattanawaha's (1978) findings raised the question—are some written tasks more likely to demand a strong internal visual response, and are some students more likely than others to give visual responses? If someone answers either of those questions in the affirmative, then a further question naturally arises—how can we know that to be the case?

One of the questions on Wattanawaha's (1977) *Monash Spatial Thinking* test is presented in Fig. 11.1.

In 1981, when presenting a paper to a meeting of the American Educational Research Association on the influence of spatial abilities and visual imagery on mathematical thinking, I asked the 100 or so people present at my talk to attempt the H-task in Fig. 11.1. After about a minute, I asked those present to indicate whether they had used either of the two following methods to find the label for "Corner 1."

1. Did you imagine the shape in Fig. 1 being rotated and turned, so that it matched the shape in Fig. 2 (and then work out that the letter *J* would be at Corner 1)? Or, . . .
2. Did you avoid imagining the transformation by thinking along the following lines: "In Fig. 2, Corner 1 is at the far opposite corner from Corner *D*, and in

Fig. 1, Corner J is at the far opposite corner to Corner D. Thus, Corner 1 must be Corner J.

One-fourth of those present indicated that they had used Method 1, and about one-third indicated that they had used Method 2 (Clements 1983).

Of interest is the fact that in 1983 when I had the privilege of working with the then 7-year-old mathematical prodigy, Terence Tao (Clements 1984), I asked him to solve the task shown in Fig. 11.1. He gained the right answers by analytic reasoning. He told me that he did not attempt to rotate the H-shape mentally, because that method was "not as good as" the method he used.

Suwarsono's Work in Establishing a Foundation for Researching Learner Preferences for Verbal-Analytic or Visual Processing

Suwarsono's (1982) Research

Stephanus Suwarsono (1982) pushed my thinking away from spatial ability toward visual imagery (Clements 1982). He wondered how a researcher might legitimately *measure* whether someone had a preference for visualizing. He also wondered whether such a preference might relate to the person's spatial abilities, and whether the preference might influence mathematical problem-solving performance.

Suwarsono (1982) developed his *Mathematical Processing Instrument* (hereafter MPI) using item response theory (IRT). MPI had two parts: the first consisted of 30 mathematics word problems developed with seventh- to ninth-grade Australian students in mind; the second contained written descriptions of different methods commonly used by students attempting the word problems in Part I. Usually, between three and five possible methods were described for each problem. Students were asked to attempt the problems in Part I and then to indicate, retrospectively, which (if any) of the methods described in Part II they had used. If a student believed that her method for solving a problem was different from any described in Part II, then she was instructed to say so, and to describe her method in writing.

In constructing his MPI, Suwarsono was guided by the following criteria:

1. The questions should range in difficulty from "very easy" to "moderately difficult" for the students. Very difficult questions were to be avoided.
2. No diagram was given, or requested, in any question.
3. For each question it could be expected that a variety of methods would be used by lower-secondary school students. In particular, it was expected that in a large group of, say, 200 students, some would think in verbal-logical ways, and others in visual ways.

Suwarsono not only measured the extent to which a person preferred thinking visually when attempting mathematics tasks; he also measured the "visualities" of the tasks themselves. That point requires explanation.

Scoring the MPI

Suwarsono scored students' responses to Part II of his instrument in the following manner:

+1 If reasoning was based on a diagram (drawn by a student) or on some visual image (constructed by the student).
0 If no answer was given, or the student could not decide which method she used.
−1 If reasoning was based on a verbal-logical method which did not involve a diagram or the construction of a visual image.

Using IRT, Suwarsono calibrated each student on an analytic-visual (ANA-VIS) dimension, and also each item on the *same* ANA-VIS dimension.

John Eliot, of the University of Maryland, an authority on how spatial abilities influence learning, served as an external assessor for Suwarsono's PhD dissertation. Eliot liked Suwarsono's ideas, and his students used Suwarsono's MPI—with due acknowledgement (see, e.g., Sheckels and Eliot 1983). This got Suwarsono's ideas known in the United States, and they are now used by many scholars interested in relationships between processing of tasks by visual and verbal means (see, e.g., Cheetham et al. 2012; Kozhevnikov et al. 2005; Lowrie and Kay 2001).

Here are 3 of the 15 questions from Part I of the MPI. According to Suwarsono's analyses, one of the three is a very visual task, and another is very verbal-analytic. See if you can decide which is which, before looking at the next paragraph.

- Two years ago Mary was 8 years old. How old will she be five years from now?
- One morning a boy walked from home to school. When he got half-way he realized that he had forgotten to bring one of his books. He then walked back to get it. When he finally arrived at school, he had walked 4 km altogether. What was the distance between his home and school?
- A girl had 11 plums. She decided to swap the plums for some apples. Her friend, who had the apples, said: "For every 3 plums, I will give you an apple." After the swap, how many plums did the girl have?

Table 11.2 indicates where each of the three appeared on Suwarsono's ANA-VIS scale. On the scale, a very visual item would have a measure of about 40, a very analytic question, a measure of about 60, and an "average" analytic-visual question, a measure of about 50. Would the measures shown in Table 11.2 be independent of the sample of students chosen to develop the MPI instrument? Before you answer that question, note that Suwarsono used a highly sophisticated form of item response theory which claimed to generate sample-free measures.

What, do you think the ANA-VIS measure would be for the following "balloon task"?

> A balloon first rose 200 m from the ground, then moved 100 m to the east, and then dropped 100 m. It then travelled 50 m to the east, and finally dropped straight to the ground. How far was the balloon from its starting point? (Suwarsono 1982, p. 292)

Suwarsono's (1982) analysis indicated it was: 40.6—which made it a "highly visual" item.

Table 11.2 ANA-VIS Measures of three questions from Suwarsono's (1982) MPI

Question	ANA-VIS measure	Classification
Two years ago Mary was 8 years old. How old will she be in five years from now?	60.6	Highly analytic
One morning a boy walked from home to school. When he got half-way he realized that he had forgotten to bring one of his books. He then walked back to get it. When he finally arrived at school, he had walked 4 km altogether. What was the distance between his home and school?	38.8	Very visual
A girl had 11 plums. She decided to swap the plums for some apples. Her friend, who had the apples, said: "For every 3 plums, I will give you an apple." After the swap, how many plums did the girl have?	51.2	Average (neither analytic nor visual)

Of course, a student's response to a mathematics task can be influenced by many factors—like, for example, teaching received from an instructor who is somewhere on the ANA-VIS scale; or the approach adopted in a textbook written by an author who is located somewhere on the same scale; or by a student herself, or himself, who is located somewhere on the same scale. It is also obvious that a student's mathematical understanding can be affected by a range of other, contextual variables.

When, in 1980, Glen Lean and I used MPI with first-year engineering students in Papua New Guinea (PNG), we got four intriguing results (Lean and Clements 1981). First, we found that MPI clearly identified visual and non-visual students. Second, there was no significant correlation between students' scores on a highly-regarded spatial visualization test (demanding mental rotation of cubes) and their ANA-VIS scores. Third, highly visual students tended to use visual methods even on highly verbal-logical (analytic) tasks, and highly verbal-analytic students tended to use verbal-logical methods, even on highly visual tasks. And, fourth, students who preferred to use verbal-analytic approaches scored higher on a mathematics test dependent variable than students who preferred to use visual processing approaches. Figures 11.2 and 11.3 from Lean and Clements (1981) show unedited responses by two PNG students to the problems shown in Fig. 11.2.

On Suwarsono's ANA-VIS scale, the first student's score indicated that he was highly analytic, and the second student's score indicated he was highly visual. Incidentally, the first student gained the highest score of the 116 engineering students on a rigorous test of mathematics. The claim that item and person calibration was sample-free was tested by calibrating MPI items' analytic-visual measures with both PNG engineering students and Australian lower-secondary school students. The Spearman-rho rank correlation between the measures obtained on the "visuality" of the items exceeded 0.9.

Suwarsono's (1982) and Lean and Clements' (1981) analyses established that there are students who do well on pencil-and-paper "visualization" tests who prefer **not** to use visual methods when attempting mathematical tasks. I often hear mathematics students refer to themselves as "visual thinkers," and have noticed that

PROBLEM 1: Tau has more money than Dilli and Mike has less money than Dilli. Who has the most money?
PROBLEM 2: In an athletics race, Johnny is 10 m ahead of Peter, Tom is 4 m ahead of Jim, and Jim is 3 m ahead of Peter. How many metres is Johnny ahead of Tom?
PROBLEM 3: Jack, Luke and Kuni all have birthdays on the 1st January, but Jack is 1 year older than Luke and Jack is three years younger than Kuni. If Kuni is 10 years old, how old is Jack?

(1) Tau $3x$
Dilli $2x$
Mike x
$\therefore$ Tau has more money.

(3) Kuni
$5x + 2x + x$
If Kuni = 10 yr old
$\therefore 5x - 2x = 3x$
$\therefore 10x - 3x = 7x$
$\therefore$ Jack is 7 years old.

(2) John 10 m to Peter x m
Tom 4 m to Jim x m
Jim 3 m to Peter x m
John to Tom?
Jo P Jo − P
10 m to 0 $10 - 0 = 10$
Jim P Jo − Jim
3 to 0 $10 - 3 = 7$
Tom 4 m to 5 $= 10 - 3 - 4 = 3$
$\therefore$ John 3 metres ahead of Tom.

Fig. 11.2 Solutions by a non-visual student (unedited) to the three problems

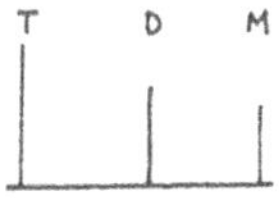

Tau has the most money.

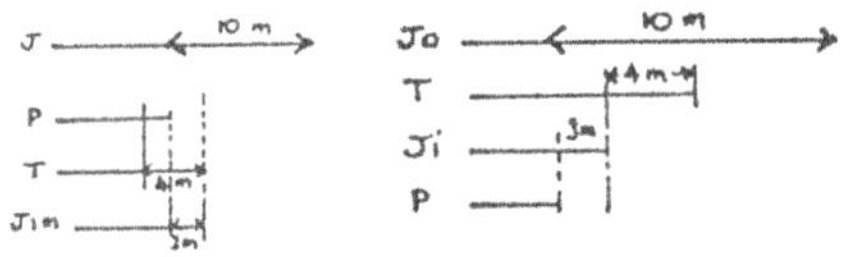

"If Jim is 3 m ahead of Peter then Tom is 7 m ahead of Peter, and if John is 10 m ahead of Peter, then John is 3 m ahead of Tom."

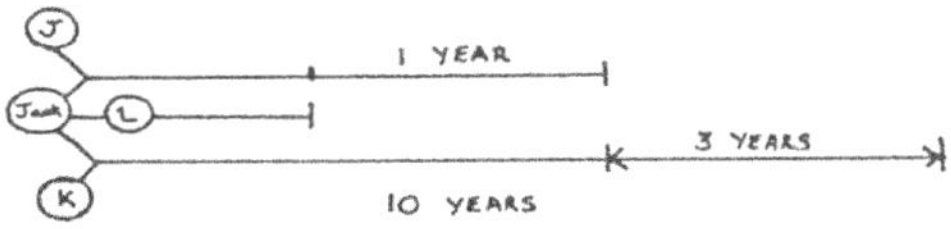

"Jack is 7 years old."

Fig. 11.3 Solutions by a visual student (unedited), to the three problems

although these students attempt to use "visual methods" (sometimes appropriately, but often inappropriately) when attempting to solve mathematics tasks, they tend to be among mathematically "weaker" students. On the other hand, Eisenberg (1994) claimed that "a vast majority of students do not like thinking in terms of pictures—

and their dislike is well documented in the literature" (p. 110). Eisenberg and Dreyfus (1991) commented that students are reluctant to use visual methods, preferring "algorithmic over visual thinking" (p. 25).

A Closing Example

Eisenberg and Dreyfus (1991) were right to call for greater use of forms of teaching that make "higher demands than thinking algorithmically" (p. 25). I can speak first-hand about what happens to students brought up on a heavy regime of algorithms that make little or no sense to them. For the past 10 years I have taught an elementary algebra course for prospective elementary and middle-school teachers who have chosen to specialize in mathematics. At the beginning of the course most of these students think they know the mathematics they learned at school well—but in fact most of them think in very wooden, mechanical ways about algebra (Ellerton and Clements 2011).

The only graphs about which they know anything much at the start of the course, are those with equation $y = mx + b$—almost all of them can parrot that this "cuts the y-axis at b, and has a slope of m." One of my aims during the course, is to seek to develop the students' visual imagery, so that they can "see," in their minds' eyes, graphs corresponding to $y = x^2$, $y = x^3$, $y = 1/x$, $y = \sqrt{x}$, $y = \sqrt{x^2}$, $y = -2(x - 1)^2$, etc. I want them to learn to approach mathematics problem solving in both analytic and visual ways, and toward that end I involve all the students in project work for which, working in pairs, they develop solutions to interesting mathematics problems, present the solutions to the whole class, and then create and solve similar problems. The students are invited to be as creative as possible, and to seek elegant, rather than pedestrian, solutions.

Here is an example of a "problem" that, over the past four years, I have asked students to solve: "Suppose, on a circular clockface the time shows exactly 12 noon. Assuming the clock is working well, how many minutes (to 1 decimal place) will it take before the minute hand and the hour hand are pointing in exactly the same direction again?" Most students immediately set about drawing a circular clockface. Another common starting point with the students is to try to set up an equation involving x (although the meaning of this variable is not usually well defined). Most cannot estimate what the answer might be. Quite a few say "one hour," and others say "65 minutes". Some simply write "1.05" and do not show a diagram. In interviews these students often say they did not think of hands moving around a circular clockface.

Invariably, pairs of students who are asked to do this problem initially find it extremely difficult. When they come to talk with me about it, I immediately point to a circular clockface on a wall and ask them where they think the two hands would be when they next point in the same direction. The two students usually generate a statement like "5 past 1." If that statement is made, I ask them: "Would it be exactly 5 past 1?" After further discussion, they usually tell me: "Well, actually, it would be a little bit more than 5 past 1." I then ask: "How could you work it out *exactly*?"

At that stage, relevant imagery has been developed. But there is still a way to go. Indeed, they have now reached what Newman (1983) called the point of "transformation", from words and imagery, to a symbolic representation of the problem. There is a fair bit of hard thinking required before they arrive at an equation like:

$$6t = 360 + \frac{1}{2}t, \quad \text{where } t \text{ minutes is the required amount of time after 12 noon.}$$

From this, the solution 65.5 minutes can easily be obtained.

The next phase is for the pair of students to work with the whole class for about 20 minutes, with the aim of getting the 25 (or so) other students in the class to the point where they can solve the problem confidently. At this *pedagogical* stage, I talk to the pair of students about developing relevant visual imagery—nearly all of them decide to borrow, or construct, a circular clockface for which they can rotate the hands. I then challenge them to think of how they would handle the transformation, or mathematization, stage of the problem. After a lot of talk, and reflection, they begin to say things like: "Well, the big hand rotates 360 degrees in 60 minutes, so it'll rotate 6 degrees every minute, and hence in t minutes it will rotate $6t$ degrees. The little hand will move at one-twelfth of that rate, that is to say, it will go $\frac{1}{2}t$ degrees when the big hand goes $6t$ degrees." Finally, they make sense of the equation $6t = 360 + \frac{1}{2}t$.

The next stage is the *problem-posing* stage. The pair of students have to pose two related problems suitable for other members of the class to solve. Here are some of the problems which students have posed:

1. If the time is 3:10, how long would it take before the two hands will be pointing in exactly the same direction?
2. If the time is 9:50, how long would it take before the two hands will be pointing in exactly the same direction?
3. How many times in any 12-hour period will the two hands point in exactly the same direction?
4. Suppose after noon the hour-hand moved backwards (i.e., counter-clockwise) at its normal rate, but the minute-hand moved forward at its normal rate. How long would it take before the two hands were pointing in exactly the same direction?
5. On a long straight road, *Car A*, which is traveling at 60 mph, is 5 miles behind *Car B*, which is travelling at 40 mph. Assuming that they continue to travel at those speeds, how long will it take for *Car A* to catch *Car B*?
6. On a long straight road, *Car A*, which is traveling at a mph, is d miles behind *Car B*, which is travelling at b mph. Assuming that $a > b$, and that they continue to travel at those speeds, how long will it take for *Car A* to catch *Car B*?
7. Suppose Train A leaves City A and travels at a mph toward City B, which is d miles from City A. At the same time, Train B, began traveling at b mph toward City B. If the two trains continue to travel at those speeds, how long will it be before Train A and Train B meet?

The two members of each pair of students who worked on the initial "clockface algebra task" not only posed problems like Questions 5, 6, 7, but also explained why those problems were "like" the first problem. Their imagery was enriched, and they began to think in images and structures. And yet, at no stage, was visual thinking an end in itself. These students began to draw diagrams, and to reflect on what the relevant variables were, and how they should be defined. There can be no doubt that they began to represent the problems, internally, in multiple ways (Amit and Fried 2005).

Presmeg (1985) suggested that teachers need to be assisted to help visualizer-students in their mathematics classes to overcome their difficulties and exploit their strengths in order that they will become more confident in planning learning environments in which students will visualize their way toward deeper knowledge of important mathematics. I feel the discussion of the clockface-algebra example above shows a pedagogical approach that will help to achieve the ends desired.

References

Amit, M., & Fried, M. (2005). Multiple representations in eighth-grade algebra lessons. In H. L. Chick & J. L. Vincent (Eds.), *Proceedings of the 29th conference of the international group for the psychology of mathematics education* (Vol. 2, pp. 57–64). Melbourne: PME.

Bishop, A. J. (1973). The use of structural apparatus and spatial ability—a possible relationship. *Research in Education, 9*, 43–49.

Bishop, A. J. (1980). Spatial abilities and mathematics education—a review. *Educational Studies in Mathematics, 11*, 257–269.

Burin, D. I., Delgado, A. R., & Prieto, G. (2000). Solution strategies and gender differences in spatial visualization tasks. *Psicológica, 21*, 275–286.

Cheetham, J. M., Rahm, B., Kaller, C. P., & Unterrainer, J. M. (2012). Visuospatial over verbal demands in predicting Tower of London planning tasks. *British Journal of Educational Psychology, 103*, 98–116.

Clements, M. A. (1982). Visual imagery and school mathematics. *For the Learning of Mathematics, 2*, 2–9; *3*, 33–39.

Clements, M. A. (1983). The question of how spatial ability is defined, and its relevance to mathematics education. *ZDM. Zentralblatt für Didaktik der Mathematik, 15*, 8–20.

Clements, M. A. (1984). Terence Tao. *Educational Studies in Mathematics, 15*, 213–238.

Clements, M. A., & Wattanawaha, N. (1978). The classification of spatial tasks suitable for the classroom. In D. Williams (Ed.), *Learning and applying mathematics* (pp. 432–447). Melbourne: Australian Association of Mathematics Teachers.

Dreyfus, T., & Eisenberg, T. (1982). Intuitive functional concepts: a baseline study on intuitions. *Journal for Research in Mathematics Education, 13*, 360–380.

Dreyfus, T., & Eisenberg, T. (1983). The function concept in college students: linearity, smoothness and periodicity. *Focus on Learning Problems in Mathematics, 5*, 119–132.

Dreyfus, T., & Eisenberg, T. (1986). On the aesthetics of mathematical thought. *For the Learning of Mathematics, 6*, 2–10.

Eisenberg, T. (1994). On understanding the reluctance to visualize. *ZDM. Zentralblatt für Didaktik der Mathematik, 26*, 109–113.

Eisenberg, T. (2008). Flaws and idiosyncrasies in mathematicians: food for the classroom? *The Montana Mathematics Enthusiast, 5*, 3–14.

Eisenberg, T., & Dreyfus, T. (1991). On the reluctance to visualize in mathematics. In W. Zimmermann & S. Cunningham (Eds.), *Visualization in teaching and learning mathematics* (pp. 25–37). Washington: Math. Assoc. of America.

Ellerton, N. F., & Clements, M. A. (2005). A mathematics education ghost story: Herbartianism and school mathematics. In P. Clarkson, A. Downton, D. Gronn et al. (Eds.), *Building connections: research, theory and practice* (pp. 313–320). Sydney: Mathematics Education Research Group of Australasia.

Ellerton, N. F., & Clements, M. A. (2011). Prospective middle-school mathematics teachers' knowledge of equations and inequalities. In J. Cai & E. Knuth (Eds.), *Early algebraization* (pp. 379–408). New York: Springer.

Gagné, R. M., & White, R. T. (1978). Memory structures and learning outcomes. *Review of Educational Research, 48*, 187–222.

Hadamard, J. (1945). *Psychology of invention in the mathematical field*. Princeton: Princeton University Press.

Hilbert, D. (1999). Preface. In D. Hilbert & S. Cohn-Vossen (Eds.), *Geometry and the imagination (pp. i –iii)*, Washington: Am. Math. Soc. [Original Work 1932].

Kosslyn, S. M. (1980). *Image and mind*. Cambridge: Harvard University Press.

Kozhevnikov, M., Kosslyn, S., & Shephard, J. (2005). Spatial versus object visualizers: a new characterization of cognitive visual style. *Memory & Cognition, 33*, 710–726.

Krutetskii, V. A. (1976). The psychology of mathematical abilities in school children. J. Kilpatrick & I. Wirszup (Eds.). Chicago: University of Chicago Press (trans. J. Teller).

Lean, G., & Clements, M. A. (1981). Spatial ability, visual imagery and mathematical performance. *Educational Studies in Mathematics, 12*, 267–299.

Lohman, D. F. (1979). *Spatial ability, individual differences in speed and level*. Stanford: Stanford University Press.

Lohman, D. F. (1996). Spatial ability and G. In I. Dennis & P. Tapsfield (Eds.), *Human abilities: their nature and assessment* (pp. 97–116). Hillsdale: Erlbaum.

Lowrie, T., & Kay, R. (2001). Relationship between visual and nonvisual solution methods and difficulty in elementary mathematics. *The Journal of Educational Research, 94*, 248–255.

McGee, M. G. (1979). Human spatial abilities: psychometric studies and environmental, genetic, hormonal, and neurological influences. *Psychological Bulletin, 86*, 889–918.

Newman, M. A. (1983). *Strategies for diagnosis and remediation*. Sydney: Harcourt Brace Jovanovich.

Owens, K. (2005). Substantive communication of space mathematics in upper primary school. In H. L. Chick & J. L. Vincent (Eds.), *Proceedings of the 29th conference of the international group for the psychology of mathematics education* (Vol. 4, pp. 33–40). Melbourne: PME.

Owens, K., & Outhred, L. (2006). The complexity of learning geometry and measurement. In A. Gutierrez & P. Boero (Eds.), *Handbook of research on the psychology of mathematics education* (pp. 83–115). Rotterdam: Sense Publishers.

Paivio, A. (1971) *Imagery and verbal processes*. New York: Holt, Rinehart & Winston.

Piaget, J., & Inhelder, B. (1971). *Mental imagery in the child*. New York: Basic Books.

Presmeg, N. C. (1985). *The role of visually mediated process in high school mathematics: a classroom investigation*. Unpublished Ph.D. dissertation, University of Cambridge, England.

Presmeg, N. C. (1986a). Visualization and mathematical giftedness. *Educational Studies in Mathematics, 17*, 297–311.

Presmeg, N. C. (1986b). Visualization in high school mathematics. *For the Learning of Mathematics, 6*(3), 42–46.

Presmeg, N. C. (2006). Research on visualization in learning and teaching mathematics: emergence from psychology. In A. Gutierrez & P. Boero (Eds.), *Handbook of research on the psychology of mathematics education* (pp. 205–235). Rotterdam: Sense Publishers.

Pylyshyn, Z. W. (1973). What the mind's eye tells the mind's brain: a critique of mental imagery. *Psychological Bulletin, 80*, 1–25.

Richardson, A. (1977). The meaning and measurement of memory imagery. *British Journal of Psychology, 68*, 29–43.

Sheckels, M. P., & Eliot, J. (1983). Preference and solution patterns in mathematics performance. *Perceptual and Motor Skills, 57*, 811–816.

Shepard, R. N., & Metzler, J. (1971). Mental rotation of three-dimensional objects. *Science, 171*, 701–703.

Skemp, R. R. (1972). *The psychology of learning mathematics*. London: Penguin.

Suwarsono, S. (1982). *Visual imagery in the mathematical thinking of seventh-grade students*. Unpublished doctoral dissertation, Monash University, Melbourne, Australia.

Tall, D., & Vinner, S. (1981). Concept image and concept definition in mathematics with particular reference to limits and continuity. *Educational Studies in Mathematics, 12*, 151–169.

Thomas, N. J. T. (1989). Experience and theory as determinants of attitudes toward mental representation: the case of Knight Dunlap and the vanishing images of J.B. Watson. *The American Journal of Psychology, 102*, 395–412.

Thurstone, L. L. (1938). *Psychometric monographs: Vol. 1. Primary mental abilities*.

Vinner, S., & Dreyfus, T. (1989). Images and definitions for the concept of function. *Journal for Research in Mathematics Education, 20*, 356–366.

Vinner, S., & Hershkowitz, R. (1980). Concept images and some common cognitive paths in the development of some simple geometric concepts. In R. Karplus (Ed.), *Proceedings of the fourth international conference on the psychology of mathematics education* (pp. 177–184). Berkeley: PME.

Wattanawaha, N. (1977). *Spatial ability and sex differences in performance on spatial tasks*. Unpublished M.Ed. thesis, Monash University, Melbourne, Australia.

Wattanawaha, N., & Clements, M. A. (1982). Qualitative aspects of sex-related differences in performances on pencil-and-paper spatial questions, grades 7–9. *Journal of Educational Psychology, 74*, 878–887.

Zimmermann, W. & Cunningham, S. (Eds.) (1991). *Visualization in teaching and learning mathematics*. Washington: Math. Assoc. of America.

Chapter 12
Reflections on Visualization in Mathematics and in Mathematics Education

Elena Nardi

Abstract Mathematics education research is far from consensus on the roles visualization can play in the teaching and learning of mathematics. This chapter offers similarly diverse perspectives: *Kupferman* illustrates a university teacher's endeavour to integrate visualization in teaching with an example of introducing the formal definition of limit to Year 1 students. He concludes that the benefits of a visually rich approach cannot be taken for granted, especially when students are not yet accustomed to it. To bring visualization into students' mathematical 'custom' *Presmeg* calls for teaching *visuality*, recognising that the relationship between logical and visual thinking in mathematics is not polarized but orthogonal, and reminding us that effective teaching of *visuality* originates in teachers whose own preferences are mixed and flexible. Analogously, *Nardi* calls for a new didactical contract that makes the rules about visualization explicit to learners, while recognising that a deliberate 'fuzziness' of this contract can also allow the manoeuvring that is often so potent in mathematics. Much like Kupferman, and in support of Presmeg's call for teaching *visuality*, *Hershkowitz*, through examples, acknowledges visualization as one of the languages of mathematics and as one of several ways of thinking mathematically. To be expressed, visual thinking needs a language, visual or other; and visual language, to be meaningful, needs to be attached to some conceptual entity. Finally, *Yerushalmy* picks up Hershkowitz's cue for meaningful integration of visualization into teaching with examples, such as interactive diagrams in algebra, that illustrate the challenges, affordances and profound epistemological shifts inherent in visually sensitive curriculum design.

Keywords Visual · Visualization · Visuality · Reasoning · Proof · Representation · Diagram · Picture · Image · Limit · Function · Infinity · Delta-epsilon · Algebra ·

With contributions by
Rina Hershkowitz, Weizmann Institute of Science, Revohot, Israel
Raz Kupferman, Hebrew University of Jerusalem, Jerusalem, Israel
Norma Presmeg, Illinois State University, Normal, USA
Michal Yerushalmy, University of Haifa, Haifa, Israel

E. Nardi (✉)
University of East Anglia, Norwich, UK
e-mail: E.Nardi@uea.ac.uk

M.N. Fried, T. Dreyfus (eds.), *Mathematics & Mathematics Education: Searching for Common Ground*, Advances in Mathematics Education,
DOI 10.1007/978-94-007-7473-5_12,

Geometry · Digital geometry environment · Algebraic · Geometric · Iconic · Symbolic · Linguistic · Kinaesthetic

Introduction

In recent years debate about the role of visual representations in mathematics and mathematics education has intensified on several fronts. For example, the potential contribution of visual representations in mathematical proof has been much discussed (e.g., Mancosu et al. 2005) and the multidisciplinary community of diagrammatic reasoning (e.g., Stenning and Lemon 2001) has been steadily growing. Central to this debate is whether visual representations should be treated as adjuncts to proof, as an integral part of proof or as proofs themselves (e.g., Byers 2007; Giaquinto 2007; Hanna and Sidoli 2007). For example, Giaquinto (2007) argues that visual means are much more than a mere aid to understanding and can be resources for discovery and justification, even proof.

Analogous attention to visualization has been apportioned also within mathematics education. Its richness, the many different roles it can play in the learning and teaching of mathematics—as well as its limitations—are increasingly being written about (e.g., Arcavi 2003). The foci of these works are diverse—Presmeg (2006a, 2006b) offers a substantial review of these as well as highlights ones that may need to take priority. Overall, we still seem far from a consensus on the many roles visualization can play in mathematical learning and teaching, as well as in pre- and post-formal aspects of mathematical thinking more generally. So, while many works clearly recognise these roles, several also recommend caution with regard to 'the 'panacea' view that mental imagery only benefits the learning process' (Aspinwall et al. 1997).

In the light of above ongoing debates and developments it is of little surprise that Ken Clements, in his chapter in this book, puts forward a challenge for clear terminology concerning visualization. The contributions to this chapter go some way in responding to this challenge.

- *Kupferman* takes us through the endeavour of integrating visualization in university mathematics teaching with urgency and immediacy. He does so through an illustrative example from introducing the formal definition of limit to incoming university students. His text brims with observations that ring bells of familiarity to mathematics education researchers. His focus on the concept of limit as a milestone in the students' early experiences in Calculus is in fine resonance with numerous studies of advanced mathematical thinking at least since the 1980s and 1990s (e.g. Tall 1991); as is his proposed visual approach to the formal definition—Roh's (2010) ε-strip is a recent example. The caveats about visualization in his closing statement also chime well with those raised by the participants in the studies referred to in the contributions by Nardi and Presmeg: students, especially if unaccustomed to a visual approach, may find it not so helpful and the benefits of this approach cannot be taken for granted. In this sense his contribution is an apt stepping stone to those that follow: to bring visualization into students' mathematical 'custom' Presmeg, for example, calls for teaching *visuality*; and Nardi calls for a new didactical contract that makes the rules about visualization explicit to learners.

- Much like Kupferman *Hershkowitz* acknowledges visualization as one of the languages of mathematics. Her working definition of *visualization* includes a consideration of it as one of the several ways of thinking mathematically; and, a group of signs and relationships among these signs, which she terms a language, by which these several ways of mathematical thinking, including the visual one, might be developed, limited, expressed and communicated. These two perspectives are interweaved: to be expressed, visual thinking needs a language, visual or other; and visual language, to be meaningful, needs to be attached to some conceptual entity. To illustrate the above she embeds her contribution to the debate in two evocative examples: one from the work with young children of the *Agam Programme* in the Weizmann Institute; and another from classroom vignettes of learners engaged with solving the visually stimulating *Matches Problem*. Across her account the significant challenges that pertain to the meaningful integration of visual thinking in teaching also emerge strongly. In a way the studies she illustrates from serve as examples of what Presmeg calls teaching *visuality*.
- *Yerushalmy* picks up Hershkowitz's call for the meaningful integration of visualization into teaching in new and dynamic perspectives. Her contribution takes us at the heart of the challenges of visually sensitive curriculum design. To this purpose she draws on examples from Algebra in order to: (i) to highlight technological affordances with noticeable impact on the way we visualize and understand mathematical objects and mathematical actions; (ii) demonstrate the potentially profound epistemological change that is inherent in efforts to design curricula in which visualization holds a central position; and, (iii) to argue that implementing ensuing curricular changes requires in-depth review of hitherto taken for granted research findings and recommendations.
- *Nardi* quotes Whiteley's (2004) 'learning to see like a mathematician' in order to explore a particular aspect of the pedagogical role of the mathematician: to foster a fluent interplay between analytical rigour and (often visually based) intuitive insight. As in Kupferman's observation fostering this fluency is much needed as students' relationship with visual reasoning is often turbulent. She attributes this turbulence to unclear didactical contracts of university and school mathematics with regard to visualization. She also recognises that the 'fuzziness' of these contracts can also allow the manoeuvring that is often so potent in mathematics. From her data some terms of a flexible but clearer didactical contract emerge—as do some of the pedagogical challenges that the honouring of such a contract implies.
- *Presmeg* closes the section of individual contributions with a challenge to the analytical-visual dichotomy by reminding us of the Krutetskiian tenet that the relationship between logical and visual thinking in mathematics is orthogonal, rather than polarized. She also reminds us of: the instrument she has used in her work and its capacity to identify individual preferences with regard to visuality; and, some of the difficulties that are inherent in the handling and generation

of visual imagery in mathematics. She continues with a brief recapitulation of how dynamic imagery and metaphorical thinking can facilitate the overcoming of these difficulties; and a revisiting of her findings that the most effective teaching of visuality originates in teachers whose own preferences are mixed and flexible. Her contribution concludes with the poignant list of questions—as in Presmeg (2006a, 2006b)—that the field ought to consider as central in future research.

We close with a few thoughts on visualization in mathematics and mathematics education taking cue from the list of questions that conclude the contribution by Presmeg and from the symposium discussion that followed the presentations on which the above contributions are based.

Visualization in First Year Calculus

Raz Kupferman

First year calculus has always been a serious stumbling block for our undergraduate students. Within my institution in recent years the percentage of failure has reached new records: 70 %–80 % failing in final exam, compared to about 50 % in the past. Amongst faculty the feeling has been that this decline reflects an increasing gap between high school and university levels.

I have chosen to focus on our Introductory Calculus course because it is the one that involves the highest degree of failure, but also because colleagues and I have always viewed it as our "display window". Its syllabus consists of the following topics: real number axioms, functions, limits, continuity, derivatives, sequences, and series. Our experience (based on bi-weekly quizzes) is that many students lose contact with the material at a very early stage.

What is it that makes Calculus such a terrifying experience for many (but also a great source of excitement for others...)? Our students are expected to learn a new language and reach a certain level of proficiency in working with abstract objects within 3 months. Within the many new concepts that the students learn, it seems as if their success hinges on their ability to understand at an early stage the notion of a *limit*.

Many of the topics learned in high school are accompanied by visual means (the only one of the five senses that can be used). When it comes to university mathematics we want our students to develop abstractions, and the natural question is to what extent we should use visual aids. Despite the stereotypical view of how mathematicians view this matter, I think that the general perception is that we should help our students assimilate new concepts and new materials by any means that helps.

Functions Before defining limits of functions we need to define functions.

A "gentle" definition of a function $f : A \to B$ is a machine that, given a number that belongs to a set A, returns a number that belongs to a set B according to a

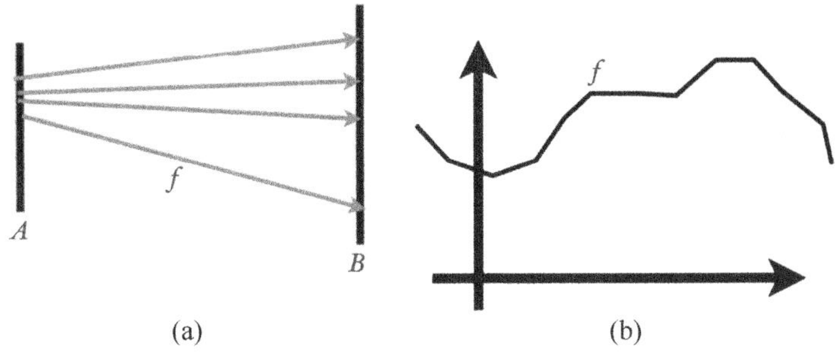

Fig. 12.1 Representations of a function as (**a**) a mapping between two sets and (**b**) as a graph

fixed assignment rule. On the other hand, and to the students' horror, the "formal" definition of a function $f : A \to B$ is a subset R of the Cartesian product set $A \times B$ that satisfies:

$$\forall x \in A\ \exists! y \in B \quad \text{such that } (x, y) \in R.$$

Interestingly, when it comes to visualizing a function, the students are used to a representation that corresponds to the "formal" definition—as a graph. The graph of the function (Fig. 12.1b), which is a visual representation they work with since middle school, is precisely a subset of the Cartesian product $A \times B$. A visual representation of the "machine" like operation of a function is shown in Fig. 12.1a. This representation is in fact more convenient when it comes to visualize more abstract functions, and for the visualization of the composition of functions.

Limits What does it mean that the limit of a function f at a point a exists and is equal to L, which we denote by

$$\lim_{x \to a} f(x) = L?$$

The formal definition, written in cryptic mathematical notation is:

$$\forall \varepsilon > 0\ \exists \delta > 0 \quad \text{such that } \forall x : |x - a| < \delta \quad \Rightarrow \quad \big| f(x) - L \big| < \varepsilon,$$

or in words: for every $\varepsilon > 0$ there exists a $\delta > 0$ such that for every $|x - a| < \delta$ it holds that $|f(x) - L| < \varepsilon$.

This definition is very convoluted to anyone unused to the nesting of logical clauses, but yet, its understanding is crucial. One of the means I am trying to use to help the students is visualization. A typical sequence of sketches is shown in Fig. 12.2. I am presenting the concept of the limit as a game: you, my audience, have to challenge me by picking a number ε; this number is represented by a shaded horizontal strip of width ε around the value L (Fig. 12.2a). I have to respond by choosing an appropriate number δ, which is represented by a shaded vertical strip of width δ around the value a (Fig. 12.2b). The value of δ has to be such that for every x within the vertical strip (Fig. 12.2c), the value of the function $f(x)$ is within the horizontal strip (Fig. 12.2d). If I can find such a δ for every ε, then indeed, the limit of f at a is L.

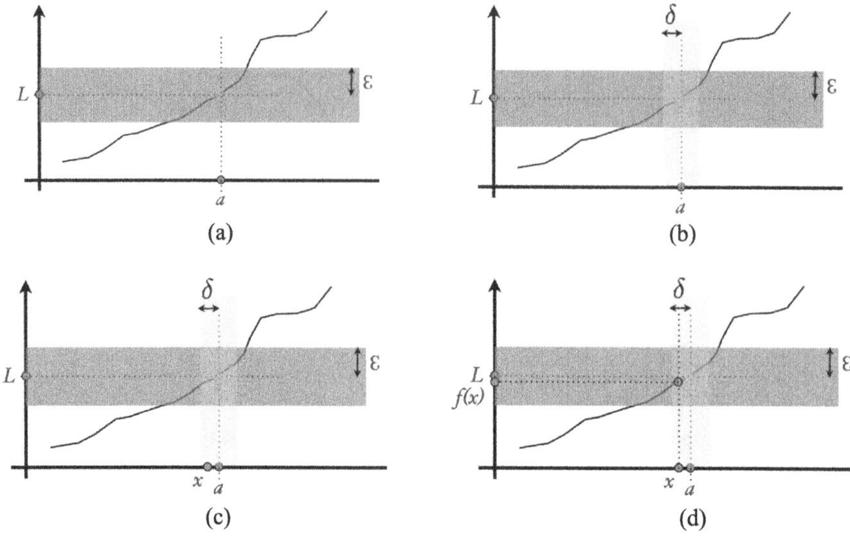

Fig. 12.2 Steps in the visualization of the defining properties of the limit of a function

While I think of such visualization as enlightening, my experience is that maybe for many students it is more confusing than helpful; they are not used to this visual language. Visualization is also a form of language that needs to be learned before one can take advantage of its strength. I encourage lecturers to use visualizations wherever they find appropriate, but the bottom line of my argument is that the students' ability to benefit from it should not be taken for granted.

Looking at Visual Thinking and Visual Communicating in Mathematics Learning Through the Lens of Examples

Rina Hershkowitz

I relate here to *visualization* mainly as:

(1) A mode of mathematical thinking.
(2) A group of *signs* and relationships among them ("a language"), by which mathematical thinking, including the visual one, might be developed, limited, expressed and communicated to oneself and to others.

Neither of these two perspectives, which may weave together, has a meaningful existence by itself: visual thinking needs "a language" (either visual or another language) to be expressed; and visual language, when it does not represent a thought, is just a group of signs without a meaning.

In the following I draw some features of the above two perspectives and raise some global questions concerning visualization in learning mathematics, by discussing two different learning topics in which visualization is involved:

Topic 1: Patterns in the "Agam Programme for Developing Visual Thinking and Visual Language" (Hershkowitz and Markovits 1992)

The Agam programme is an example of an effort to interweave the development of a visual language with a process of developing the visual thinking of young children. The programme is a vision of the artist Yaacov Agam, and it has become an educational reality through the ongoing work of a team of researchers and educators of the Science Teaching Department in the Weizmann Institute. The development and investigation of the programme focuses on several groups of students some beginning with three-to-four-year-olds in the nursery school and continuing with the same groups to the third grade. This activity was followed by research that showed that the "Agam children" can apply visual abilities and visual thinking in learning tasks more successfully than children in the control groups (Razel and Eylon 1990).

Some of the programme's 36 curriculum units introduce children to such basic visual concepts as the main geometric figures, directions, colours, and size relationships. These units make up a "visual alphabet" that forms the basis for more advanced concepts, such as symmetry, ratio and proportion, numerical intuition, dimensions, that serve as building blocks in scientific and mathematical thinking.

The first unit is on the circle and the second on the square. From circles and squares as "visual letters", patterns (the third unit in the programme), which are "visual words", or "visual sentences", can be created. A pattern is a *visual periodic series* whose elements, at this stage in the curriculum, are: squares, circles, different colours, different sizes, the figures' orientation and the interval between the figures (see, for example, Fig. 12.3a). When children create patterns they are seen as problem solvers with high-level visual thinking. They analyze the main characteristics of patterns that are to be used in their creation—for example, the building blocks for the periodic theme and the length of the period—and choose those that they would like to have in their own special pattern. Finally, they synthesize all the above in the reproduction of their pattern. This creation of patterns is not always obvious and straightforward as we can see in Fig. 12.3b, where the child adds a second dimension and creates "periodic themes" which keep the periodicity of form and size but not of colour. All the purposeful activities in the unit are linear, but children's creativity is unbounded (see the "sun" in Fig. 12.3c and the matrix in Fig. 12.3d).

It is worth noting that the above patterns differ from those students typically encounter when learning early algebra (Radford 2012). While "algebraic patterns" are visual sequences that can be generalized by means of a quantitative-numerical rule-of-change, the "periodic patterns" present the principles of a theme, its length, and its repetition, and therefore can be expressed and described by visual language

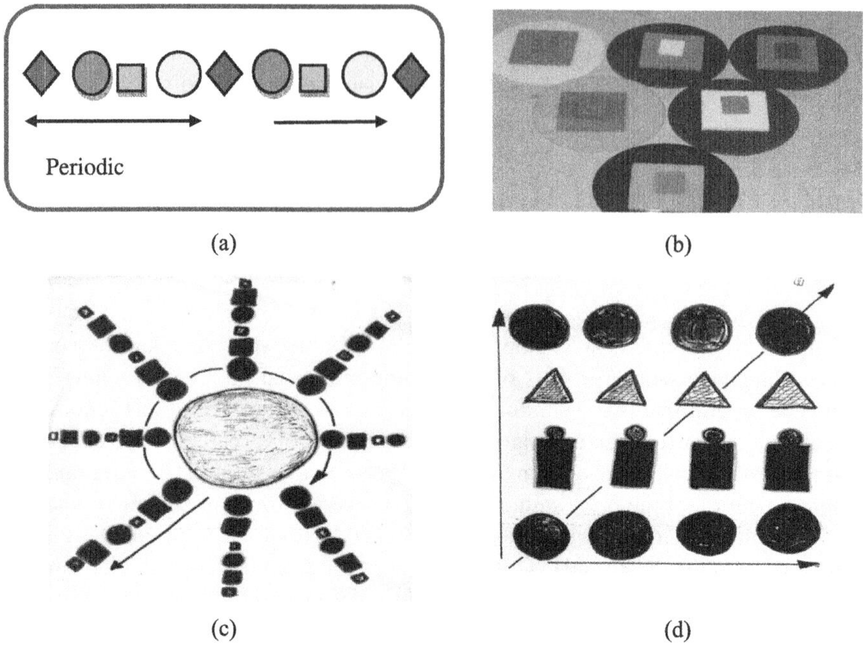

Fig. 12.3 Patterns: (**a**) Linear pattern; (**b**) Vertical incomplete patterns; (**c**) The "sun"; (**d**) The "matrix"

(or verbal which is quite heavy in this case). It seems that the difference between the two types of "patterns" is mainly in the fact that the "algebraic-visual patterns" express the accumulation of the repeated elements' quantities in the pattern (see the next example), where the visual patterns of the *Agam programme* do not. In this sense the later are similar to periodic phenomena in science.

Topic 2: Visual-Quantitative-Patterns in Two Dimensions and Their Algebraic-Symbolic Generalizations—The Matches Problem (Hershkowitz et al. 2001)

This is a "story" of a problem, borrowed from a verbal communication with Prof. I. Weinzweig, which I implemented in many teachers' courses in various countries. The problem is presented visually (see Fig. 12.4) and when different individuals solve it by visual thinking it affords and triggers a diversity of visual solution processes. In the following I show a few examples of such solution processes which uncover "mechanisms" of visualization as a mathematical way of thinking (e.g. the mechanism of *composing and de-composing*, Duval 1998).

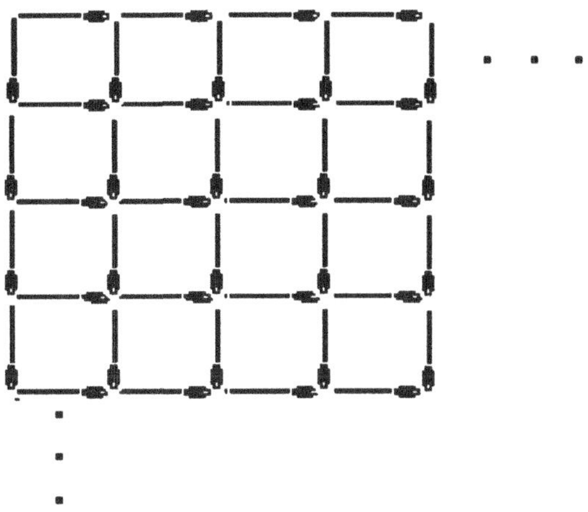

Fig. 12.4 The Matches Problem

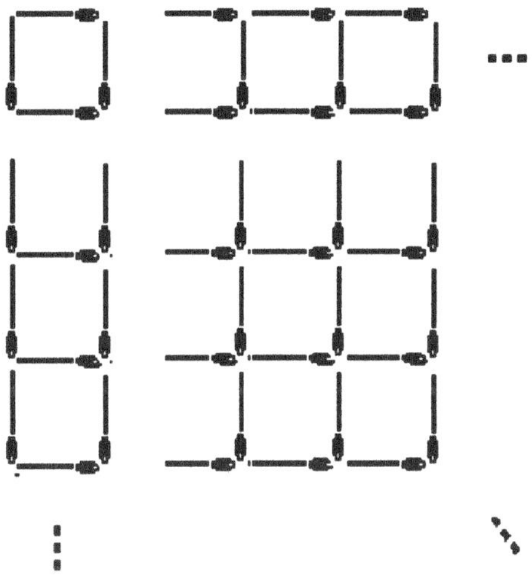

Fig. 12.5 From One Square on...

The Matches Problem There are squares of "small squares" made of matches. How many matches are needed for a square with n matches in its side? Explain how you reached your conclusion. Try to find more than one way.

Example Solution 1 (Fig. 12.5): *From One Square on...* A common strategy was to start from a corner-square of 4 matches, to continue by counting two chains of **U**'s (three matches each) sideways and downwards to obtain $3(n-1) \times 2$. The counting then focuses on the $(n-1)(n-1)$ L's (with 2 matches each) opposite the four initial matches. The sum of all the components taken together is therefore:

$$4 + 3(n-1) \times 2 + 2(n-1)(n-1)$$

The problem solver made a visual transformation (here decomposition and composition) and obtained a new pattern, which affords systematic counting. The "prod-

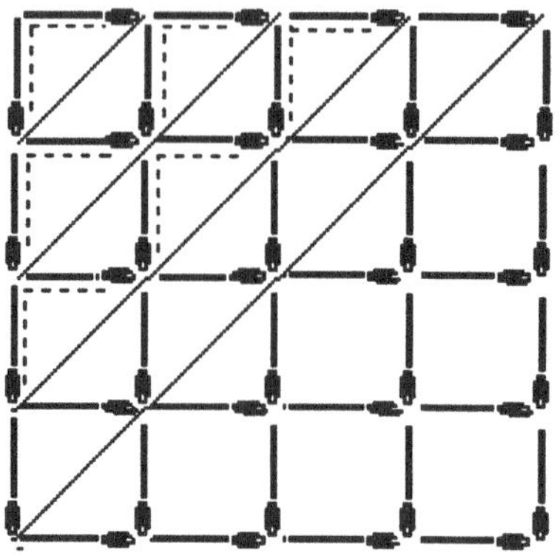

Fig. 12.6 Looking Above the Diagonals

uct", as a mathematical object, is signified in a meaningful way (only?) by the symbolic "language" of Algebra.

Example Solution 2 (Fig. 12.6): *Looking Above the Diagonals* Here again the problem solver made a visual transformation in her way of looking at the combination of matches which enables her to count the matches systematically:

The number of matches in half a square is

$$2(1+2+3+4+\cdots+n)$$

So in the whole square there are

$$2\times 2(1+2+3+4+\cdots+n).$$

Example Solution 3 (Fig. 12.7): *Decomposition of the Square Into Two "Equal Triangles"* The number of horizontal matches "as we descend the staircase" is $1+2+3+4+\cdots+n$. The same is true for the vertical matches "as we ascend the staircase". Thus in one triangle there are $2(1+2+3+4+\cdots+n)$. Since there are two triangles, the final count is: $2\times 2(1+2+3+4+\cdots+n)$.

Many more example solutions are presented in Hershkowitz et al. (2001).

It is worth noting that I saw very few teachers who did not use the power of visualization in solving this problem. The teachers who did not hold in the visualization way, "translated" the situation into numbers organized in tables, but failed to generalize a solution numerically.

Discussion

It has been advocated (see, for example, Steen 1988) that, the search for patterns and their organization in mathematical language, is a central component of mathematical thinking.

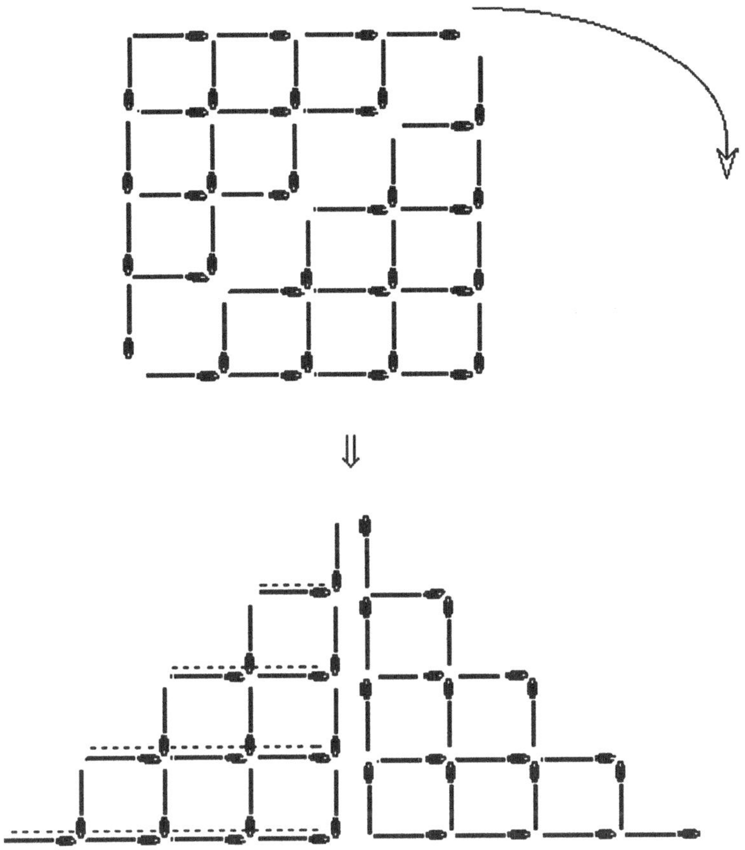

Fig. 12.7 Decomposition of Square Into Two "Equal Triangles"

The two learning topics described above show that visualization can play a crucial role in this. This supports Fischbein's claim that visualization "not only organizes data at hand in meaningful structures, but it is also an important factor guiding the analytical development of a solution." (Fischbein 1987, p. 101). From analyzing the above two cases, it seems that visualization can be even more than that: it can be the analytical process itself which concludes with a general solution. As such, its analytical components may include: (a) decomposition of a structure into substructures and units, (b) creation of auxiliary constructions, (c) transformation of the whole structure into another structure (in the second case), and (d) re-composition and synthesis.

In the first topic above, thinking and language are both visual. Design experiments in the context of the *Agam programme* showed that by implementing the purposefully designed activities children construct visually quite compound mathematical and scientific concepts.

In the second case, the *Matches Problem* elicited many different solution strategies, lively discussions and a non-negligible amount of enthusiasm on each and every occasion we proposed it to teachers and students. A few points are worth noting:

- While the typical solution processes of this problem are visual, the best (and might be the only) presentation ("language") to express a generalized solution of this problem is a symbolic one.
- Different visual solution processes (visual mathematical thinking) give rise to different (but of course equivalent) algebraic expressions.
- Hence the problem affords symmetrical (reflexive) relationships between visualization and the symbolic representation, rather than the asymmetrical "classic" way of considering visualization as the intuitive support of a higher level of reasoning.
- Do the groups of teachers who failed to use a visual strategy to solve the *Matches Problem* represent a different mathematical culture? It might be speculated that their persistence with numerical approaches is because: (a) their mind's eye was not used to visual analysis, and/or (b) visual means were not highly regarded and not considered as a legitimate mathematical way to produce a general and formal solution.

Epilogue

There is a need to design (through design experiments) learning trajectories that target visualization for its own sake. The excellent example "Visual Math" of Michal Yerushalmy, (in the following section of this chapter), as well as the two cases discussed above show ways in which this can be achieved for specific topics, and/or especially designed programmes. All this suggests that the time is ripe to unify existing research results in order to create an expanded learning trajectory that includes a strand of visual learning alongside other mathematical strands in the mathematics syllabus.

What Do We Know About Visualization in the Age of Rapid Technological Change?

Michal Yerushalmy

Summarizing trends of current interest of research on visualization in mathematics education, Norma Presmeg (2006a, 2006b) identifies curriculum development and research of effective teaching of mathematical visualization mainly in dynamic computer environments to be of importance and interest. Presmeg also points into the newer theoretical directions that visualization research is looking at, including visualization as related to the embodied nature of mathematical learning and to the semiotic analysis of inscriptions, signs and gestures. The new foci of interest, Presmeg argues, mark a change from previous research challenges such as the reluctance to visualize in mathematics and the status of visual reasoning in mathematics; and, the diversification of terms and intentions concerning visualization (ibid. p. 209).

I would like to extend Presmeg's call to make specific aspects of thinking with technology an important direction of the study of visualization. To do that I would use examples from algebra in order (i) to highlight technological affordances that are found to have noticeable impact on the way we visualize and understand mathematical objects and mathematical actions, (ii) to demonstrate the possible depth of the epistemological change that should be considered when designing a curriculum that assumes visualization to be central to mathematical reasoning and, (iii) to argue that implementing such curricular changes requires in-depth review of earlier finding and recommendation of research.

New Visual Landscapes

For over two decades educators study the impact of Dynamic Geometry Environments (DGEs) on learning and mainly the impact of dragging. The terms *figure* and *drawing* coined by Parzysz (1988) became the true core of the study of geometric reasoning with technology. It is only recently that we begin to understand the visual catalog in other subject matters such as algebra and calculus. Earlier studies of students' interaction and uses of diagrams in its static mode focus on the explanatory ways of visual diagrams and its functions related to problem solving. These functions are changing when mathematical text embeds interactive diagrams and when digital multimodal communication channels also serve mathematical conversations (Yerushalmy and Botzer 2011). Issues such as personalization of a diagram, direct access modification of visuals, randomness, generality and representativeness of visual examples become the noticeable processes to study (Naftaliev and Yerushalmy 2009). As the actions performed with a visual sign and the motivations to take these actions are two critical semiotic aspects, the visual functions of interactive technology change the semiotic landscape.

To demonstrate the implication of the shift from static to dynamic diagram let us view a diagram (Fig. 12.8) that shows a visual solution of an equation presented symbolically and graphically in the form of $f(x) = g(x)$. Graphically, the representation of a solution is the intersection between the graphs of the two functions. This visual indication may act as feedback for the correctness of the symbolic operation. But if the practice of solving equations is part of interaction with the interactive diagram that Fig. 12.8 demonstrates (developed by Schwartz 2011), more important occasion for learning algebraic skills with understanding occur. Interacting with the diagram would mean transforming (e.g., scaling, translating) the graphs of given functions' comparison to produce new equivalent comparisons. Simultaneous transformation of both graphs would keep the solution. Transforming only one side of

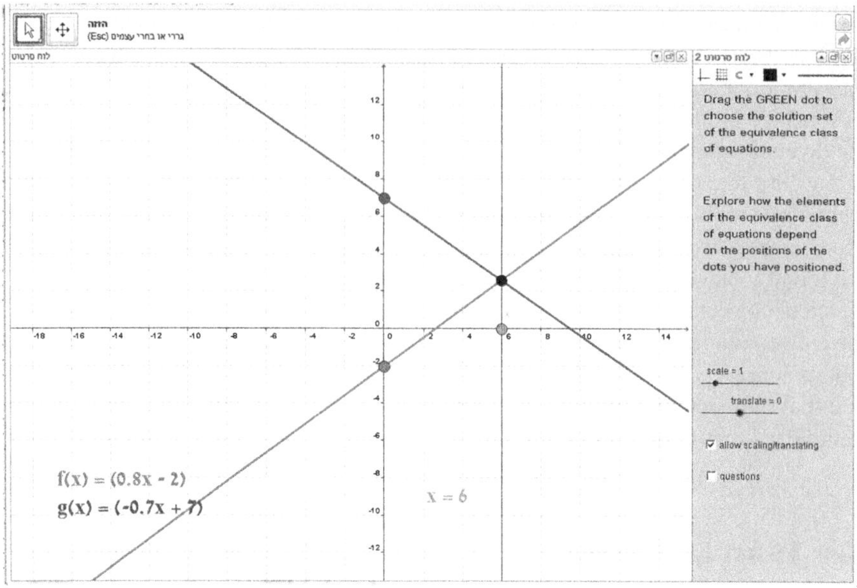

Fig. 12.8 Transformations of equation that yield equivalent equations (Schwartz 2011)

the equation and still keeping the solution is obvious when treating the solution as the intersection of two graphs, but quite surprising for all of us who learned that the only way to correctly arrive to solution is "to do the same" to both sides of the equation.

Setting up learning occasions of this type represents a change of focus and possible change of order relative to the traditional assumption about algebraic skills being the foundation for the understanding of functions.

What Does It Take to Change the Lenses?

The dynamic setting and the operations on visual objects in the equation task were designed to raise conjectures that then can be generalized and proved as part of learning and practicing symbols' manipulations. It represents the type of algebra, the skills and the pedagogical approach of Visual Math (1995/2005). *Visual Math* is a function-based algebra curriculum designed to use functions in multiple representations early to be the foundation for mastering algebraic skills with understanding, by all students. This is different from strategies that append traditional algebra problems with graphs as it seeks ways to introduce mathematical concepts by visual objects upon which symbolic understanding of variables, expressions and equations are built. The design principles implemented in *Visual Math* attempt to respond to the challenge by adopting a view of the domain and representing it in a structure that is visible to its users. Technology then becomes necessary as it supports direct

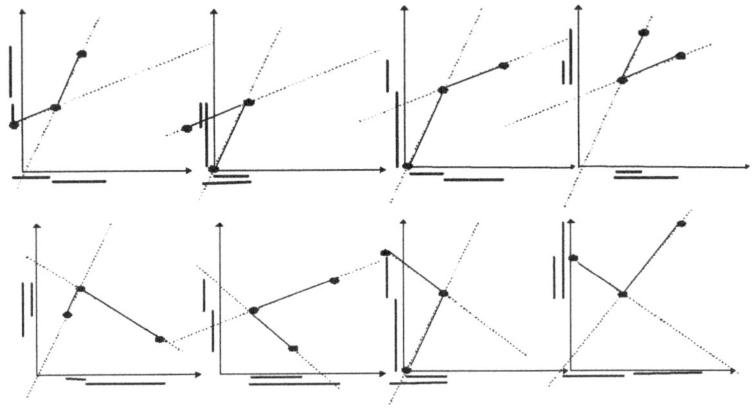

Fig. 12.9 Eight problem situations (adopted from Yerushalmy and Gilead 1999, p. 194)

construction and manipulations of visual objects. The traditional curricular order aiming at conceptual development of knowledge of algebra has to be changed; symbols such as letters, parameters, equal sign and equation now acquire their meaning only after graphs of functions, geometrical transformations, comparisons of processes modeled by two functions graphs and the rate of change of a given process are learned.

Designing learning opportunities for the inquiry of linear, single-variable word problems in context illustrates the research and development process involved in the unpacking of the traditional school-mathematics known as "linear word-problems" to the new visual curricular agenda and the learning sequences. Traditionally, word problems are organized into a learning unit, ordered by the complexity of the equations to be solved. A *Visual Math* sequence is based on the centrality of comparison of two processes and the algebraic meaning of equation and solution that can be derived from this comparison to algebra beginners. Attempting to map this subdomain of school algebra, we identified (Yerushalmy and Gilead 1999) eight main problem situations derived from this representation: two types of comparisons of two functions; four intersecting lines that have the same inclination; and, four that have opposite inclinations (Fig. 12.9). With some limitations, this organization allows the mapping of most linear word problems in algebra that involve a situation of comparison of two functions.

This visual mapping idea makes the design of the tasks for inquiry manageable because the number of problem types is reduced and ordering by complexity is based on the structure of the situations as represented visually by linear functions. The tasks involve practicing the required skills and provide opportunities that are mathematically interesting and manageable. On another level, the tasks are aimed at promoting the heuristics of problem solving (e.g., the study of the differences and similarities between problems) and strategies of inquiry (e.g., generalizations or counter-examples). Viewing the eight problem situations sequence through dynamic lenses help to understand the mathematical similarity: linear transformation of scaling, translations and reflections (in horizontal or vertical lines) would turn a given model (one of the 8) to any of the other models. The technology acts here as a cognitive tool offering a way to experiment the generality of an example and its representativeness of the domain, and to learn similarity among problems and the solution strategies derived from that.

This example illustrates the challenge of redesigning the pedagogy that would make the new epistemological assumptions relevant to algebra teaching and learning. While dynamic visual actions with new tools are assumed, it certainly takes more than a tool and a task to change the traditional symbolic lenses of school algebra.

New Challenges for the Study of Visualization

In "*Challenging known transitions: Learning and teaching algebra with technology*" (Yerushalmy 2005) I speculate about the degree to which new technologies will lead to the replacement of current curricula with new content. I also ask how the use of a new curriculum that is based upon new epistemological assumptions changes our capability to anticipate students' difficulties and strengths. We should wonder whether the hope of educational systems that digital environments would motivate change is realistic and likely to succeed unless they understand more about what they are up against. In this regard I suggested and hopefully illustrated that advances in this area would benefit from in-depth study of curricular approaches. I would challenge the reliability of earlier research findings regarding students' visual thinking and suggest that curricular research could benefit from systematic studies that reexamine visualization as cognitive challenge and as pedagogical preferences, especially those that concern the semiotic potential of technological tools, for teaching school algebra.

Learning 'To See like a Mathematician'

Elena Nardi

The mathematicians whose interviews form the evidence base for (Nardi 2008) offer a response to Clements' challenge quoted in the Introduction (that mathematics educators need to tighten their working definition of "visualization") that is much less rhetorical than it may seem at first. Much like Kupferman and Hershkowitz in earlier sections of this chapter, these mathematicians describe the role that they hope visualization has in their students' thinking as follows:

> ...the diagram is used almost as a third type of language—where the other two are words and symbols—as an extension of their power to understand. (Nardi 2008, p. 145)

The origins of these mathematicians' perspectives often lie in their own practice. Many—e.g. as reported in Dreyfus et al. (2012)—make a relatively straightforward point: consider "what mathematicians often do" (Whiteley 2009, p. 258, also citing Brown 1997), how "mathematicians work" (Nardi 2009, p. 117), what constitutes 'expert behaviour in doing mathematics' (Iannone 2009, p. 224) as one criterion for deciding pedagogical priorities for university mathematics teaching. For example, Whiteley (2009) reflects on his own practices as a mathematician and highlights the pedagogical importance of exemplification. His support for visual arguments fits squarely within this emphasis: illustrations and gestures can be close to the cognitive processes students need to carry out in order to develop understanding, sometimes even of the 'purest' mathematical idea. The usual critique against visual reasoning, that visuals 'are "merely" examples', 'too specific to be used in general proofs' should not deter us, he stresses: "Visuals are strong particularly because they are examples' and they can indeed 'carry general reasoning as symbols for the general case, provided the readers bring a range of variation to their cognition of the figure" (p. 260). Furthermore, not only there is nothing wrong with a 'partial' perception of a mathematical idea but also this very 'partiality', and any work students may do towards developing conventions and expressions for it, can be instructive. Pedagogical practice that deprives students of these instructive opportunities is impoverished.

The mathematicians I and colleagues interviewed (2008, 2009) elaborate Whiteley's "learning to see like a mathematician" (2004, p. 279) further. Part of the pedagogical role of the mathematician, they state, is to foster a fluent interplay between analytical rigour and (often visually based) intuitive insight. The need to foster this fluency is particularly pronounced as students' relationship with visual reasoning is often turbulent. Even when students overcome resistance to employing visualization, their reliance on it can be somewhat fraught: pictures may appear unaccompanied by any explanation of how they came to be, or they may appear disconnected from the rest of the students' writing. In fact, students' reticence about employing visualization has often been attributed (Nardi 2008) to what they perceive as the 'fuzzy' didactical contract (Brousseau 1997) of university mathematics: a contract that allows them to employ only previously proven statements but does not clarify which parts of their prior knowledge, or ways of knowing, count as proven or acceptable. Like Whiteley (2009), the mathematicians colleagues and I interviewed stress the potentially creative aspects of this 'fuzziness'. They argue that: a picture provides evidence, not proof; pictures are natural, not obligatory elements of mathematical thinking; pictures are "a third type of language" (as quoted in the opening paragraph of this section and the contributions in this chapter by Hershkowitz and Yerushalmy). From these views emerges a didactical contract in which students are allowed to use facts that have not been formally established; later, they are expected to establish those facts formally. The students are encouraged to make use of the power that visualization allows them. However, they are required to do so in a sophisticated way—for example, through including articulate accounts of their thinking in their writing and through acknowledging the support (e.g., of a graphic calculator) that facilitated the emergence of an insight.

The 'contract' described above poses certain challenges to the teacher, whether at university or school levels. I elaborate some of these challenges by drawing on the perspective put forward by one such teacher (of secondary school mathematics), Spyros (as reported in the analyses of task responses and interview data examined in (Biza et al. 2009 and Nardi et al. 2012). We asked Spyros to elaborate on whether he would accept an argument based on a graph and his answer was firm: 'No, first of all it is not an adequate answer in exams' (especially those requiring formal proof explicitly). We asked him to let aside the examination requirements for a moment and consider whether an argument based on a graph would be adequate mathematically. He replied: 'Mathematically, in the classroom, I would welcome it at lesson-level and I would analyse it and praise it, but not in a test'. When we asked him to elaborate he said: 'Through [the graph-based argument] I would try to lead the discussion towards a normal proof... with the definition, the slope, the derivative, etc.'. And when we asked him to justify he said: 'This is what we, mathematicians, have learnt so far. To ask for precision.... we have this axiomatic principle in our minds.... And this is what is required in the exams. And we are supposed to prepare the students for the exams.'

Spyros's statement is clear: while he cannot accept a graph-based argument as proof, he recognises graph-based argumentation as part of the learning trajectory towards the construction of proof. He seems to approach visual argumentation from three different and interconnected perspectives: the restrictions of the current educational setting, in this case the Year 12 examination; the epistemological constraints with regard to what makes an argument a proof within the mathematical community; and, finally, the pedagogical role of visual argumentation as a means towards the construction of formal mathematical knowledge. These three perspectives reflect three roles that a mathematics teacher needs to balance: educator (responsible for facilitating students' mathematical learning), mathematician (accountable for introducing the normal practices of the mathematical community) and professional (responsible for preparing candidates for one of the most important examinations of their student career). Spyros' awareness of these roles, and their delicate interplay, is evidence of the multi-layered didactical contract he seems able to offer to his students. As Nardi et al. (2012) propose, Spyros' views are underlain by a nexus of priorities (pedagogical, curricular, professional and personal/evaluative) that need to be considered, and handled, concurrently and with equal urgency.

Visualization in the Learning and Teaching of Mathematics

Norma Presmeg

In response to Ken Clements' injunction to be precise about terminology concerning visualization, I want to put forward a model that challenges the analytical-visual dichotomy that has been used in some research studies. Following Krutetskii's (1976) formulation, the strength of logic (and analysis) determines the effectiveness of mathematical problem solving, whereas the presence or absence of visualization determines its type. That is, *all* mathematical thinking involves logic (which could

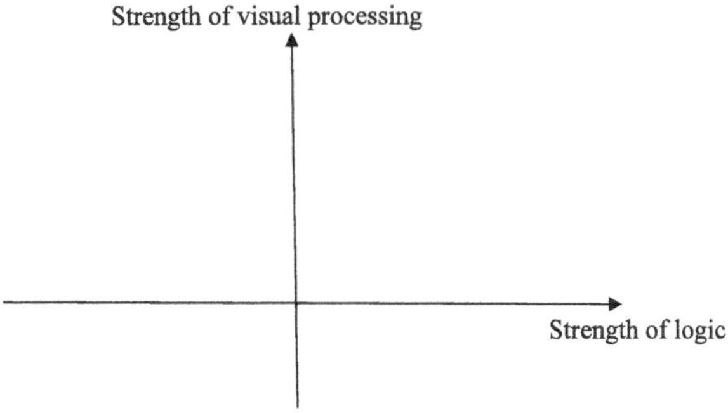

Fig. 12.10 Orthogonal axes for logic and visualization in mathematics

be depicted on the x axis), but mathematical visualization is orthogonal to it (on the y axis) and could be present or absent. My characterization of visualization from the 1980s went beyond Bishop's (1980) distinction between Interpreting Figural Information (IFI) and Visual Processing (VP), although these provided a starting point. Krutetskii's (1976) theoretical formulation was central in my research. For him, strength of logic determines the effectiveness of mathematical thinking, whereas visualization is optional (Fig. 12.10). There is no duality between logical analysis and visualization in an either-or sense.

My research identified individuals in all four quadrants of this model (Presmeg 1985) according to their mathematical logic and preference. In my work, visualization could be of the form of mental visual imagery (internal representations)—but it could also be of the form of inscriptions of various kinds (external representations). Following Piaget and Inhelder's (1971) claim that visual imagery underlies the creation of a drawing or spatial arrangement, the distinction between external and internal representations will not be pursued further as a theoretical issue. In keeping with the Peircean semiotic framework I used in my later research (Presmeg 2006a, 2006b), my working definition is that a visual image is a mental sign vehicle involving visual or spatial information, whereas inscriptions are the external sign vehicles.

Preference for Visualization in Mathematics

In mathematics, sign vehicles are often of a visual nature; even algebraic symbolism has a structure and needs to be seen, either mentally or in written form. However, one might talk more broadly about individual preferences for visualization in mathematics, and guided by the work of Suwarsono (1982) who worked with seventh graders in Australia, I constructed an instrument to measure the mathematical visuality of high school students (grades 11 and 12) and their mathematics teachers. Parts A (6 items: word problems without any figural content) and B (12 items) were designed for students in the last years of high school; Parts B (the same 12 items) and C (6 more difficult items) were intended for their mathematics teachers. After standardization and checks for validity and reliability, the instrument was used to select teachers of a range of styles, and visualizers in their grade 12 mathematics classes. A *visualizer* is a person who prefers to use visual methods (including visual imagery) to solve problems that are capable of solution by visual and nonvisual means, as in my instrument. The frequency distribution graphs of visuality scores indicated that for most populations this frequency follows a normal, Gaussian, distribution. But there are people at both ends of the scale: some who seldom, if ever, feel the need to visualize, and others for whom it is not an option, they always feel the need. Those whose visuality scores were above the median were taken to be the visualizers.

Tasks given to students may be more or less visual—whether or not a diagram is presented. Ken Clements notes that Suwarsono used item response theory to assign visuality scores to the items (all without diagrams) in his instrument, using the same scale as he used for the preference for visuality of students. It is noteworthy that in the development of my instrument, too, based on the responses of an initial group of high school mathematics students, an item analysis was performed, assigning a visuality score to each of the 32 items in the original instrument, and also a degree of difficulty score. As a result of this analysis, 3 items that were judged to be too difficult (none or only one of the students in the test sample could solve each of these problems) and 5 that were too visual or too nonvisual, were discarded, resulting in the 24 items that were retained, some of which were rearranged according to three criteria (sections should not start with a highly visual or nonvisual problem, problems should be varied, and the problems should be arranged in roughly ascending degree of difficulty).

For a mathematical task, the way the task is done depends on the following: the task itself; instructions to do the task in a certain way (Paivio 1971); individual preferences (as manifest in the scores on my instrument); and also the culture of the learning environment. This last aspect may underlie the claim by Dreyfus (1991) and Eisenberg (1994) that students are reluctant to visualize in mathematics. On the contrary, some visualizers do not have an option. But they may hide their preference if the culture does not value and encourage this mode of thought, as was the case of visualizers in classes of my non-visual group of teachers. Tasks and teaching are topics in a later section.

Difficulties and Affordances of Use of Visual Imagery in Mathematics

In keeping with the theme of this book, *Searching for Common Ground* amongst the components involved in mathematics research and mathematics education research, with mathematics squarely at the centre, I shall concentrate on a topic that is relevant to both fields. One result of my research on visualization, starting in the early 1980s, was that all of the *difficulties* experienced by the 54 visualizers in high school in my initial main study, related in one way or another to the abstraction and generalization that are essential aspects of doing mathematics.

- The one-case concreteness of an image may be tied to irrelevant details, or introduce false information.
- A prototypical image may induce inflexible thinking.
- An uncontrollable image may persist, thus preventing more fruitful avenues of thought.
- Imagery needs to link with rigorous analytical thought processes to be effective.

Implicit in these difficulties is *compartmentalization*. The damaging effect of compartmentalization in mathematics education has long been noted by several authors (Duval 1999; Nardi et al. 2005; Vinner et al. 1981). My later research on the learning of trigonometry (Presmeg 2006b) examined this aspect in more detail.

There were two basically different ways that these difficulties could be overcome: firstly, a visual image or a diagram of one concrete case could be the bearer of abstract information—a sign vehicle for an abstract object. *Dynamic imagery* was useful in this case, as was *pattern imagery*. Secondly, *metaphorical thinking* could link the domain of abstract mathematical objects with visual imagery or inscriptions in a different domain. The intricacies of each of these cases are apparent in the research data. Mathematical thinking that alternates between visualization and logical deduction is particularly effective. Finally, visual images of all types have mnemonic advantages: It is easier to remember a "picture".

The interaction of visual styles of mathematical learning with different forms of teaching was a central component of my research on visualization.

Teaching Visuality

From the literature, augmented by a full year of classroom observations, twelve aspects of teaching that were facilitative of visualization were identified and refined by triangulation of three viewpoints (those of the teachers themselves, their students, and the observer). From these classroom aspects, it was possible to assign a *teaching visuality* (TV) score to each of the thirteen teachers in the initial research. These teachers had been chosen to provide a range of *mathematical visuality* (MV) scores according to the preference instrument. However, one strong result of the research was that TV and MV scores were only weakly correlated (Spearman's rho $= 0.404$). It made sense that some teachers who themselves did not require visualization in doing mathematics, nevertheless recognized their students' need for such means, and taught accordingly. According to their TV scores, the thirteen teachers fell neatly into three groups: a visual group (5 teachers), a middle group (4 teachers), and a non-visual group (4 teachers). Analysis of the qualitative data from 178 transcribed clinical interviews with the 54 student visualizers in the study revealed that the group of teachers whose pedagogy was optimal for the mathematical achievement of these visualizers was not the *visual* group, but, counter-intuitively, the *middle* group. The visual group of teachers encouraged pictorial thinking and inscriptions, and valued these means, but they were unaware of the difficulties relating to abstraction and generalization that could be experienced by these visual students. The middle group of teachers, while encouraging visualization, also stressed general thinking, systematization, and logical analysis. In the classes of the non-visual teachers, visualizers were like fishes out of water. I gained the impression that all the visualizers would have benefitted if their mathematics teachers had been more aware of the pitfalls as well as the positive potential of visual thinking in teaching and learning mathematics.

In the light of the intricacies involved in these relationships between student learning preferences and teaching visuality, including the *stance* towards visualization in mathematics that is taken by the teacher, I want to say a bit more about the nature of tasks presented to students by a teacher. It is commendable that Raz Kupferman (this chapter) presents a diagram illustrating the ε–δ definition of limits of functions, when he teaches this concept to university undergraduates. However, he is right that students' ability to benefit from such a diagram should not be taken for granted. Many of the visualizers in my research would have attempted to make such a diagram for themselves, or at least to entertain a mental image of it: they could make sense of such a definition—if they made sense of it—in no other way. But potential difficulties tied to generalization need to be taken into account. Not all students may be benefited by a "visual example": The research of Elena Nardi and colleagues with mathematicians (this chapter) confirms that although the introduction of visual "examples" in teaching college-level mathematics is a positive step, in itself it is not enough. Nardi quotes Whiteley (2009), "Visuals are strong particularly because they are examples, and they can indeed carry general reasoning as symbols for the general case, provided the readers bring a range of variation to their cognition of the figure" (p. 260). This is the core issue: mathematics teachers—even those with high *teaching visuality* scores—do not usually have trouble with abstraction and generalization in mathematics. However, unless they stress these elements and the logic of the processes involved (as did my *middle* group of teachers), the presentation of a visual example alone may not take into account the difficulties associated with the "one-case concreteness" of such a visual presentation (Presmeg 1985, 2006a). The stance of the teacher is also important: and *teaching visuality* was measured according to 12 criteria, not only the presentation of a picture. My research showed over and over again, the effectiveness of alternating holistically visual and rigorously sequential logical processing in mathematics. Nardi (this chapter) points out that some of the mathematicians she and her colleagues interviewed were aware of this issue: "Part of the pedagogical role of the mathematician, they state, is to foster a fluent interplay between analytical rigour and (often visually based) intuitive insight."

The tasks presented by Rina Hershkowitz (this chapter), and the way they are addressed in teaching, illustrate the power of activities that admit a range of flexible methods in their implementation. Curriculum development is suggested, and this aspect comes out even more specifically in the research on visual algebra of Michal Yerushalmy and her colleagues. Yerushalmy (this chapter) underscores the social and semiotic elements of such work: "I would challenge the reliability of earlier research findings regarding students' visual thinking and suggest that curricular research could benefit from systematic studies that reexamine visualization as cognitive challenge and as pedagogical preferences, especially those that concern the semiotic potential of technological tools, for teaching school algebra."

Thirteen Significant Questions

At the end of my Handbook chapter (Presmeg 2006a) I put forward a list of thirteen questions in this field that seem to be significant, as follows. As suggested in previous sections, some of these issues have started to be addressed by the participants in this panel on mathematical visualization.

1. What aspects of pedagogy are significant in promoting the strengths and obviating the difficulties of use of visualization in learning mathematics?
2. What aspects of classroom cultures promote the active use of effective visual thinking in mathematics?
3. What aspects of the use of different types of imagery and visualization are effective in mathematical problem solving at various levels?
4. What are the roles of gestures in mathematical visualization?
5. What conversion processes are involved in moving flexibly amongst various mathematical registers, including those of a visual nature, thus combating the phenomenon of compartmentalization?
6. What is the role of metaphors in connecting different registers of mathematical inscriptions, including those of a visual nature?
7. How can teachers help learners to make connections between visual and symbolic inscriptions of the same mathematical notions?
8. How can teachers help learners to make connections between idiosyncratic visual imagery and inscriptions, and conventional mathematical processes and notations?
9. How may the use of imagery and visual inscriptions facilitate or hinder the reification of processes as mathematical objects?
10. How may visualization be harnessed to promote mathematical abstraction and generalization?
11. How may the affect generated by personal imagery be harnessed by teachers to increase the enjoyment of learning and doing mathematics?
12. How do visual aspects of computer technology change the dynamics of the learning of mathematics?
13. What is the structure and what are the components of an overarching theory of visualization for mathematics education?

I started to work on "overarching theory" for a presentation on this topic at Topic Study Group 20 of the 11th meeting of the *International Congress on Mathematical Education* (Presmeg 2008a), and the problem of compartmentalization was addressed in research on the learning of trigonometry (Presmeg 2006b), along with further research on metaphors (Presmeg 2008b). However, much more needs to be learned in answer to all of these questions.

Concluding Reflections

These reflections aim to integrate into this book chapter part of the symposium discussion that followed the presentations on which the above contributions are based. Quotation marks indicate (here anonymised) verbatim excerpts from the audio recording of this discussion.

Visualization can be used as a bridge between more and less familiar aspects of mathematics. One example of this is introducing the horrendously complex, multi-quantified formal definition of limit through graphical illustrations of the convergence of functions that are familiar to students, such as x^2 or $\sqrt{x}$. While acknowledging that students who take advanced mathematics courses need to understand, endorse and employ mathematical formalism, much of this understanding can be built through visual means. This is not to say that visual means do not have inherent constraints. Diagrams can be confusing: for example, a diagram cannot be *of* infinity the way a diagram can be *of* a triangle. Or, within Analysis and the aforementioned example, students need to be able to see that diagrams—such as the one in Kupferman's contribution in this chapter—illustrate how the epsilon-delta expression defines the limit always in terms of *finite* quantities, succeeding thus to remove infinity. A key question that emerges then is what pedagogical practices can help students benefit most from visualization, while in full awareness of its limitations—and in full awareness that students do not 'see' in the same way their (much more experienced) teachers do. Some of these limitations include the ambiguous status of visual arguments in mathematics and their lack of transparency: 'sometimes visualization will cheat you. You draw a picture, you see something but you do not see something invisible, behind the picture.'

A perspective of visualization as an approach to mathematics that has to be judged by the criteria that apply to the analytical approach is problematic. Visualization is often seen as a 'secondary school' approach that students need to steer clear of when they arrive at university, and certainly later on. IT can help avoid this unnecessary dichotomy: 'visual signs can be the first channel of making mathematics, and not necessarily the additional one, [to] what we call symbolic.' 'Visual signs are symbols. As long as we live it as a secondary category, we will not encourage students to do this.' The issue therefore becomes what kind of pedagogical practice can make visualization a legitimate approach for students.

One approach is to allow the curriculum to explore 'visualization for its own sake': this should not be taken to mean 'teach visualization' as such but explore the power of visualization, acknowledge that 'it has a logic of itself' and do so explicitly: we cannot simply 'expect [students] to be visual' upon arrival at university, for example. And the communities of mathematics and mathematics education need to work harder towards making the substantial mathematics education research on these matters known to practitioners of mathematics teaching across all levels. Often excellent intuition, accumulated over years of experience, is but one source of pedagogical insight.

At the heart of recommendations that can be put forward is alertness to the dangers of 'opposing visualization with rigour' and of not acknowledging that there is a 'radical change in the didactic contract between secondary and university mathematics regarding graphical representations'; and of the need to prioritise work towards making 'the change of the didactical contract possible to the student and understandable to the student'. 'Reluctance [to visualise] depends on the culture and the environment' and therefore endorsing the legitimacy of visual thinking in mathematics needs to be 'part of the explicit intended curriculum'.

Fostering in students the learning of the 'grammar and syntax of diagrams, the same way they learn the grammar and syntax of written mathematics' would then be an inextricable part of this curriculum. The assumption that 'because something is visual, it is [therefore] transparent, and you understand it' is incorrect when there is substantial intermediate work that students need to engage with in order to achieve this understanding. And this work includes learning to distinguish between visualization in a general sense and mathematical visualization, and learning to engage not just with the symbolic and the visual registers of mathematics but also the linguistic and kinaesthetic. The complexity of this task is formidable.

Another inextricable part of this explicit intended curriculum on visualization would be deploying IT that fosters students' acquiring of the 'new lenses' and 'rethinking the mathematics that we do'. We simply cannot continue with the same curriculum, purporting to do the same mathematics and merely having visualization as an add-on. The mathematics we do needs to be reorganized: the content, the order of difficulty of the problems presented to students, their structure.

The design of this curriculum also needs to consider that classical hierarchies in psychology have been seriously challenged in recent years. While few would argue against Bruner's enactive, iconic, and symbolic *classification*, its *hierarchical* structure, its proposed *progression* are being doubted in very substantial ways. These are 'just different means, different kinds of thinking', seems to be the proposed more contemporary approach. 'We cannot say that visualization gives way to generalisation and abstraction' and the 'representation development hypothesis' is put again to the test.

There is another issue that this curriculum would need to engage with: visualization can 'take a lot of time, in comparison to working with symbolism'. Its time-consuming nature is however compensated by its power to give insight and to support our power for abduction. This is one more reason why the opposition of logic and proof to visualization is unproductive.

The optimal approach therefore appears to be to 'have them [logic/proof and visualization] simultaneously and working together, sometimes alternating' and to acknowledge them as 'different ways of thinking'. Again new developments in technology, particularly work on Digital Geometry, are key to achieving this acknowledgement. 'It's not a question of either-or, it's a question of both and throughout'.

References

Arcavi, A. (2003). The role of visual representations in the learning of mathematics. *Educational Studies in Mathematics*, *52*, 215–241.

Aspinwall, L., Shaw, K. L., & Presmeg, N. C. (1997). Uncontrollable mental imagery: graphical connections between a function and its derivative. *Educational Studies in Mathematics*, *33*, 301–317.

Bishop, A. J. (1980). Spatial abilities and mathematics education—a review. *Educational Studies in Mathematics*, *11*, 257–269.

Biza, I., Nardi, E., & Zachariades, T. (2009). Teacher beliefs and the didactic contract on visualization. *For the Learning of Mathematics*, *29*(3), 31–36.

Brousseau, G. (1997). *Theory of didactical situations in mathematics*. Dordrecht: Kluwer Academic.

Byers, W. (2007). *How mathematicians think: using ambiguity, contradiction, and paradox to create mathematics*. Princeton: Princeton University Press.

Dreyfus, T. (1991). On the status of visual reasoning in mathematics and mathematics education. In F. Furinghetti (Ed.), *Proceedings of the 15th conference of the international group for the psychology of mathematics education* (Vol. 1, pp. 33–48).

Dreyfus, T., Nardi, E., & Leikin, R. (2012). Forms of proof and proving. In G. Hanna & M. de Villiers (Eds.), *Proof and proving in mathematics education—the 19th international commission for mathematics instruction study* (pp. 111–120). New York: Springer.

Duval, R. (1998). Geometry from a cognitive point a view. In C. Mammana & V. Villani (Eds.), *Perspectives on the teaching of geometry for the 21st century* (pp. 37–52). Dordrecht: Kluwer Academic.

Duval, R. (1999). Representation, vision and visualization: cognitive functions in mathematical thinking. Basic issues for learning. In F. Hitt & M. Santos (Eds.), *Proc. of the 21st conf. of the North American chapter of the int. group for the psychology of mathematics education* (Vol. 1, pp. 3–26). Cuernavaca: PME-NA.

Eisenberg, T. (1994). On understanding the reluctance to visualize. *ZDM. Zentralblatt für Didaktik der Mathematik*, *26*, 109–113.

Fischbein, E. (1987). *Intuition in science and mathematics: an educational approach*. Dordrecht: Reidel.

Giaquinto, M. (2007). *Visual thinking in mathematics*. New York: Oxford University Press.

Hanna, G., & Sidoli, N. (2007). Visualisation and proof: a brief survey of philosophical perspectives. *ZDM. The International Journal on Mathematics Education*, *39*, 73–78.

Hershkowitz, R., & Markovits, Z. (1992). Conquer math concepts by developing visual thinking. *The Arithmetic Teacher*, *39*, 38–41.

Hershkowitz, R., Arcavi, A., & Bruckheimer, M. (2001). Reflections on the status and nature of visual reasoning—the case of matches. *International Journal of Mathematical Education in Science and Technology*, *32*, 255–265.

Iannone, P. (2009). Concept usage in proof production: mathematicians' perspectives. In F. L. Lin, F. J. Hsieh, G. Hanna, & M. de Villiers (Eds.), *ICMI study 19: proof and proving in mathematics education* (Vol. 1, pp. 220–225). Taipei: National Taiwan Normal University.

Krutetskii, V. A. (1976). *The psychology of mathematical abilities in schoolchildren*. Chicago: University of Chicago Press.

Mancosu, P., Jorgensen, K. F., & Pedersen, S. A. (Eds.) (2005). *Visualization, explanation and reasoning styles in mathematics*. Dordrecht: Springer.

Naftaliev, E., & Yerushalmy, M. (2009). Interactive diagrams: alternative approach to the design of algebra inquiry. In M. Tzekaki, M. Kaldrimidou, & H. Sakonides (Eds.), *Proceedings of the 33rd conference of the international group of psychology of mathematics education* (Vol. 4, pp. 185–192). Thessaloniki: PME.

Nardi, E. (2008). *Amongst mathematicians: teaching and learning mathematics at university level*. New York: Springer.

Nardi, E. (2009). 'Because this is how mathematicians work!' 'Pictures' and the creative fuzziness of the didactical contract at university level. In F. L. Lin, F. J. Hsieh, G. Hanna, & M. de Villiers (Eds.), *ICMI study 19: proof and proving in mathematics education* (Vol. 2, pp. 112–117). Taipei: National Taiwan Normal University.

Nardi, E., Jaworski, B., & Hegedus, S. (2005). A spectrum of pedagogical awareness for undergraduate mathematics: from 'tricks' to 'techniques'. *Journal for Research in Mathematics Education, 36*, 284–316.

Nardi, E., Biza, I., & Zachariades, T. (2012). 'Warrant' revisited: integrating teachers' pedagogical and epistemological considerations into Toulmin's model for argumentation. *Educational Studies in Mathematics, 79*, 157–173.

Paivio, A. (1971). *Imagery and verbal processes*. New York: Holt, Rinehart & Winston.

Parzysz, B. (1988). 'Knowing' vs. 'Seeing': problems of the plane representation of space geometry figures. *Educational Studies in Mathematics, 19*, 79–92.

Piaget, J., & Inhelder, B. (1971). *Mental imagery and the child*. London: Routledge/Kegan Paul.

Presmeg, N. C. (1985). *The role of visually mediated process in high school mathematics: a classroom investigation*. Unpublished Ph.D. dissertation, University of Cambridge, England.

Presmeg, N. C. (2006a). Research on visualization in learning and teaching mathematics: emergence from psychology. In A. Gutierrez & P. Boero (Eds.), *Handbook of research on the psychology of mathematics education* (pp. 205–235). Rotterdam: Sense Publishers.

Presmeg, N. C. (2006b). A semiotic view of the role of imagery and inscriptions in mathematics teaching and learning. In J. Novotna, H. Moraova, M. Kratka, & N. Stehlikova (Eds.), *Proceedings of the 30th annual meeting of the international group for the psychology of mathematics education* (Vol. 1, pp. 19–34). Prague: PME.

Presmeg, N. C. (2008a). An overarching theory for research on visualization in mathematics education. Plenary paper. In *Proceedings of topic study group 20, visualization in the teaching and learning of mathematics 11th international Congress on mathematics education (ICME-11)*, Monterrey, Mexico, July 6–13, 2008. Published on the ICME-11 web site: http://tsg.icme11.org (TSG20). Accessed 6 June 2013.

Presmeg, N. C. (2008b). The power and perils of metaphor in making internal connections in trigonometry and geometry. In *Proceedings of the 5th conference of the European society for research in mathematics*, Larnaca, Cyprus, February 22–26, 2007 (pp. 161–170) (CD published March 2008).

Radford, L. (2012). Early algebraic thinking: epistemological, semiotic, and developmental issues. In *12th international Congress on mathematical education*, 8–15 July 2012, COEX, Seoul, Korea.

Razel, M., & Eylon, B. (1990). Development of visual cognition: transfer effects of the Agam program. *Journal of Applied Developmental Psychology, 11*, 459–485.

Roh, K. (2010). An empirical study of students' understanding of a logical structure in the definition of limit via the ε-strip activity. *Educational Studies in Mathematics, 73*, 263–279.

Schwartz, J. L. (2011). *Unsolving linear and quadratic equations*. Poincare Institute for Mathematics Education, Tufts University, Medford, MA.

Steen, L. A. (1988). The science of patterns. *Science, 240*, 611–616.

Stenning, K., & Lemon, O. (2001). Aligning logical and psychological perspectives on diagrammatic reasoning. *Artificial Intelligence Review, 15*, 29–62.

Suwarsono, S. (1982). *Visual imagery in the thinking of seventh grade students*. Unpublished Ph.D. dissertation, Monash University, Australia.

Tall, D. (Ed.) (1991). *Advanced mathematical thinking*. Dordrecht: Kluwer Academic.

Vinner, S., Hershkowitz, R., & Bruckheimer, M. (1981). Some cognitive factors as causes of mistakes in addition of fractions. *Journal for Research in Mathematics Education, 12*, 70–76.

Visual Math: Algebra (1995/2005). Center for educational technology, Ramat-Aviv. http://www.cet.ac.il/math-international/first.htm. Accessed 6 June 2013. http://www.cet.ac.il/math/function/english. Accessed 6 June 2013.

Whiteley, W. (2004). To see like a mathematician. In G. Malcolm (Ed.), *Multidisciplinary approaches to visual representations and interpretations* (Vol. 2, pp. 279–291). London: Elsevier.

Whiteley, W. (2009). Refutations: the role of counter-examples in developing proof. In F. L. Lin, F. J. Hsieh, G. Hanna, & M. de Villiers (Eds.), *ICMI study 19: proof and proving in mathematics education* (Vol. 2, pp. 257–262). Taipei: National Taiwan Normal University.

Yerushalmy, M. (2005). Challenging known transitions: learning and teaching algebra with technology. *For the Learning of Mathematics*, *25*(3), 37–42.
Yerushalmy, M., & Botzer, G. (2011). Teaching secondary mathematics in the mobile age. In O. Zaslavsky & P. Sullivan (Eds.), *Mathematics teacher education series: Vol. 6. Constructing knowledge for teaching secondary mathematics tasks to enhance prospective and practicing teacher learning* (pp. 191–208). New York: Springer.
Yerushalmy, M., & Gilead, S. (1999). Structures of constant rate word problems: a functional approach analysis. *Educational Studies in Mathematics*, *39*, 185–203.

Justification and Proof

Chapter 13
Making Sense of Mathematical Reasoning and Proof

David Tall

Abstract This chapter charts the growth of proof from early childhood through practical generic proof based on examples, theoretical proof based on definitions of observed phenomena, and on to formal proof based on set theoretic definitions. It grows from human foundations of perception, operation and reason, based on human embodiment and symbolism that may lead, at the highest level, to formal structure theorems that give new forms of embodiment and symbolism.

Increasing sophistication in mathematical thinking and proof is related to earlier experiences, called 'met-befores' where supportive met-befores encourage generalisation and problematic met-befores impede progress, causing a bifurcation in the perceived nature of mathematics and proof at successive levels of development and in different communities of practice. The general framework of cognitive development is offered here to encourage a sensitive appreciation and communication of the aims and needs of different communities.

Keywords Mathematical proof · Crystalline concepts · Met-before · Generic proof · Van Hiele theory · Structure theorems

This article is a product of personal experience, working with colleagues such as Shlomo Vinner who gave me the insight into the notion of concept image, Eddie Gray, whose experience with young children led me to grasp the essential ways in which children develop ideas of arithmetic and to build a theoretical framework for the different ways in which mathematical concepts are conceived, Michael Thomas who helped me understand more about how older children learn algebra, the advanced mathematical thinking group of PME who broadened my ideas about the different ways that undergraduates come to understand more formal mathematics, many colleagues and doctoral students who I celebrate in Tall (2008) and, more recently, the working group of ICMI 19 who focused on the cognitive development of mathematical proof (Tall et al. 2012).

D. Tall (✉)
University of Warwick, Coventry, UK
e-mail: davidtall@mac.com

M.N. Fried, T. Dreyfus (eds.), *Mathematics & Mathematics Education: Searching for Common Ground*, Advances in Mathematics Education,
DOI 10.1007/978-94-007-7473-5_13,

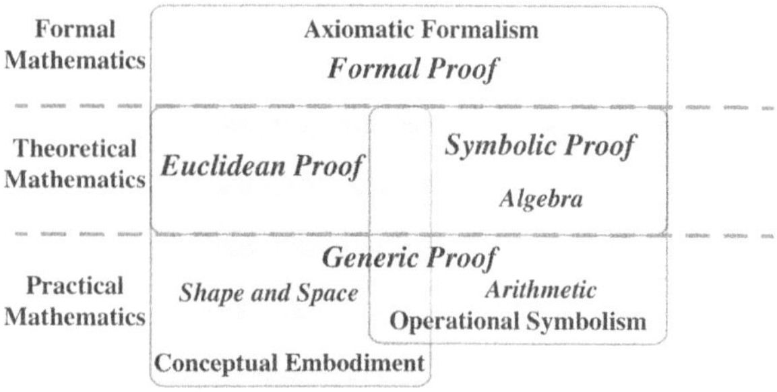

Fig. 13.1 Outline of long-term development of proof

Mathematical Thinking in Terms of Human Perception, Operation and Reason

The cognitive development of mathematical thinking and proof is based on fundamental human aspects that we all share: human perception, action and the use of language and symbolism that enables us to develop increasingly sophisticated thinkable concepts within increasingly sophisticated knowledge structures. It is based on what I term the *sensori-motor language of mathematics*, blending together perception, operation and reason (Tall 2013).

Mathematical thinking develops in the child as perceptions are recognised and described using language and as actions become coherent operations to achieve a specific mathematical purpose. According to Bruner (1966), these may be communicated first through enactive gestures, then iconic images, then the use of symbolism, including not only written and spoken language but also the operational symbolism of arithmetic and the axiomatic formal symbolism of logical deduction.

The theoretical framework proposed here follows a similar path enriched by the experience over time, building from *conceptual embodiment* that combines the enactive and iconic modes of human perception and action, developing into the mental world of perceptual and mental thought experiment (Fig. 13.1). Embodied operations, such as counting, adding, sharing, are symbolised as manipulable concepts in arithmetic and algebra in a second mental world of *operational symbolism*. As the individual matures, there is a further shift into a focus on the *properties* of mental objects as in Euclidean geometry, the blending of visual and symbolic modes of thought and the properties of arithmetic operations recast as 'rules' that underlie the generalized operations and expressions in algebra. Each of these leads to different forms of mathematical proof: *Euclidean proof* in geometry, *symbolic proof*, based on the 'rules of arithmetic', and *blending embodied and symbolic reasoning* using language.

Embodiment and symbolism develop alongside each other and interact with each other. The early stages of *practical mathematics* begin with experience of shape and space, and of operations in arithmetic, in which properties of specific examples are seen to offer *generic proof*, such as realising that $2 + 3 = 3 + 2$ holds not just for the numbers 2 and 3, but for *any* pair of whole numbers. This develops into the *theoretical mathematics* of definition and deduction in Euclidean and symbolic forms of proof.

Properties in both embodiment and symbolism develop into the *formal mathematics* of set-theoretic definition and proof in the *axiomatic formal* world of pure mathematics. While theoretical mathematics is based on embodied and symbolic experiences, formal mathematics guarantees that all the properties proved from given set-theoretic axioms and definitions will also hold in any new context that satisfies the given axioms and definitions.

Embodiment and symbolism continue to play their part in axiomatic formalism, not only in imagining new possibilities that may be defined and proved formally, but also in an amazing turnaround in which certain theorems (called structure theorems) prove that axiomatic systems have embodied and symbolic structures established by formal proof. This reveals mathematical thinking at the highest level, and mathematical proof in particular, as an intimate blend of embodiment, symbolism and formalism where individual mathematicians develop a preference for different aspects.

The Evolution of Theories of Mathematical Thinking and Proof

Pierre Van Hiele (1986) focused on *structure* and insight, seeing a succession of levels that may be described as *recognition* and *description* of figures, leading to *definition* and *deduction* of properties through Euclidean proof.

Ed Dubinsky and others (Asiala et al. 1996) took an apparently different path, following Piaget's idea of reflective abstraction to focus on operations that are seen first as *actions*, routinized as *processes*, then *encapsulated* as mental *objects* within knowledge *schemas*.

Anna Sfard (1991) proposed a framework that alternated between *operational* and *structural* ways of thinking in which operations are *condensed* as *processes*, and then *reified* as mental *objects* that now have a certain structure. She suggested at the time that an operational approach inevitably precedes structural mathematics. However, her examples involve operational symbolism being reified as mental objects, without any reference to the van Hiele development of the properties of objects.

This led to a three-part analysis in Tall et al. (2000) through parallel developments of conceptual embodiment (broadly following van Hiele) and operational symbolism (using process-object theories) in school, leading much later to the axiomatic formal framework of set-theoretic definition and proof in university pure mathematics (Tall 2004a, 2004b).

Following the recent death of Van Hiele in 2011 at the grand old age of one hundred, I revisited his ideas of structure and insight, which he applied to geometry, but not to the symbolism of arithmetic and algebra (Van Hiele 2002). I realised that the term *operation* should not be restricted to the symbolic operations in arithmetic and algebra. Operations occur in the constructions of Euclidean geometry. For instance, we may operate on an isosceles triangle by joining the vertex to the midpoint of the base to cut the triangle into two parts that are congruent (with three corresponding sides). This proves that the base angles must be equal, and various other properties follow, such as the property that the line from the vertex to the midpoint of the base is at right angles to the base.

The operations of construction in geometry and the various operations in arithmetic and algebra have a common definition: they consist of 'a coherent sequence of actions and decisions performed to achieve a specific purpose.' A geometric operation is a construction that focuses on the *object* (the figure) and results in enabling us to see relationships concerning *the properties of the object*. A symbolic operation performs a calculation or manipulation, focusing more on *the properties of the operations themselves* as the operations lead to a symbolic output.

Furthermore the compression of operation into mental object in symbolism begins for the child as *embodied operations* on objects such as counting, adding, sharing, and is compressed into *symbolic operations* on whole numbers, fractions, signed numbers and so on. This reveals two distinct forms of compression from operation to mental object that I termed *embodied compression* and *symbolic compression* (Tall 2013, Chap. 7).

Embodied compression focuses on the effect of the operations on the objects, such as counting a collection to find the number of objects, such as 'six'. Focusing on the way that the objects are placed leads to a realisation of the fundamental properties of whole number arithmetic. For instance, the set of six objects may be subdivided, say, into subsets of 'four' and 'two' and, by rearranging the sets, it may be seen that 'two' and 'four' is also 'six'. Reorganizing the subsets as two rows of 'three' allows them to be seen as three columns of 'two' so that 'two threes' is the same as 'three twos'. Embodied compression enables us to see *at a glance* the flexible properties of arithmetic. 'Proof' at this early stage is a form of reasoning based on our interpretation of the coherence of our own perceptions and actions. This form of proof, in which a specific example is seen to be typical of a whole category of examples, is termed *generic proof* (Mason and Pimm 1984; Harel and Tall 1991).

Symbolic compression involves performing a counting operation to obtain a number concept, for instance, the operation of 'count-on' calculates 'two and eight' as counting on eight to get 'three, four, five, six, seven, eight, nine, *ten*' while 'eight and two' is the short count 'nine, *ten*'. Here the two operations are very different, one is a long count, and the other is short. The general properties of the symbolic compression are therefore not as self-evident as they are with embodied compression.

A gifted child may grasp the flexible properties of arithmetic as part of a coherent knowledge structure in which symbols operate dually as process or concept (which we termed a 'procept') that may be used as an organising principle to simplify operations. A child who focuses on procedural operations of counting taking place in time will find arithmetic operations to be far more difficult to cope with. Eddie Gray and I called this bifurcation 'the proceptual divide' between those fixed in increasingly complicated counting procedures and those who develop flexible ways to derive new facts from known facts (Gray and Tall 1994).

This bifurcation between those who find mathematics 'easy' and those who find it impossibly difficult begins at a very early age. It should be taken into account in seeking to explain and predict how each individual attempts to make sense of mathematics, building on increasingly sophisticated perception, operation and reason.

Long-Term Pleasure and Pain

Emotions play a vital role in mathematical thinking and have a profound effect on how individuals make sense of mathematical proof. As my supervisor, Richard Skemp used to say: 'pleasure is a signpost, not a destination.' His goal-oriented theory of learning (Skemp 1979) saw children starting out with the goal of seeking to make sense of the world. Successfully linking together ideas in coherent ways gives pleasure, success breeds more success, so that a child with a history of success builds up a positive feed-back loop where an encounter with a problematic situation is often met with the determination to conquer the difficulty. However, lack of success leads to an anti-goal, to avoid a sense of stress. Further encounters with stress may lead to a negative feed-back loop in which the desire to avoid failure leads to less engagement with the mathematics and less technical proficiency that causes even more difficulty and greater mathematical anxiety (Baroody and Costlick 1998). As a result of the negative feedback, students may seek the comfort of learning procedures by rote to succeed in examinations and prefer to learn proofs procedurally rather than seek to grasp deeper meanings that do not seem to make sense.

An analysis of the development of mathematical thinking reveals the surprising conclusion that mathematics is not a system that builds logically on previous experience at each stage, even though every mathematics curriculum in the world is intent on presenting topics in a coherent sequence, carefully preparing the necessary pre-requisites at each stage for the more sophisticated stages that follow. On the contrary, an experience that has been 'met before' may be supportive in some new situations yet problematic in others.

The concept of 'met-before' was introduced by de Lima and Tall (2008) and McGowen and Tall (2010) to describe 'a structure we have in our brains *now* as a result of experiences we have met before.' Some ideas that work in one situation such as 'addition makes bigger' or 'take away makes smaller' in whole number arithmetic are supportive in the context of fractions yet problematic in the context of signed numbers. This recalls the concept of 'epistemological obstacle' developed by Bachelard (1938) and Brousseau (1983) and the need for accommodation by Piaget (see, for example, Baron et al. 1995) or reconstruction by Skemp (1971).

However, the notion of met-before refers to the effect of previous experience on new learning. A particular met-before is not in itself supportive or problematic, it *becomes* supportive or problematic in a new situation when the learner attempts to make sense of the new ideas. For instance, 'take away leaves less' is supportive in some contexts (e.g. everyday situations where something is removed, in the postulates of Euclidean geometry, or taking one whole number from another) but it is problematic in others (such as taking away a negative number or in the theory of infinite cardinals).

A problematic met-before arises not only in the individual learner, *it is a widespread feature of the nature of mathematics itself.* In shifting to a new context, say from whole numbers to fractions, or from positive numbers to signed numbers, or from arithmetic to algebra, *generalization is encouraged by supportive met-befores* (ideas that worked in a previous context and continue to work in the new one) *and impeded by problematic met-befores* (that made sense before but do not work in the new context).

For instance, properties such as commutativity, associativity, distributivity are supportive as number systems are broadened through whole numbers, integers, real numbers, complex numbers, but other aspects such as 'take away gives less' or 'the square of a non-zero number is positive' become problematic.

Crystalline Concepts

Given this increasing difficulty of problematic aspects that occur in generalization, I sought a unifying principle that is supportive in mathematical thinking and binds mathematical ideas together in any given context. In Tall (2011), I formulated a working definition of a *crystalline concept* as 'a mathematical concept that has an internal structure of relationships that cause it to have specific properties in the given mathematical context.' Such concepts include:

- *platonic objects in geometry*, such as points, lines, triangles, circles, congruent triangles, parallel lines that have properties arising through Euclidean proof;
- *operational symbols as flexible procepts* in arithmetic, algebra and symbolic calculus that have necessary properties through calculation and manipulation;
- *set-theoretic concepts* in axiomatic formal mathematics whose properties are deduced by formal proof.

Not only do crystalline concepts occur at the highest levels of mathematical thinking, they emerge in the thinking of a young child who sees the flexible proceptual structure of arithmetic through embodied compression rather than the procedural step-by-step counting procedures of arithmetic that operate in time.

They enable flexible thinkers to see mathematical ideas in astonishingly simple ways. It is not that the fractions $\frac{4}{8}, \frac{7}{14}, \frac{101}{202}$ are all *equivalent* to each other and to the simplest possible canonical form $\frac{1}{2}$, it is that they are all manifestations of a single crystalline concept—the rational number one half—also represented as a unique point on the number line.

It is not that the expressions $2(x + 7)$ and $2x + 14$ are equivalent but different, where the first can be turned into the second by 'multiplying out the brackets' and the second can be turned into the first by 'factorization', it is that *both expressions are different ways of writing the same crystalline concept* as an algebraic expression. Indeed, the functions $f(x) = 2(x + 7)$ and $g(x) = 2x + 14$ are not simply equivalent, *they are precisely the same function*. Students who think flexibly in terms of crystalline concepts have much more powerful means of relating mathematical ideas than those who see equivalent ideas that are changed from one form to another by carrying out procedures.

Likewise, in axiomatic formal mathematics, an axiomatic system such as 'a group' is a crystalline concept with rich interconnections between its properties. We may not know what specific group we are dealing with, but we *do* know that it has an identity that we may denote by e, and that if x is any element, we can define the power x^n for any positive or negative integer and prove that $x^{m+n} = x^m x^n$ for any integers m, n.

A crystalline concept may be defined formally and then its properties may be deduced as theorems to build up a knowledge structure where relationships are tightly interconnected by formal proof. For example, we can prove that if we begin with the axiomatic definition of an ordered field F, then in this context we may formulate any of the equivalent definitions for completeness, to prove that a complete ordered field is not only unique up to isomorphism, it is also unique as a crystalline concept.

At the highest level of pure mathematical research, it is the compression of structural properties of defined formal concepts into crystalline concepts that gives gifted mathematicians a simplicity of thought that is beyond the mere proving of theorems of equivalence. An ordered field not only contains a subfield *isomorphic* to the rational numbers, it can be conceived as a crystalline concept that *contains* the crystalline concept of the rational numbers.

I recall the ideas that I encountered as a graduate student when theoreticians spoke of the identification of one structure with another structure as 'an abuse of notation'. On the contrary, it is this way of thinking that gives the biological brain of the mathematician a level of flexibility to conceive mathematical ideas in more simple and insightful ways.

Formal constructions building up more general systems—for example, from natural numbers, to integers, to rational numbers, to real numbers, and beyond—all involve equivalence relations of ordered pairs in one structure to construct the next. At each stage we get an isomorphism between equivalence classes of ordered pairs and a substructure of the larger system. This development involves supportive met-befores that encourage generalization and problematic met-befores that impede progress. Yet once we have the larger system, we no longer need to speak of isomorphisms, we can simply refer to the subsystem as a subset given by specified properties. Being able to move flexibly between seeing subsystems as subsets or as isomorphic copies leads naturally to the cognitive notion of crystalline concept. It offers the human brain a simpler way to think of strictly formulated isomorphic systems as a single underlying crystalline concept that can occur in different contexts yet operate in the same coherent way in every representation.

The Transition from Proof in Embodiment and Symbolism to Formal Proof

The overall framework for cognitive development from the newborn child to the frontiers of mathematical research was further developed in the *ICMI Study 19 on Proof and Proving* (Tall et al. 2012), and has been extended in *How Humans Learn to Think Mathematically* (Tall 2013).

The Van Hiele levels (1986) have been variously reconsidered by a range of authors, may now be seen in as four successive levels which I term

- *Recognition* of basic concepts such as points, lines, and various shapes;
- *Description* of observed properties;
- *Definition* of concepts to test new examples to see if they satisfy the definition and to use the definitions to formulate geometric constructions;
- *Deduction* in the form of Euclidean proof in plane geometry.

Each of these is a form of *structural abstraction* in which the structure of the objects under consideration and their relationships shift to successive new levels of sophistication. This begins first with observations of geometric objects whose structures are recognised and described. At this point the foundations of Euclidean proof are laid down by formulating definitions for figures that not only allow them to be categorised and constructed but also to use ideas such as congruent triangles and parallel lines to construct Euclidean proof.

Van Hiele also described a fifth level of *rigour* that may be seen as shifting in two directions, the first is to different embodied contexts such as projective geometry or spherical geometry, the second is in terms of the more sophisticated world of *axiomatic formalism* as prescribed by Hilbert.

Van Hiele (2002) saw these levels apply to geometry and not to the symbolic development from arithmetic to algebra. The calculation with numbers and manipulation of algebraic symbols involve quite different mental activities from those in Euclidean proof. However, once operations are encapsulated as number concepts and generalized as algebraic expressions, these too have properties that can be *recognised* and *described*, then *defined* as 'rules of arithmetic' to be used in algebraic proofs to *deduce* theorems. Thus the sequence of structural abstraction also occurs in the higher levels of operational symbolism to provide definitions of whole numbers, such as even, odd, prime and to deduce theorems such as the uniqueness of factorization into primes.

Exactly the same structural abstraction arises in the axiomatic formal world of set-theoretic definition and formal proof. This builds on our experience of conceptual embodiment and operational symbolism, beginning with the *recognition* and *description* of mathematical situations and then the *definition* of axiomatic systems and of defined concepts within those systems, and *deduction* of properties of systems and defined concepts using formal proof.

Experienced mathematicians have flexible knowledge structures that they wish to pass on to their students. However, by the time students pass through school to enter university, they will have already developed in very different ways based on how they have managed to make sense of previous experiences.

Krutetskii (1976) produced significant evidence that the most gifted children are more likely to develop a strong verbal-logical basis to mathematical thinking than a visual-pictorial foundation. Out of over a thousand students, the most gifted nine were classified with five analytic (verbal logical), one geometric (visual-pictorial), two combining both (one more visual, the other more verbal) and one who was not classified. Presmeg (1986) found that the most outstanding senior school mathematics students in her study (7 pupils out of 277) were almost always non-visualizers. Of 27 'very good' students (10 % of the sample), eighteen were non-visualizers and five were visualizers.

This suggests that a small number of those students who enter university are powerful verbal-analytic thinkers who may benefit from making sense of set-theoretic definitions, an even smaller number base their thinking on visual-pictorial representations, and others who may have a blend of visual embodied thinking and operational symbolism or who prefer to learn procedurally by rote.

Some students seek a *natural* approach based on a blend of previous experiences of embodiment and symbolism from school mathematics. Some with a more verbal-logical basis may seek to use a *formal* approach based on set-theoretic definitions and the deduction of properties using formal proof. Others seek to learn proofs procedurally to reproduce in examinations. All of these approaches may involve supportive and problematic aspects, which have been detailed in the literature (e.g. Pinto and Tall 1999; Weber 2004; Tall 2013).

As students become more experienced and shift to graduate studies, Weber (2001) produced evidence that research graduates are more likely to respond flexibly to problems by making links between concepts in a sophisticated knowledge structure while undergraduates in their early studies, have yet to develop such flexibility.

This is consistent with the lack of aesthetic appreciation of mathematical ideas noted by Dreyfus and Eisenberg (1986) and also with the relationship noted by Koichu et al. (2007) between "aesthetical blindness" of students and factors such as self-esteem that affect their aesthetic judgement.

The theoretical framework presented here traces the development of cognitive and emotional aspects throughout the lifetime of the individual. A few students, characterized as being 'gifted' develop verbal-analytic skills that enable them to build formally from set-theoretic definitions to construct highly connected crystalline concepts that may have embodiments and operations linked to underlying formally proved structure theorems. But many others, who focus on 'maximising their mark on the exam' to 'get a good degree' to move on in their lives, have good reasons for doing so. The mathematics is *problematic* for them and *it doesn't make sense*.

	Problems (recognition)	Possibilities (description)	Conjectures (definition)	Proof (deduction)
Atiyah	Thinking about a vague and uncertain situation	Trying to guess what may be found out	Reaching definitions	And definitive theorems and proofs
MacLane	Getting and understanding needed theorems	Working with them to see what could be calculated	What might be true	Come up with new theorems

Fig. 13.2 Van Hiele-like developments in mathematical research

Structure Theorems

Some theorems based on formal axioms and definitions prove formal structures that enable the ideas to be reconsidered in embodied and symbolic terms. For example, a finite dimensional vector space over a field F is isomorphic to F^n, so that its elements may be represented symbolically as n-tuples and its linear maps as matrices, and in the case where F is the field of real numbers and $n = 2$ or 3, it may be embodied in two or three dimensional space. In the same way a finite group is isomorphic to a subgroup of a group of permutations, which allows it to be operated on symbolically and embodied as the transformations of a geometric object.

Structure theorems enrich formal mathematics with new forms of embodiment and symbolism, to enable mathematicians to recognise problems, imagine possibilities, to formulate conjectures and to prove new theorems. Mathematicians of different persuasions see proof as their main research goal, but achieve it in different ways, as the algebraist Saunders MacLane observed when comparing his approach with that of the geometer Michael Atiyah:

> For MacLane it meant getting and understanding the needed definitions, working with them to see what could be calculated and what might be true, to finally come up with new 'structure' theorems. For Atiyah, it meant thinking hard about a somewhat vague and uncertain situation, trying to guess what might be found out, and only then finally reaching definitions and the definitive theorems and proofs. (MacLane 1994, pp. 190–191)

Both strategies follow the same format—becoming aware of a problem, considering possibilities, formulating conjectures and seeking proof—and this follows the broad van Hiele format of recognition, description, definition and deduction (Fig. 13.2).

The Overall Development of Proof

The long-term growth of mathematical thinking of proof begins with the perceptions and actions of young children, and develops through three successive levels:

- *practical mathematics* exploring shape and space and developing experience of the operations of arithmetic. This involves the *recognition* and *description* of properties, such as the observation that the sum of numbers is not affected by the order of operation and proof is often formulated as *generic proof*.
- *theoretical mathematics* of *definition* and *deduction*, as exhibited by Euclidean proof in geometry, and of the definition of the 'rules of arithmetic' and properties such as even, odd, prime, composite, and the theoretical deduction of theorems such as uniqueness of factorization into primes.

Theoretical mathematics is appropriate for most applications of mathematics, while those who go on to study pure mathematics change meaning once more to

- *formal mathematics* based on set-theoretic *definition* and *deduction*.

In mathematical research, mathematicians use various combinations of embodiment, symbolism and formalism to imagine possible theorems and to formulate conjectures to seek proof and to shift to ever more sophisticated levels using structure theorems.

This framework offers mathematicians, mathematics educators, teachers and learners the opportunity to share an overall development of proof based on the fundamental sensori-motor bases of human thinking that becomes increasingly sophisticated through the use of language and symbolism. It offers an integration of the cognitive and affective development of mathematical knowledge and mathematical proof.

To enable different communities of practice to come together for mutual benefit, it is essential to develop a common context of discourse that enables different communities to speak meaningfully to each other. In the final chapter of *How Humans Learn to Think Mathematically* (Tall 2013), I consider the problems encountered in communication between different communities. It becomes clear that each community has its own ways of working that may be highly appropriate in its own context but that the shift to another context involves met-befores that may impede the possibility of an expert in one community making sense of the needs of another community. This suggests the need for a sense of openness and willingness to listen to other points of view and to see the relevance of various viewpoints in different contexts. It should be possible for a community to realise that viewpoints that may be essential in their own context may not be appropriate in others. For instance, a formal mathematician could become more sensitive to the practical needs of mathematics in the everyday community, or recognise the theoretical requirements of applied mathematicians, who build on natural modelling of real situations rather than formal set-theoretic definitions and proof. In the other direction, it should be possible for those involved with practical mathematics to develop some insight into more technical requirements, or for technical mathematicians to have a sense of the power of the greater generality of axiomatic mathematics. The goal should surely be a more respectful understanding between various communities of practice involved in mathematics, including pure and applied mathematicians, mathematics educators and a range of other communities of practice in science, sociology, psychology, philosophy, history, cognitive science, constructivism and so on.

The theory presented here focuses on the fundamental ideas of proof that occur as humans use their perception, operation and reason to build increasingly sophisticated mathematical knowledge. It begins with practical experiences in which specific examples may be seen as *generic* examples of proof. Then these experiences lead to *theoretical* proof based on Euclidean definition and proof in geometry, definitions based on the symbolic 'rules of arithmetic' in arithmetic and algebra, or a blending of embodied thought experiment and symbolic proof. At a formal level, definitions are given as quantified set-theoretic definitions and *formal* proof that apply in any context where the axioms and definitions are satisfied.

The long-term development is affected by supportive and problematic met-befores that apply not only to developing students, but also to the historical evolution of mathematics and to the competing views of differing communities of practice. Experts with sophisticated knowledge structures are subject to personal conceptions of mathematics that they may share with other experts in their community but perhaps not with other communities. The framework given here offers an opportunity to evolve theoretical ideas into the future by blending differing viewpoints to grasp the fundamental basis of the long-term development of mathematical thinking and proof by building on the fundamental ideas of perception, operation and reason.

References

Asiala, M., Brown, A., DeVries, D., Dubinsky, E., Mathews, D., & Thomas, K. (1996). A framework for research and curriculum development in undergraduate mathematics education. In *CBMS issues in mathematics education: Vol. 6 Research in collegiate mathematics education II* (pp. 1–32).

Bachelard, G. (1938). *La formation de l'esprit scientifique*. Paris: Vrin (reprinted 1983).

Baron, R., Earhard, B., & Ozier, M. (1995). *Psychology (Canadian edition)*, Scarborough: Allyn & Bacon.

Baroody, A. J., & Costlick, R. T. (1998). *Fostering children's mathematical power. An investigative approach to K-8 mathematics instruction*. New Jersey: Erlbaum.

Brousseau, G. (1983). Les obstacles épistémologiques et les problèmes en mathématiques. *Recherches En Didactique Des Mathématiques*, *4*, 164–198.

Bruner, J. S. (1966). *Towards a theory of instruction*, New York: Norton.

de Lima, R. N., & Tall, D. O. (2008). Procedural embodiment and magic in linear equations. *Educational Studies in Mathematics*, *67*, 3–18.

Dreyfus, T., & Eisenberg, T. (1986). On the aesthetics of mathematical thought. *For the Learning of Mathematics*, *6*(1), 2–10.

Gray, E. M., & Tall, D. O. (1994). Duality, ambiguity and flexibility: a proceptual view of simple arithmetic. *Journal for Research in Mathematics Education*, *26*, 115–141.

Harel, G., & Tall, D. O. (1991). The general, the abstract, and the generic in advanced mathematics. *For the Learning of Mathematics*, *11*(1), 38–42.

Koichu, B., Berman, A., & Katz, E. (2007). What is a beautiful problem? An undergraduate students' perspective. In J. H. Woo, H. C. Lew, K. S. Park, & D. Y. Seo (Eds.), *Proceedings of the 31st conference of the international group for the psychology of mathematics education* (Vol. 3, pp. 113–120). Seoul: PME.

Krutetskii, V. A. (1976). *The psychology of mathematical abilities in schoolchildren*. Chicago: University of Chicago Press.

MacLane, S. (1994). Responses to "Theoretical mathematics". *Bulletin of the American Mathematical Society*, *30*, 190–191.

Mason, J. H., & Pimm, D. (1984). Seeing the general in the particular. *Educational Studies in Mathematics*, *15*, 277–289.

McGowen, M. C., & Tall, D. O. (2010). Metaphor or met-before? The effects of previous experience on the practice and theory of learning mathematics. *The Journal of Mathematical Behavior*, *29*, 169–179.

Pinto, M. M. F., & Tall, D. O. (1999). Student constructions of formal theory: giving and extracting meaning. In O. Zaslavsky (Ed.), *Proceedings of the 23rd conference of the international group for the psychology of mathematics education* (Vol. 4, pp. 65–73). Haifa: PME.

Presmeg, N. C. (1986). Visualisation and mathematical giftedness. *Educational Studies in Mathematics*, *17*, 297–311.

Sfard, A. (1991). On the dual nature of mathematical conceptions: reflections on processes and objects as different sides of the same coin. *Educational Studies in Mathematics*, *22*, 1–36.

Skemp, R. R. (1971). *The psychology of learning mathematics*. London: Penguin.

Skemp, R. R. (1979). *Intelligence, learning, and action*. London: Wiley.

Tall, D. O. (2004a). The three worlds of mathematics. *For the Learning of Mathematics*, *23*(3), 29–33.

Tall, D. O. (2004b). Thinking through three worlds of mathematics. In M. J. Joines, M. Johnsen, & A. E. Fuglestadt (Eds.), *Proceedings of the 28th conference of the international group for the psychology of mathematics education* (Vol. 4, pp. 281–288). Bergen: PME.

Tall, D. O. (2008). A life-time's journey from definition and deduction to ambiguity and insight. *Mediterranean Journal for Research in Mathematics Education*, *7*, 183–196.

Tall, D. O. (2011). Crystalline concepts in long-term mathematical invention and discovery. *For the Learning of Mathematics*, *31*(1), 3–8.

Tall, D. O. (2013). *How humans learn to think mathematically*. New York: Cambridge University Press.

Tall, D. O., Thomas, M. O. J., Davis, G., Gray, E. M., & Simpson, A. P. (2000). What is the object of the encapsulation of a process? *The Journal of Mathematical Behavior*, *18*, 1–19.

Tall, D. O., Yevdokimov, O., Koichu, B., Whiteley, W., Kondratieva, M., & Cheng, Y. H. (2012). Cognitive development of proof. In G. Hanna & M. de Villiers (Eds.), *Proof and proving in mathematics education: the 19th ICMI study* (pp. 14–39). New York: Springer.

Van Hiele, P. (1986). *Structure and insight*. Orlando: Academic Press.

Van Hiele, P. (2002). Similarities and differences between the theory of learning and teaching of Skemp and the Van Hiele levels of thinking. In D. O. Tall & M. O. J. Thomas (Eds.), *Intelligence, learning and understanding—a tribute to Richard Skemp* (pp. 27–47). Flaxton: Post Pressed.

Weber, K. (2001). Student difficulty in constructing proofs: the need for strategic knowledge. *Educational Studies in Mathematics*, *48*, 101–119.

Weber, K. (2004). Traditional instruction in advanced mathematics courses: a case study of one professor's lectures and proofs in an introductory real analysis course. *The Journal of Mathematical Behavior*, *23*, 115–133.

Chapter 14
Reflections on Justification and Proof

Justification and Proof in Mathematics and Mathematics Education

Keith Weber

Abstract In this chapter, we explore how investigations into mathematicians' practice can inform instruction on justification and proof. Each co-author of this practice presents an investigation of how mathematicians use justification and proof in their professional practice and suggests pedagogical implications based upon insights from their investigations.

Keywords Justification · Mathematicians' practice · Proof

Introduction

In 2000, Eric Knuth noted that there was a rebirth of proof in mathematics classrooms. In the 1990s, many researchers noted that the role of proof in mathematics was limited (e.g., Schoenfeld 1994; Wu 1996). But in the last decade, mathematics educators have advocated that justification and proof was expected to play an important role in all aspects of students' mathematical learning (e.g., Knuth 2000). If we accept the premise that proof should be taught to all students of mathematics, this begs the questions: what role should proof play in the mathematics classroom and how should it be taught? Often proof is taught independently from other mathematical content, leading students to view proof as a pedantic ritual that they are required to engage in rather than as a tool for facilitating communication and advancing mathematical knowledge (e.g., Harel 1998; Schoenfeld 1994)—consequently students often see little value in the proof they observe or produce (e.g., Harel 1998; Healy and Hoyles 2000; Schoenfeld 1989).

With contributions by
Gila Hanna, Ontario Institute for Studies in Education, University of Toronto, Toronto, Canada
Guershon Harel, University of California-San Diego, La Jolla, CA, USA
Ivy Kidron, Jerusalem College of Technology, Jerusalem, Israel
Annie Selden and John Selden, New Mexico State University, Las Cruces, NM, USA

K. Weber (✉)
Graduate School of Education, Rutgers University, 10 Seminary Place, New Brunswick, NJ 08901, USA
e-mail: keith.weber@gse.rutgers.edu

M.N. Fried, T. Dreyfus (eds.), *Mathematics & Mathematics Education: Searching for Common Ground*, Advances in Mathematics Education,
DOI 10.1007/978-94-007-7473-5_14,

Addressing these shortcomings involves an important opportunity for mathematicians and mathematics educators to collaborate. An important source of insight into what justification and proof should mean and how these constructs should be used for students is mathematicians' practice with justification and proof. Toward this end, mathematics educators have explored how justification and proof have been used in mathematical practice. This is based on the assumption that what motivates mathematicians to engage in justification and proof might similarly motivate students to do the same and that students may gain similar benefits from the kind of proving mathematicians do. Narrowing the gap between mathematical and classroom practice with respect to justification and proof has the potential also to bridge the gap between mathematicians and mathematics educators.

In this chapter, we explore different ways that mathematicians treat justification and proof in their professional practice and explore the implications for mathematics education. We also explore different ways that educational researchers have used to investigate mathematicians practice, including different sources of data from mathematicians, to form these implications. Gila Hanna's contribution examines recent philosophical theses with respect to mathematical proof and their implications for mathematics education. Notably, philosophers have stressed the communicative functions of (informal) proofs, including providing explanation and illustrating methods, and Hanna believes proof can have similar functions in mathematics classrooms. Guershon Harel's contribution explores the different types of intellectual need that have the potential to motivate all individuals (notably mathematicians and students) to engage in justification. Harel argues that creating these types of need in the classroom environment can help students view justification and proof as meaningful mathematical activities. Ivy Kidron's contribution is based on her fine-grained analyses of the opportunities to construct mathematical knowledge in the process of forming justifications, illustrating how many of the insights gained from constructing a proof are obtained by linking aspects of an individual's knowledge that were previously disconnected. Annie Selden and John Selden's contribution explores how proofs are typically written by mathematicians, contrasting this both with other genres of writing and the ways in which students write proofs. They base their conclusions on their reading of the mathematical literature, as well as their own experience as research mathematicians. Keith Weber's contribution uses insights gained from laboratory experiments with mathematicians to explore the heterogeneity in mathematicians' practice with respect to proof, and he suggests that a variety of practices with regard to justification may be beneficial to include in mathematics teaching.

Proof in the Curriculum: Reflecting Modern Mathematical Practice

Gila Hanna

My starting point is the view that educational researchers and curriculum developers cannot foster the use of reasoning and proving in mathematics teaching without understanding what it means to reason and to prove in mathematics itself. I firmly believe that an appreciation of current thinking in mathematical practice and in the philosophy of mathematics stands to provide mathematics educators with new perspectives on questions crucial to the teaching of mathematics, as well as to open up new avenues for research in mathematics education.

My research interests focus on epistemological issues related to proof and proving in mathematics education. The focus of my research is not on cognition. Its objective is to direct the attention of educational researchers to the newest developments in the philosophy and practice of mathematics and to their relevance for effective mathematics education.

Over the past thirty years or so, philosophers of mathematics have shifted their interest markedly, away from a preoccupation with the logical foundations of mathematics and towards a detailed study of mathematical practice. This new focus brings with it a greater relevance to mathematics education, because of the light it throws on applications of mathematics, on the role of new technologies in mathematics, on the ways in which today's mathematicians actually devise and judge proof, on the reasoning styles they use, and on the ways in which they present and weigh evidence.

At the same time, philosophers of mathematics have come to a greater recognition of the central importance of mathematical understanding, and so have looked more closely at how that understanding is best conveyed and thus at what counts as explanation in mathematics. As might be expected with these two shifts in focus, these philosophers have turned their attention more and more from the justificatory to the explanatory role of proof (Corfield 2003; Kitcher 1981; Mancosu 2001; Rav 1999; Sandborg 1997; Tappenden 2005).

Proof in Mathematical Practice

There is a standard definition of mathematical proof which is universally accepted as the ideal: A mathematical proof is a finite sequence of propositions, each of which is either an axiom or follows from preceding propositions by the rules of logical inference. Now, this is actually the definition of a formal proof, a Hilbertian view of proof, and it is what students learn when they first encounter the concept of deductive proof. But in mathematical practice the vast majority of proofs are not purely syntactical derivations. Proofs do not necessarily conform to this standard definition simply because in many contexts mathematicians either cannot afford to live up to the demands of formal proof or simply choose to employ informal proofs that provide more insight and understanding.

In addition, many of the rather informal proofs that mathematicians routinely produce are nevertheless considered by other mathematicians to be sufficiently rigorous and thus reliable. In fact, one could say that there are informal standards of rigor, or standards of informal rigor, that are acceptable to experts in the field (Azzouni 2009; Marfori 2010).

It is informative to examine what the mathematician Yu. Manin (1998) has to say about such proofs. He perceives the practice of mathematics as both imprecise and informal, and describes proof as having three levels: (1) An individual level, where proving depends on the preferences of individual mathematicians and their inclinations for various styles of reasoning, such as geometric, visual, or algebraic calculations; (2) A social level, where mathematicians have to rely on the work of others and on the authority of accepted proofs and reasonings; (3) An epistemological level, where mathematicians know that a rigorous proof has an ideal (formal) representation, but consider this ideal to be an "imaginary text."

Manin goes on to explain that the standard definition of mathematical proof (as stated above) is an "ideal representation" which might arouse either "a strong aversion" or a high degree of "enthusiasm." However, he also concedes that because mathematicians must maintain high standards of proof, ultimately "we have to resort to the ideal of mathematical proof as an ultimate judge of our efforts" (Manin 1998, pp. 154–155).

Terrence Tao (2010) points out mathematicians are expected to adhere to "the highest standards of rigor that are practical" (rigor not being the same as logical formality) when they devise proofs. But he also states that the same level of rigor is not necessarily appropriate to every part of mathematics.

From Mathematical Practice to Mathematics Education

The Use of Proof to Promote Understanding

Incorporating reasoning and proof in the curriculum to reflect its importance in mathematical practice is good as far as it goes. But it has become clear that there is much to be gained by also looking closely at *how* mathematicians reason and prove. In other words, mathematical practice has lessons for instructional practice. Since in mathematical practice proofs that provide insights and understanding are highly valued, it behooves mathematics educators to pay great attention to the explanatory role of proof, and perhaps even to assign more weight to its explanatory than to its justificatory role. This suggests that a somewhat more "liberal" view of proof, encompassing the use of computer graphics and other forms of experimentation and visualization, would have beneficial effects on the effectiveness of mathematics teaching. Also the ability to understand a proof or to provide a proof depends on a good grasp of the specific mathematical topic under consideration and on one's familiarity with the concepts involved. Thus it is crucial that students be taught more mathematics.

The Use of Proof to Teach Mathematical Methods

This approach is inspired by mathematical practice, in which proofs do much more than provide warrants for mathematical statements (Rav 1999). In the course of proving a mathematical proposition, a mathematician brings to bear not only other mathematical truths, but also many powerful tools of mathematical practice—resources that encompass strategies and techniques and may best be referred to as methods. But proofs do not just make use of these methods. They also serve as a test bed, an opportunity for mathematicians to verify the validity of methods, adapt them to new contexts, and extend them. Often they also provide the incentive to construct entirely new ones (Avigad 2006; Corfield 2003; Dawson 2006; Hanna and Barbeau 2008; Rav 1999). Recently the mathematician Terrence Tao has recommended "learning the power of other mathematicians' tools", and adding them to one's bag of tricks (Tao 2001, Blog).

Using proof to teach mathematical methods is not intended or expected to replace or compromise in any way the teaching of proof itself or of logical derivation in the Euclidean sense. What it could do is provide mathematics educators with another very important reason to accord to proof a position of significance in the mathematics curriculum.

Formal Proofs and Proof Assistants

In ordinary mathematical practice, as mentioned above, many informal proofs are considered to be an appropriate standard of proving, and suitable for publication. Nevertheless, mathematicians would certainly prefer a higher level of certainty than these proofs afford. It so happens that mathematicians now have access to a very promising way of gaining confidence in a proof, with the development, over the past twenty years, of several computer programs known as "automatic proof checkers" and "proof assistants." Checking the correctness of a proof has reached a level that no ordinary proof can match. According to Wiedijk (2008), these programs have been successful in checking the validity of the proofs of several well-known theorems, such as the Fundamental Theorem of Algebra (2000), Jordan's Curve Theorem (2005), the Fundamental Theorem of Calculus (1996), the Four Colour Theorem (2004), and the Prime Number Theorem (2008).

Some mathematicians (Calude and Müller 2009; Hales 2008; Harrison 2008; Wiedijk 2008) envisage the use of proof assistants in daily mathematical practice, to provide an objective criterion for the correctness of a proof and thus to supplement or even partially replace the process of peer review. As Wiedijk (2008) put it, "When the part of refereeing a mathematical article that consists of checking its correctness takes more time than formalizing the contents of the paper would take, referees will insist on getting a formalized version before they want to look at a paper" (p. 1414).

Theorem-Provers for Education

Mathematics education has greatly benefitted from the intensive use of several educational software packages, such as Dynamic Geometric Software (DGS), spreadsheets, and Computer Algebra Systems (CAS). As reported by Maric and Neuper

(2011) there is now a fully functional version of Theorem-Prover System (TPS) appropriate for the school and undergraduate levels and dubbed eduTPS. Unfortunately, it has not yet been sufficiently noticed by mathematics educators, and thus has not been tested by educational researchers or tried in the classroom. A group of researchers will be presenting this new software at the upcoming Conference on Intelligent Computer Mathematics, CICM 2012. It is still an open question whether the use of eduTPS would be beneficial to the teaching of mathematics.

Categories of Intellectual Need

Guershon Harel

The goal of this paper is to define *intellectual need* and outline its different manifestation in mathematical practice. The definition is oriented within a larger conceptual framework called *DNR-based instruction in mathematics* (*DNR*). *DNR* can be thought of as a system consisting of three categories of constructs: *premises* (explicit assumptions underlying the *DNR* concepts and claims), *concepts* oriented within these premises, and *instructional principles*. The latter are claims about the potential effect of teaching actions on student learning justifiable in terms of these premises and empirical observations. The initials D, N, and R stand for the three foundational instructional principles of the framework: *Duality, Necessity, and Repeated reasoning*. (For a more comprehensive discussion of *DNR*, see Harel 2008a, 2008b). Here we only present four of the eight premises of *DNR*—those that are needed for the definition of *intellectual need*:

1. The *knowledge of mathematics premise:* Mathematics consists of two related but different categories of knowledge: all the ways of understanding and ways of thinking that have been institutionalized throughout history.
2. The *knowing premise:* Knowing is a developmental process that proceeds through a continual tension between assimilation and accommodation, directed toward a (temporary) equilibrium.
3. The *knowledge-knowing linkage premise:* Any piece of knowledge humans know is an outcome of their resolution of a problematic situation.
4. The *subjectivity premise:* Any observations humans claim to have made are due to what their mental structure attributes to their environment.

Definition of Intellectual Need

If K is a piece of knowledge possessed by an individual or community, then, by the *Knowing-Knowledge Linkage Premise*, there exists a problematic situation S out of which K arose. S (as well as K) is subjective, by the *Subjectivity Premise*, in the sense that it is a perturbational state resulting from an individual's encounter with a situation that is incompatible with, or presents a problem that the individual recognizes as unsolvable by, her or his current knowledge. Such a problematic situation S, prior to the construction of K, is referred to as an individual's *intellectual need*: S is the need to reach equilibrium by learning a new piece of knowledge. Thus, intellectual need has to do with disciplinary knowledge being created out of people's current knowledge through engagement in problematic situations conceived as such by them. One may experience S without succeeding to construct K. That is, intellectual need is only a necessary condition for constructing an intended piece of knowledge.

Categories of Intellectual Need

DNR offers five categories of intellectual needs: (1) *need for certainty*, (2) *need for causality*, (3) *need for computation*, (4) *need for communication*, and (5) *need for structure*.

Need for Certainty When an individual (or a community) considers an assertion, he or she conceives it either as a *fact* or as a *conjecture*—an assertion made by a person who has doubts about its truth. The assertion ceases to be a conjecture and becomes a fact in her or his view once the person becomes certain of its truth. The *need for certainty* is the natural human desire to know whether a conjecture is true—whether it is a fact. When the person fulfills this need, through whatever means deemed appropriate by her or him, the person gains new knowledge about the conjecture.

Need for Causality Certainty is achieved when an individual determines (by whatever means he or she deems appropriate) that an assertion is true. Truth alone, however, may not be the only aim for the individual, and he or she may desire to know *why* the assertion is true—the cause that makes it true. Thus, the *need for causality* is one's desire to *explain*, to determine a cause of a phenomenon. "Mathematicians routinely distinguish proofs that merely demonstrate from proofs which explain" (Steiner 1978, p. 135). For many, the role of mathematical proofs goes beyond achieving certainty—to show that something is true; rather, "they're there to show... why [an assertion] is true," as Gleason, one of solvers of Hilbert's Fifth Problem (Yandell 2002, p. 150), points out. Two millennia before him, Aristotle, in his *Posterior Analytic*, asserted that "we suppose ourselves to possess unqualified scientific knowledge of a thing, as opposed to knowing it in the accidental way in which the sophist knows, when we think that we know the cause on which the fact depends as the cause of the fact and of no other."

Need for Computation The *need to compute* refers to one's desire to quantify, determine a missing object or construct a mathematical object, determine the property of an object or relations among objects, etc. by means of symbolic algebra. It also includes the need to find more efficient computational methods, such as one might need to extend computations to larger numbers in a reasonable "running time." Often the need to compute reflects two inseparable abilities: (a) the ability to represent a situation symbolically and manipulate the representing symbols as if they have a life of their own, without necessarily attending to their reference; and (b) the ability to pause at will during the manipulation process in order to probe into the referential meanings for the symbols involved in the manipulation.

Need for Communication In mathematics, the *need for communication* refers collectively to two reflexive acts: *formulating* and *formalizing*. Formulating is the act of transforming strings of spoken language into algebraic expressions (i.e., expression amenable to computation by means of symbolic algebra as discussed in the preceding section). Formalization is the act of externalizing the exact intended meaning of an idea or a concept or the logical basis underlying an argument. In modern mathematics the acts of formulation and formalizations are reflexive in that as one formalizes a mathematical idea it is often necessary to formulate it, and, conversely, as one formulates an idea one often encounters a need to formalize it.

Need for Structure The *need for structure* is the need to re-organize the knowledge one has learned into a logical structure. A critical element in this definition is the verb "to re-organize," and, by implication, its source verb "to organize." The verb "to organize" implies an action on something that already exists, and the verb "to re-organize" implies that something has already been organized. Accordingly, the need for structure is not a forward need; that is, one does not feel intellectually compelled to learn new knowledge in a particular order and from that fit a predetermined structure; rather, one assimilates knowledge into one's existing structure, and reorganizes it if and when one perceives a need to do so. The nature of the structure into which one organizes one's own knowledge is idiosyncratic and depends entirely on one's past experience. Such a structure is unlikely to be logically hierarchical, and even mathematicians are unlikely to involuntarily organize their knowledge into a systematic logical structure. Thus, the term "re-organize" in the above definition recognizes that individual learners or communities of learners first organize the mathematical knowledge they learn in a form determined by their existing cognitive structures; later they may meet the need to *re-organize* what they have learned into a logical structure.

Summary

In this brief paper, we have identified five categories of intellectual needs. Collectively, these five needs are ingrained in all aspects of mathematical practice—in forming hypotheses, proving and explaining proofs, establishing common interpretations, definitions, notations, and conventions, describing mathematical ideas unambiguously, etc. DNR-based instruction is structured so these same needs drive student learning of specific topics, *and* by realizing the different needs that drive mathematical practice, students are likely to construct a global understanding of the epistemology of mathematics as a discipline. These needs, thus, are common to students and mathematicians, and so they are real components of the common ground this volume is striving for.

Justification and Construction of Knowledge

Ivy Kidron

In the following contribution, I use my work with Dreyfus (Dreyfus and Kidron 2006; Kidron and Dreyfus 2009, 2010a, 2010b) to elucidate the intricate relationships between processes of justification and construction of knowledge from a cognitive perspective. I organize this issue around three questions: What is justification for the mathematician and the student? How do mathematicians construct justifications? How can justification lead to new knowledge?

What Is Justification for the Mathematician and the Student?

My position is that justification can lead to the construction of knowledge because *justification provides insights into the connections underlying the statement to be* justified (Kidron and Dreyfus 2010a). This position is consistent with the view of Rota (1997), who wrote:

> Verification alone does not give us a clue to the role of a statement within the theory; it does not explain the *relevance* of the statement… the logical truth of a statement does not *enlighten* us as to the *sense* of the statement. …every teacher of mathematics knows that students will not learn by merely grasping the formal truth of a statement. Students must be given some enlightenment as to the *sense* of the statement. (pp. 131–132)

The term enlightenment has been introduced by Rota (1997) to refer to "insight into the connections underlying the statement to be justified". Rota pointed out that enlightenment is not easily formalized and that unlike mathematical proofs, it admits degrees: "Mathematical proof does not admit degrees. A sequence of steps in an argument is either a proof, or it is meaningless. Heuristic arguments are a common occurrence in the practice of mathematics. However, heuristic arguments do not belong to formal logic. … Proofs given by physicists admit degrees. In physics, two proofs of the same assertion have different degrees of correctness. … A great many characteristics of mathematical thinking are neglected in the formal notion of proof" (ibid., pp. 134–135).

We might learn from Rota two important ideas: the first is that even though heuristic (non-deductive) arguments are not proofs, they nevertheless deserve more of a place in mathematics. Since heuristic arguments have a place in mathematics they ought to have a place in mathematics teaching as well. Moreover, heuristic arguments provide different levels of conviction. The second is that, to mathematicians, proofs can provide insights. Thurston (1994) also expressed his view of justification when he wrote that mathematicians should pay much more attention to communicating not just definitions, theorems and proofs, but also mathematical ideas. He added that there are certain theorems that are generally accepted and as long as people in the field are comfortable that the idea works, it does not need to have a formal written source. Thurston observed that when people are doing mathematics, the flow of ideas and the social standard of validity are much more reliable than formal documents. An implication is that when presenting proofs to students, the instructor should be transparent about the ideas that motivated the proof.

How Do Mathematicians Construct Justifications?

We may ask: Are there regularities in the processes of constructing justifications that should be common to mathematicians and students?

Selden (2012, p. 398) investigated the issue of how tertiary students deal with various aspects of proof and proving? Selden points out that upon being given a statement to prove, students' first job is to understand both the statement's structure and its content. Then the students have to interpret and use previous theorems and definitions in proving. Tall (2006) also emphasizes the role of incorporating previous constructs toward the building of justifications, as appears in his concept of "met before": a structure we have in our brains *now* as a result of experiences we have met before. In the new situation, the "met before" becomes supportive or problematic. Tall (2006) maintains that "a teacher can do a great deal by adopting a connectionist viewpoint to help each learner to address a problem by building on current knowledge" (p. 211); hence teachers should understand and address not only the mathematical structure but also the role of the learner's prior knowledge in constructing a justification. Kidron and Dreyfus (2010a) described justification as a process of combination of selected previous constructs. They analyzed the justification construction of one mathematician who was attempting to understand the family of solutions to the differential equation, $dx/dt = rx(1 - x)$ and sought to justify how changes in the parameter r led to different final state solutions. In this case study, significant advances in the justification process came from viewing the same diagram in different ways. For instance, she noticed that the bifurcation map of the family of solutions, which she used as an aid for her reasoning, branched when the discriminant of the differential equation vanished (see Kidron and Dreyfus 2010a, for more details). Noticing this connection led to an enlightenment (in the sense of Rota).

How Can Justification Lead to New Knowledge?

The mathematician in the study by Kidron and Dreyfus (2010a) constructed a justification which constitutes a complex learning process. Her attempts at justification gave rise to several interweaved modes of thinking that develop in parallel and interact. Kidron and Dreyfus observed a combination of numerical and graphical modes of thinking as well as a close approach of an algebraic mode to an analytical mode of thinking. Finally, they observed an integration of a dynamic graphical view with an algebraic/analytic one incorporating all four modes of thinking, the numerical, the algebraic, the dynamic graphical and the analytic one into a single consistent image.

These different modes of thinking refer to different constructions of knowledge. Kidron and Dreyfus' analysis refers not only to what justification means for the mathematician in the study but also to the relationships of this meaning of justification for the patterns of knowledge construction. They analyzed the interaction pattern of combining constructions of knowledge and show that combining constructions is associated with the construction of justification. They observed that "enlightenment," to use Rota's term, occurs when constructions combine—at the integration of different modes of thinking. An important question might be: is every justification a construction of new knowledge? it is not easy to answer this question. Nevertheless, if we observe the dynamics of the process of constructing a justification: it goes from a bunch of disconnected selected previous constructs to an interlinked network of previous constructs that provide a justification. The establishment of new connections requires construction of new knowledge towards a more formal reasoning.

Similarities and Differences Between Justification Processes of Students and Mathematicians

The relationship between the process of constructing a justification and the phenomenon of combining modes of thinking were observed by Kidron and Dreyfus in a few case studies including mathematicians and young learners (Kidron and Dreyfus 2009). The combining points indicated the integration of different knowledge constructs and different modes of thinking. An important question for which I do not have yet an answer is: Are combining modes of thinking an indicator for certain types or forms of justification? Which types and forms?

Studying how mathematicians construct mathematical knowledge via the process of justifying can inform mathematics instruction. The main idea is to learn how expert mathematicians bring together different types of mathematics resources while combining rigorous and intuitive thinking. This idea is well developed in Wilkerson-Jerde and Wilensky (2011). They point out that experts are likely to refer to specific examples or instantiations when making sense of an unknown aspect of a mathematical idea. They claim that for some experts, specific instantiations of the mathematical object being explored serve a central role in the process of constructing a proof.

This is well expressed by de Villiers (2004) who argues that "mathematicians are often convinced by the truth of their results (usually on the basis of quasi-empirical evidence) long before they have proofs" (p. 402). Quasi-empirical evidence might be provided by means of generating examples. We may ask: What is the aim of generating examples? Is it only for empirical evidence and conviction? Or maybe the aim is a better insight in the sense of the statement to be proved, helping to judge the probable truth of the conjectures? Indeed, Selden (2012, p. 403) points out that generating examples and counter-examples can help students judge the truth of conjectures. But students are often reluctant to generate examples (Watson and Mason 2005). Moreover, Alcock and Weber (2010) have noted that students who attempt to use examples are often not successful.

Enlightenment and Aesthetics

Analyzing similarities and differences between justification processes of students and mathematicians especially in connection to the notion of "enlightenment" which accompanies some of the justification processes, we may ask: how to build in students an appreciation for the power and beauty of a mathematical argument?

Viewed this way, Rota's "enlightenment" may be compared to the "sudden illumination that underlies the aesthetics of the solution processes" (Dreyfus and Eisenberg 1986). Ted Eisenberg pointed out the importance of appreciation of the aesthetics of mathematical thought yet was concerned to find that few students derive pleasure from the beauty of mathematics.

The Genre[1] of Proof

Annie Selden and John Selden

A Study of Mathematicians' Views on Features of Proofs

Some years back, while attending the Park City Mathematics Institute,[2] we interviewed mathematicians regarding what they thought of the following seven conjec-

[1] Here we are invoking the "new view" of genre. As Friedman and Medway (1994) wrote, "... the notion of genre has been reconceived.... Genres have come to be seen as typical ways of engaging rhetorically with recurring situations." (p. 2).

[2] The Park City Mathematics Institute is a program of the Institute for Advanced Study, Princeton, NJ. It is designed for mathematics researchers, post-secondary students, and mathematics educators at the secondary and post-secondary levels.

tured features of proofs, while looking at one of their own published mathematics papers. Overall, they tended to agree.

Proofs Are not Reports of the Proving Process

For example, one does not write into a final written proof things like, "I tried this [technique or idea] and it did not work." One also does not (usually) write "I want to show [the conclusion or sub-conclusion that one will prove next], except perhaps in rather long, complicated proofs."

Mathematicians do not consider "the function of the written document as a record of the work done... and consistency with historical development is hardly considered at all as a relevant factor..." (Csiszar 2003, pp. 247–248). "False starts, mistakes, revisions—these are all part of the creative process [in discovering and proving theorems]. But when the final result is published, we seldom see the enormous effort that was necessary for the creation; we see the polished product, the correct statement with a clean proof. ... [This is] an important feature of mathematics." (John Ewing, as quoted by Csiszar 2003, p. 244).

Proofs Contain Little Redundancy

Unlike arguments in philosophical papers, one does not consider the argument from another point of view (at least in the same proof). Furthermore, "... mathematicians appear to prize brevity, conciseness, and precision of meaning." (Shepherd et al. 2012). "Mathematicians collectively take pride in their writing style for its rigor and precision." (Csiszar 2003, p. 244).

Symbols Are (Generally) Introduced into Proofs in One-to-One Correspondence with Mathematical Objects

For example, one does *not* do what one of our transition-to-proof course students did, when proving the following **Theorem**: *For integers m, n, and p, if m divides n and m does not divide p, then n does not divide p.* The attempted proof began with the following unnecessary profusion of letters: **Proof**: *Since m, n, and p are integers and m divides n and n does not divide p, let $m = j$ and $n = jk$ and $p = l$, where j, k, and l are integers*, and did not get better. "In the genre of mathematical proofs it is not permissible to let the same symbol represent two different numbers, except across independent subproofs. Perhaps this is because doing so seems very likely to cause validators confusion." (Selden and Selden 2003, p. 13).

Proofs Contain Only Minimal Explanations of Inferences, that Is, Warrants are Often Left Implicit

For example, one of our transition-to-proof students correctly proved the **Theorem**: *For all sets A and B, if* $A \cap B = A$, *then* $A \cup B = B$ as follows. **Proof**: *Let A and B be sets. Suppose* $A \cap B = A$. *Let* $x \in A$. *Since* $A = A \cap B$, *then* $x \in B$. A warrant for this step, that the student deemed unnecessary, would have been: *From* $x \in A$ *and* $A = A \cap B$, *one has* $x \in A \cap B$. *Then* $x \in A$ *and* $x \in B$, *so* $x \in B$.

The *Manual for Authors of Mathematical Papers* (MAMP) suggests that authors, "Omit any computation which is routine (i.e., does not depend on unexpected tricks). Merely indicate the starting point, describe the procedure, and state the outcome..." (as quoted by Csiszar 2003, p. 244). Also, the Associate Editor for the *Journal of Geometric Analysis*, suggests, in giving advice to authors, that while most mathematical arguments need to be justified, sometimes "the reason will be so obvious to the reader that it is actually more effective to leave it out." (Lee 2012, p. 3).

Proofs Contain only very Short Overviews or Advance Organizers

For example, one might state at the beginning of a proof by contradiction "Suppose to the contrary that..." However, one would not generally give the overall organization of the proof in advance, such as indicating that there would be a case argument, and that the first case would be proved directly.

This contravenes Leron's (1983) idea of using a hierarchical structure of levels when *presenting* proofs in the classroom. He stated that, "While the age-old and venerable method [of presenting proofs in a step-by-step, 'linear' fashion] may be well suited for securing the validity of proofs, it is nonetheless unsuitable for... presentations."

Entire Definitions (Available Outside the Proof) Are not Quoted in Proofs

For example, it is quite common for beginning undergraduate real analysis students to state the entire definition of continuity within a proof, rather than merely writing, "By definition of continuity, we have..." Also, in a point-set topology proof, one would not usually include standard definitions of compact and connected. This, and other, student proving difficulties related to the genre of proof were noted by Selden and Selden (2011).

Proofs Are "Logically Concrete" in the Sense that Quantifiers, Especially Universal Quantifiers, Are Avoided Where Possible

Students often want to argue for all x within a proof, rather than selecting a fixed, but arbitrary x and arguing about that. It is implicit that, whichever element is selected, the argument can be about that element. Although this is a simple rhetorical device, it is very powerful and it simplifies the logic required of validators, thereby perhaps avoiding some errors.

Even mathematics graduate students can take a very long time to employ, and get comfortable with, this feature of the genre of proof. Mary was a teacher, returning to graduate school in mathematics. She was taking beginning real analysis with Dr. K, who assigned 3 or 4 weekly proofs, graded them very thoroughly, and allowed them to be resubmitted. He emphasized things like writing "Let x be a number" into proofs. She recalled feeling this requirement was not particularly important or appropriate. However, she complied to get full credit. Near the middle of the course, Mary came to feel that this "made sense and it was the way to do it." She reported to us, two years later, that she could not think of any other way to write (this feature of) proofs (Selden et al. 2010).

Discussion and Conclusion

The genre of proofs has developed over considerable time, but not necessarily by conscious intent. This is probably because mathematicians have come to see this genre as having value. Csiszar (2003, p. 268) suggested that this rhetoric contributes to a sense of a "proof's inevitability" and that readers can obtain pleasure from "seeing a proof unfold as it must before one's [their] eyes." In addition, sometimes mathematicians want to get the flow of the argument without wallowing in the details. As the *MAMP* stated, readers want "to see the path—not examine it with a microscope." Finally, we suggest that one reason this genre may have developed is because it makes the validation of proofs as easy, and hence as reliable, as possible. That is, distractions are minimized, thereby maximizing the finding of errors.

On the Heterogeneity of Mathematical Practice with Respect to Proof

Keith Weber

Many mathematics educators contend that the teaching of mathematical proof should be informed by how proof is used by mathematicians. For instance, Herbst and Balacheff (2009) averred that, "since a notion of proof exists in the discipline of mathematics, it might be entitled to exist in classroom activity. And if it were to exist, it would be expected to exist in a form that was accountable to, if not compatible with, how it exists in the discipline" (p. 43). Similarly, Harel (2001) claimed that in his research program, "the goal of instruction must be unambiguous; namely to gradually refine current students' proof schemes toward the proof scheme shared and practiced by the mathematicians of today" (p. 188). Claims such as these are based on the premise that there is significant agreement amongst mathematicians with respect to their professional practice with proof. Indeed, Harel and Sowder (2007) explicitly acknowledged as much, stating that their research program "is based on the premise that a shared scheme exists" among mathematicians. In this contribution, I will report the results of several recent large-scale empirical studies that my colleagues and I have conducted with mathematicians that reveal that mathematicians often behave quite differently in how they seek conviction and read proofs. I will then suggest some consequences for mathematics educators about the goals of proof-related instruction.

Do Mathematicians Agree on what Constitutes a Proof?

Selden and Selden (2003) and others have argued that an important goal of proof-oriented mathematics classes is to teach students how to distinguish between valid and invalid arguments (e.g., Weber 2008). Such a goal would appear to be based on the premise that we can, in a sense, objectively classify an argument as valid or invalid. Indeed, many mathematics educators and mathematicians express exactly this viewpoint, contending that although one might not be able to give precise criteria for what constitutes a proof, mathematicians for the most part agree on whether or not a given argument is valid. Azzouni's (2004) article, for instance, attempted to explain why "mathematicians are so good at agreeing with one another on whether some proof convincingly establishes a theorem" (p. 84). McKnight et al. (2000) asserted that "all agree that something is either a proof or it isn't and what makes it a proof is that every assertion in it is correct" (p. 1). Selden and Selden (2003) remarked on "the unusual degree of agreement about the correctness of arguments and the truth of theorems arising from the validation process" (p. 7); they contended that validity is a function only of the argument and not of the reader: "Mathematicians say that an argument proves a theorem, not that it proves it for Smith and possibly not for Jones" (p. 11).

Inglis et al. (2013) recently conducted a study whose findings challenge the notion of uniform agreement. In this study, 109 mathematicians were shown an argument purporting to prove that $\int \frac{dx}{x} = \ln x + C$ and were told the argument was submitted for publication in the *Mathematics Gazette*, an expository mathematics publication. The argument first established that $\lim_{k \to -1} \int x^k dx = \ln x + C$ and then concluded the theorem. In other words, the argument established the theorem by commuting the limit and integral sign. When asked if the argument was valid, 29 mathematicians (27 %) agreed that it was and 80 (73 %) said it was not, with applied mathematicians significantly more likely than pure mathematicians to judge the argument to be valid (50 % vs. 17 %). To make certain that this disagreement was not due to performance error, perhaps due to a lack of content knowledge, we added the following question after the mathematicians had made their judgments. We first informed participants that a mathematician critiqued this argument by noting that the limit and integral sign were commuted, asked if such a critique was reasonable, and then asked if this alone was enough to render the argument invalid. Participants found this critique to be reasonable—82 % agreed it was with only 9 % saying it was unreasonable with no reliable statistical differences between those who originally judged the argument to be valid or invalid. Of the 29 who judged the argument to be valid, 24 (83 %) claimed this critique was not sufficient to render the argument invalid. Of the 80 who judged the argument to be invalid, only 24 % of those who initially judged the argument to be invalid made this judgment. Hence, most of those who accepted the argument as valid did so despite being aware that the limit and integral were commuted and finding this to be a reasonable critique of the argument. This suggests that mathematicians do not agree about whether particular inferences within an argument are permissible, even in a domain as basic as elementary calculus.

Is Empirical Evidence Convincing?

Suppose a student reads a claim that all integers have a mathematical property (e.g., all integers are not odd perfect squares). This student verifies this claim for a large number of integers and becomes convinced that this claim is true. Is this student's behavior desirable? Is it consistent with mathematical practice? Many in the mathematics community believe the answer to both questions is no. Balacheff (1987), for instance, refers to this type of empirical reasoning as "naïve". However, others disagree, arguing in fact that it is fairly common for mathematicians to gain conviction via empirical reasoning. de Villiers (2004) expressed this as follows: "Contrary to the belief common amongst many mathematics teachers that only proof provides certainty for the mathematician, mathematicians are often convinced by the truth of their results (usually on the basis of quasi-empirical evidence[3]) long before they have proofs" (p. 402). The following study provides evidence that both viewpoints, to some extent, are correct.

Weber (2013) presented 49 mathematicians with an argument that no term in a particular sequence b_n was a perfect square. Without informing them of what the sequence was, the mathematicians were presented with an empirical argument in support of this claim. The argument listed the first 12 terms of the sequence and verified that the odd-numbered terms in the sequence were congruent to 2 (mod 4) and the even terms in the sequence were congruent to 3 (mod 4). They were also told that a computer verified that this trend continued for the first 10,000 terms of the sequence. As no perfect square is congruent to 2 or 3 modulo 4, the argument concluded that no term in the sequence b_n was a perfect square. Mathematicians were asked, on a scale of 0 through 100, how persuasive they found the argument. They were also asked a multiple choice question on how persuasive the argument was, with their choices being: (a) completely persuasive, (b) highly persuasive, as

[3]By quasi-empirical evidence, de Villiers (2004) was including naïve empirical evidence collected with the aid of computers (see p. 398).

persuasive as some proofs that they have read, (c) somewhat persuasive, or (d) not persuasive, meaning their conviction in the statement did not increase as a consequence of reading this argument. Of the 49 mathematicians, 8 (16 %) said they found the argument completely or highly persuasive while 11 (22 %) found the argument not persuasive. This illustrates how some mathematicians might dismiss arguments that others find to be quite persuasive. These 49 participants, as well as 48 participants who were in a control condition for this study, were also asked if they were ever convinced that a claim was correct solely on the basis of empirical evidence. Of these 97 participants, 26 (27 %) answered yes, while the remaining 71 participants (73 %) answered no. In summary, some mathematicians find arguments based on empirical evidence to be highly persuasive while others do not.

Implications for Instruction

If we can make a claim of the form "most mathematicians do *X* with respect to proof", then it would arguably be good policy to set a goal for mathematics instruction for students to do *X* as well. For instance, I believe few mathematicians believe that checking a general claim for a small number of examples constitutes a mathematical proof and consequently students should learn that these types of empirical arguments are not proofs. However, the situation is more complicated if it is the case that "some mathematicians do *X* while others do not". This makes it harder to say whether students should find a particular argument to be valid or whether students should ever find empirical evidence to be convincing. One approach might be to expose students to the different viewpoints that mathematicians have. This goes against the popular view that an answer in mathematics is either correct or it is not, but it would paint a more accurate picture of mathematical practice.

Reflective Summary

Many mathematics educators believe that mathematicians' practice with justification and proof should inform how proof is taught in the mathematics classroom. This presents an opportunity for mathematics educators and mathematicians to collaborate as mathematics educators can explore mathematicians' insights about their practice. The reports in this chapter inform how this might occur. Many students find the genre of proof in mathematics to be perplexing. Selden and Selden's contribution aims to understand what characteristics the genre of proof has. The insights from this chapter can be shared with students, so they can better comprehend the proofs they read and so they have a greater understanding of the types of arguments they are asked to produce. Kidron's contribution tackles the important issue of what can be learned by the process of producing a justification. One way that producing a justification may lead to insight is by having the learner construct connections that are novel to him or her. Kidron documents how this occurs in the work of the professional mathematician and contends it can occur in the work of the student as well. Hanna explores modern trends in the philosophy of mathematics. Throughout her piece, she argues that the role that proof plays for mathematicians might be able to play analogous roles in the mathematical classroom, particularly with regard to understanding. Harel urges the mathematics education community to consider justification more broadly. Typically mathematics instructors treat concepts and their definitions as a starting point for their investigations. Harel challenges this practice, arguing that mathematicians only create concepts to fill an epistemic need as a tool for solving problems that they consider important. He illustrates how such an epistemic need can be created for students as well. Weber's contribution highlights the limits of using mathematicians' practice to inform instruction by demonstrating that mathematicians sometimes disagree on important aspects of justification and proof, including whether mathematicians find empirical evidence to be highly persuasive or if mathematicians can agree on what types of inferences are valid within a proof. If mathematicians disagree on these issues, mathematics educators must use other criteria in deciding what behavior is desirable or normatively correct on the part of their students.

References

Alcock, L., & Weber, K. (2010). Undergraduates' example use in proof construction: purposes and effectiveness. *Investigations in Mathematics Learning*, *3*, 1–22.

Avigad, J. (2006). Mathematical method and proof. *Synthese*, *153*(10), 105–159.

Azzouni, J. (2004). The derivation-indicator view of mathematical practice. *Philosophia Mathematica*, *12*, 81–106.

Azzouni, J. (2009). Why do informal proofs conform to formal norms? *Foundations of Science*, *14*(1), 9–26.

Balacheff, N. (1987). Processus de preuves et situations de validation. *Educational Studies in Mathematics*, *18*(2), 147–176.

Calude, C. S., & Müller, C. (2009). Formal proof: reconciling correctness and understanding. In L. Dixon et al. (Eds.), *LNAI: Vol. 5625. Proceedings Calculemus/MKM* (pp. 217–232). Dordrecht: Springer.

Corfield, D. (2003). *Towards a philosophy of real mathematics*. Cambridge: Cambridge University Press.

Csiszar, A. (2003). Stylizing rigor: or, why mathematicians write so well. *Configurations*, *11*(2), 239–268.

Dawson, J. W. (2006). Why do mathematicians re-prove theorems? *Philosophia Mathematica*, *14*, 269–286.

de Villiers, M. (2004). The role and function of quasi-empirical methods in mathematics. *Canadian Journal of Science, Mathematics and Technology Education*, *4*, 397–418.

Dreyfus, T., & Eisenberg, T. (1986). On the aesthetics of mathematical thoughts. *For the Learning of Mathematics*, *6*(1), 2–10.

Dreyfus, T., & Kidron, I. (2006). Interacting parallel constructions: a solitary learner and the bifurcation diagram. *Recherches En Didactique Des Mathématiques*, *26*(3), 295–336.

Friedman, A., & Medway, P. (1994). Introduction. New views of genre and their implications for education. In A. Friedman & P. Medway (Eds.), *Learning and teaching genre* (pp. 1–22). Portsmouth: Heinemann.

Hales, T. C. (2008). Formal proof. *Notices of the American Mathematical Society*, *11*, 1370–1380.

Hanna, G., & Barbeau, E. (2008). Proofs as bearers of mathematical knowledge. *ZDM. The International Journal on Mathematics Education*, *40*(3), 345–353.
Harel, G. (1998). Two dual assertions: the first on learning and the second on teaching (or vice versa). *The American Mathematical Monthly*, *105*, 497–507.
Harel, G. (2001). The development of mathematical induction as a proof scheme: a model for DNR-based instruction. In S. Campbell & R. Zazkis (Eds.), *Learning and teaching number theory* (pp. 185–212). New Jersey: Ablex.
Harel, G. (2008a). DNR perspective on mathematics curriculum and instruction, part I: focus on proving. *ZDM. Zentralblatt für Didaktik der Mathematik*, *40*, 487–500.
Harel, G. (2008b). DNR perspective on mathematics curriculum and instruction, part II: with reference to teachers' knowledge base. *ZDM. Zentralblatt für Didaktik der Mathematik*, *40*, 893–907.
Harel, G., & Sowder, L. (2007). Towards a comprehensive perspective on proof. In F. Lester (Ed.), *Second handbook of research on mathematical teaching and learning*. Washington: NCTM.
Harrison, J. (2008). Formal proof—theory and practice. *Notices of the American Mathematical Society*, *11*, 1395–1406.
Healy, L., & Hoyles, C. (2000). A study of proof conceptions in algebra. *Journal for Research in Mathematics Education*, *31*, 396–428.
Herbst, P., & Balacheff, N. (2009). Proving and knowing in public: the nature of proof in a classroom. In M. Blanton, D. Stylianou, & E. Knuth (Eds.), *Teaching and learning proof across the grades: a K-16 perspective* (pp. 40–64). New York: Routledge.
Inglis, M., Mejia-Ramos, J. P., Weber, K., & Alcock, L. (2013). On mathematicians' different standards when evaluating elementary proofs. *Topics in Cognitive Science*, *5*, 270–282.
Kidron, I., & Dreyfus, T. (2009). Justification, enlightenment and the explanatory nature of proof. In F.-L. Lin, F.-J. Hsieh, G. Hanna, & M. de Villiers (Eds.), *Proceedings of the ICMI study 19 conference: proof and proving in mathematics education* (Vol. 1, pp. 244–249). Taipei: National Taiwan Normal University.
Kidron, I., & Dreyfus, T. (2010a). Justification enlightenment and combining constructions of knowledge. *Educational Studies in Mathematics*, *74*(1), 75–93.
Kidron, I., & Dreyfus, T. (2010b). Interacting parallel constructions of knowledge in a CAS context. *International Journal of Computers for Mathematical Learning*, *15*(2), 129–149.
Kitcher, P. (1981). Explanatory unification. *Philosophy of Science*, *48*, 507–531.
Knuth, E. (2000). The rebirth of proof in school mathematics in the United States? *International Newsletter on the Teaching and Learning of Mathematical Proof*. http://www.cabri.imag.fr/Preuve/.
Lee, J. L. (2012). Some remarks on writing mathematical proofs. http://www.math.washington.edu/~lee/Writing/writing-proofs.pdf. Accessed 3 June, 2012.
Leron, U. (1983). Structuring mathematical proofs. *The American Mathematical Monthly*, *90*(3), 174–180.
Mancosu, P. (2001). Mathematical explanation: problems and prospects. *Topoi*, *20*, 97–117.
Manin, Y. (1998). Truth, rigour, and common sense. In H. G. Dales & G. Oliveri (Eds.), *Truth in mathematics* (pp. 147–159). Oxford: Oxford University Press.
Marfori, M. A. (2010). Informal proofs and mathematical rigour. *Studia Logica*, *96*, 261–272.
Maric, F., & Neuper, W. (2011). Theorem-prover based systems (TPS) for education (eduTPS). http://sites.dmi.rs/events/2012/CADGME2012/files/working%20group/Theorem-Prover%20based%20Systems.pdf.
Marks, G., & Mousley, J. (1990). Mathematics education and genre: dare we make the process writing mistake again? *Language and Education*, *4*, 117–135.
McKnight, C., Magid, M., Murphy, T. J., & McKnight, M. (2000). *Mathematics education research: a guide for the research mathematician*. Washington: Am. Math. Soc.
Nardi, E., & Iannone, P. (2006). To appear and to be: acquiring the genre speech of university mathematics. In *Proceedings of the 4th conference on European research in mathematics education*, CERME, Sant Feliu de Guixols, Spain (pp. 1800–1810).
Rav, Y. (1999). Why do we prove theorems? *Philosophia Mathematica*, *7*(3), 5–41.

Rota, G.-C. (1997). *Indiscrete thoughts* (pp. 131–135). Boston: Birkhäuser.

Sandborg, D. (1997). *Explanation in mathematical practice*. Unpublished Ph.D. dissertation, University of Pittsburgh.

Schoenfeld, A. H. (1989). Explorations' of students mathematical beliefs and behaviour. *Journal for Research in Mathematics Education, 20*, 338–355.

Schoenfeld, A. H. (1994). What do we know about mathematics curricula? *The Journal of Mathematical Behavior, 13*(1), 55–80.

Selden, A. (2012). Transitions and proof and proving at tertiary level. In M. de Villiers & G. Hanna (Eds.), *Proof and proving in mathematics education—the 19th ICMI study* (pp. 391–420). New York: Springer.

Selden, A., & Selden, J. (2003). Validations of proofs considered as texts: can undergraduates tell whether an argument proves a theorem? *Journal for Research in Mathematics Education, 34*, 4–36.

Selden, A., & Selden, J. (2011). Mathematical and non-mathematical university students' proving difficulties. In L. R. Wiest & T. D. Lamberg (Eds.), *Proceedings of the 33rd annual conference of the North American chapter of the international group for the psychology of mathematics education*. Reno, NV (pp. 675–683).

Selden, A., McKee, K., & Selden, J. (2010). Affect, behavioural schemas and the proving process. *International Journal of Mathematical Education in Science and Technology, 41*(2), 199–215.

Shepherd, M. D., Selden, A., & Selden, J. (2012). University students' reading of their first-year mathematics textbooks. *Mathematical Thinking and Learning, 14*, 226–256.

Steiner, M. (1978). Mathematical explanation. *Philosophical Studies, 34*, 135–151.

Tall, D. (2006). A theory of mathematical growth through embodiment, symbolism and proof. In *Annales de didactique et de sciences cognitives* (Vol. 11, 195–215) Strasbourg: IREM.

Tao, T. (2001). Learn the power of other mathematicians' tools. http://terrytao.wordpress.com/career-advice/learn-the-power-of-other-mathematicians-tools/.

Tao, T. (2010). Buzz-public. https://profiles.google.com/114134834346472219368/buzz/WJkEENg19Sz#114134834346472219368/buzz/WJkEENg19Sz.

Tappenden, J. (2005). Proof style and understanding in mathematics 1: visualization, unification and axiom choice. In P. Mancosu, K. F. Jørgensen, & S. A. Pedersen (Eds.), *Visualization, explanation and reasoning styles in mathematics* (pp. 147–214). Dordrecht: Springer.

Thurston, W. P. (1994). On proof and progress in mathematics. *Bulletin of the American Mathematical Society, 30*(2), 161–177.

Watson, A., & Mason, J. (2005). *Mathematics as a constructive activity: learners generating examples*. Mahwah: Erlbaum.

Weber, K. (2008). How mathematicians determine if an argument is a valid proof. *Journal for Research in Mathematics Education, 39*, 431–459.

Weber, K. (2013). On the sophistication of naïve empirical reasoning: factors influencing mathematicians' persuasion ratings of empirical arguments. *Research in Mathematics Education, 15*, 100–114.

Wiedijk, F. (2008). Formal proof—getting started. *Notices of the American Mathematical Society, 11*, 1408–1414.

Wilkerson-Jerde, M. H., & Wilensky, U. J. (2011). How do mathematicians learn math?: resources and acts for constructing and understanding mathematics. *Educational Studies in Mathematics, 78*, 21–43.

Wu, H. (1996). The role of Euclidean geometry in high school. *The Journal of Mathematical Behavior, 15*, 221–237.

Yandell, B. H. (2002). *The honors class: Hilbert's problems and their solvers*. Natick: AK Peters.

Policy: What Should We Do, and Who Decides?

Chapter 15
Mathematics and Mathematics Education Policy

Mogens Niss

> *"The first and foremost goal of a mathematics teacher is to help students learn mathematics, not to make them feel good about not knowing mathematics" (Ted Eisenberg in a dialogue with Michael Fried, Eisenberg and Fried 2009, p. 145)*

Abstract This chapter explores aspects of the relationship between mathematics education policy and mathematics. It argues that some of the differences of views and opinions encountered on the stages on which mathematics education policy is discussed and debated, are often rooted in very different views of and stances on the nature and essence of mathematics as a discipline and as a subject. After an initial attempt at introducing and clarifying some key concepts used in the chapter, the analysis is supported and illustrated by a number of concrete examples from the writings of influential organisations, mathematicians and mathematics educators who have articulated their positions with regard to mathematics education policy.

Keywords Nature of mathematics · Mathematics education · Mathematical pedagogy · Math wars · Education policy · Policy makers · Policy agents · Justification question

Introduction: The Notion of Policy

This paper deals with the relationship between three entities, "mathematics", "mathematics education", and "policy", of which policy perhaps appears to be "the odd term out", as it typically deals with issues related to decision-making, society, politics, economy, management and administration, whereas the two other may seem relatively familiar and well-defined, at least within a community of mathematicians and mathematics educators. However, in this paper I shall argue that problems with and disagreement about policy in the context of mathematics and, above all, math-

M. Niss (✉)
IMFUFA/NSM, Roskilde University, Roskilde, Denmark
e-mail: mn@ruc.dk

M.N. Fried, T. Dreyfus (eds.), *Mathematics & Mathematics Education: Searching for Common Ground*, Advances in Mathematics Education,
DOI 10.1007/978-94-007-7473-5_15,

ematics education are in large part, but of course not solely, rooted in disagreement about what mathematics and mathematics education are supposed to mean and be.

As it is a key point in this paper that the relationships amongst the three components at issue are markedly influenced by the meaning attached to each of them, it follows that we need to take a closer look of the terms.

Firstly, I propose the following definition of policy:

- A *policy* is a set of measures—decisions and actions—designed and implemented to pursue certain ends and goals that are deemed desirable by those adopting the policy.

Admittedly, this is a very wide definition as it ranges from decisions and actions undertaken by, say, members of a household in order to achieve certain goals, through to measures put in place by a government, or a trans-national body such as the UN or the EU, to pursue some general ends. Oftentimes, however, the term policy is restricted to contexts in which measures are adopted by some powerful body or agency to bring about desired changes, typically termed "reform", of a larger system. In some cases a policy may also be adopted in order to prevent certain changes from happening to a system, e.g. global warming. I submit that a wide definition of policy, such as the one proposed here, has the advantage of conceptual clarity, even though it leaves the issue of distinguishing between different kinds, levels and scopes of actual policies to the specific usage of the term in concrete situations.

The policies that preoccupy us here are predominantly to do with mathematics education at large, i.e. the teaching and learning of mathematics in schools and other educational institutions. At national and international levels there are, indeed, significant policy issues related to the position, maintenance and development of mathematics as a discipline. However such policy issues are not in focus in this paper.

Mathematics education policy is being conducted by *policy makers*, each of whom operates at a variety of different policy levels, ranging from single classrooms in single institutions, over networks of different categories of local, national or international peers, to the insides of agencies, associations or organisations. Policy makers in mathematics education are those who make the final decisions on the overall measures that are to be implemented in the teaching and learning of mathematics. They include

- The teacher or instructor
- The teaching institution
- The teacher educator
- The teacher education institution
- Politico-administrative authorities
- National official bodies
- International organisations (e.g. the OECD, The World Bank).

Some will argue that it is too far-fetched to consider the individual teacher, instructor or teacher educator a policy maker. I agree that when such a person operates in a "normal mode" within the established curricula, practices, rules, and frameworks of his or her institution, he or she is not a policy maker. In contrast, the individual teacher, instructor of teacher educator does in fact function as a policy maker when deliberately working to change his or her own approach to teaching or assessment, or when engaging in discussions with others about changing the local curriculum, or adopting new materials, or when trying to influence policies through his or her involvement in professional associations, committees and suchlike.

Policy levels and policy makers are influenced by a multitude of different *policy agents* who have their own philosophical, ideological or commercial interests and agendas to pursue. Significant policy agents include:

- Textbook authors and publishing houses
- Employers
- Testing agencies
- Teacher organisations
- Mathematics educators (including researchers)
- Mathematicians
- Representatives of other disciplines
- Individual politicians
- Lobby groups
- Media

The demarcation line between a policy maker and a policy agent is a bit blurred and not so easy to draw. For instance, the same individual or agency can have either role depending on the context. Thus a mathematician is a policy agent when pleading in a magazine article for a particular stance as regards the role of proof and proving in school mathematics, but a policy maker when serving on a department committee equipped with the power to decide on a new mathematics programme in the mathematician's university.

Amongst and across mathematics education policy levels, policy makers and policy agents, respectively, there are usually different, and in some cases even conflicting, interests, priorities, and agendas. Not only the ends and goals pursued can be different, the preferred means put forward or adopted to pursue them can be very different as well. In both respects we sometimes encounter substantive disagreement or fights, if not outright battles or even (math) wars, as in the USA, the Netherlands and Portugal (e.g., see Schoenfeld 2004). How may this be explained?

Three Interrelated Explanations

I can think of three possible, interrelated explanations.

The *first*—and most fundamental explanation—is that different policy makers and agents often hold differing, and sometimes incompatible, views of *the nature and essence of mathematics as a discipline*. These differences seem to be rooted in different experiences of what mathematics is (all about), and in different ensuing images and ideas about the role of mathematics in society.

The *second* explanation is to do with differing answers to what is sometimes called the *justification question*: To whom in society should mathematics education be provided, and with which aims?

Answers to this question—and to the ensuing question of what mathematics should then be taught—in turn depend on underlying, explicit or implicit, conceptions of mathematics as a discipline and of its role in society and culture.

The *third* explanation focuses on differing views concerning the (best) *ways to teach and to learn mathematics* with regard to different categories of students.

Such views reflect views of teaching and learning in general but become specific when focusing on the teaching and learning of *mathematics*. Views of and perspectives on what constitute teachers and learners and their respective relationships with mathematics, as well as what constitutes teaching and learning of mathematics, differ across and amongst policy makers and agents and, once again, reflect different perceptions of mathematics as a discipline.

It is now time to substantiate these attempts at explaining possible origins of disagreement and fights over mathematics education policy.

Before doing so, we have to acknowledge a methodological problem, though. Policy makers are people who make decisions, and their perceptions and views typically manifest themselves first and foremost through the very specification and implementation of those decisions, e.g. when, say, a ministry of education in a country introduces a new mathematics curriculum for some segment of the education system, or when a school board endorses some textbook systems while discarding others, or when a private or public assessment agency introduces a new set of assessment modes and instruments in mathematics. Sometimes, but not always, policy makers accompany their decisions with explanations so as to clarify their intentions or to defend themselves against actual or potential criticism. To the extent they also publicise their reasoning, if so typically in rather general terms, this is primarily for official use, and one cannot be sure that the publicised reasons are the real, underlying reasons that drive the policy makers. One also cannot be sure—in case underlying reasons, different form the publicised ones, do exist—that either set of reasons (the underlying ones, the publicised ones, or their union) constitute a consistent set of motivations and justifications. The identification of policy maker's real, underlying perceptions, thinking and reasoning therefore requires thorough, in-depth investigations, combining interpretation of documents with empirical studies, for example by way of questionnaires, interviews and suchlike. Cooper (1985) provides an excellent example of such in-depth investigations. In summary, as policy makers' perceptions and views of mathematics and mathematics education are usually not directly accessible on the surface of things, interpretive investigations—as a matter of fact research proper—are needed to uncover these perceptions and views. There are, however, exceptions to this picture. They occur when policy makers operate directly on the specific advice and recommendations produced by some group of people (typically a committee entrusted with the task of producing advice and making recommendations to policy makers), while officially buying into the explanations and reasoning offered by the group to substantiate its recommendations.

When it comes to policy agents, the situation is different. Policy agents work by articulating themselves as best they can, presenting their concerns, analyses, investigations, and recommendations, while making their underlying reasoning and justification as explicit, transparent and convincing as possible. Some policy agents operate as lobbyists, i.e. they try to influence policy makers under clandestine circumstances away from the public domain, whereas others operate in public or semi-public domains in order to contribute to shaping public opinion and, eventually, the views of the policy makers. This means that policy agents' views and perceptions of mathematics and mathematics education are often explicit and accessible, and subject to scrutiny and debate on the basis of their face value. As policy agents have interests and agendas, some more than others, it is not necessarily the case, however, that the analyses put forward to support conclusions and recommendations encountered in the open domain correspond to the real driving forces behind the conclusions and recommendations.

The Nature and Essence of Mathematics

When attempting to capture and characterise the nature and essence of mathematics, the questions one might ask represent different perspectives:

- What are the purposes and goals that mathematics as a field serves or might serve?
- What are the content and the structure of mathematics—what are its concepts, objects, results and theories?
- What are the methods and processes—the modes of operation—by which mathematics obtains its results?
- What are the means of justification of statements and claims, including results, in mathematics, and what are the bases for these means?
- What are the relationships and links between mathematics and other fields (disciplines and fields of practice)?
- What are the driving forces and mechanisms in the historical development of mathematics?
- What is the socio-cultural role and the sociology of mathematics, i.e. who are the people who make, do and use mathematics for different purposes and in different ways, and in what contexts and under what conditions do they work and operate?

As they stand, all of these questions are phrased as descriptive/analytic questions, focusing on what *is* the case. However, most of them have a dual, normative counterpart, focusing on what *ought to be* the case.

When trying to identify the views of different policy makers and policy agents in mathematics education, it appears that these views differ greatly as to which of the above perspectives are perceived as important for capturing the nature and essence of mathematics.

In order to provide some blood and flesh to the generalities we have stuck to so far, it is time to consider some examples. Let us begin by paying attention to some *policy agents* who have expressed their perceptions and conceptualisations of the nature of mathematics so as to pave the way for inferences pertaining to mathematics education.

In a regular lecture at ICME-10, the 19th International Congress on Mathematical Education, in Copenhagen 2004, Vagn Lundsgaard Hansen, a professor of mathematics at the Danish Technical University with a strong interest and frequent personal involvement in mathematics education at upper secondary and tertiary levels, spoke about what he calls *the dual nature* of mathematics:

> ... maybe today mathematics can best be understood as a framework for studying concrete real-world phenomena in terms of the underlying abstract mathematical models. (Hansen 2008, p. 1)

and

> In mathematical modelling, however, the abstract aspects are inseparably related to concrete aspects like a yin-yang relation where neither aspect can flourish without the other. (op. cit., p. 2)

So, amongst the perspectives listed above, Hansen emphasises the purposes and goals of mathematics with particular regard to its links with other fields, and relates these links with the content, structure, methods and processes of mathematics. He submits that this dual nature of mathematics should occupy a pivotal role in the teaching of mathematics: "Accordingly, the teaching of mathematics ought to include both concrete and abstract mathematics right from the beginning of the educational system" (op. cit., p. 1).

Also another mathematician with a strong interest and activity in mathematics education of a long standing spoke about the nature of mathematics at ICME-10, namely Zbigniew Semadeni of Warsaw University in Poland. The focal perspective of his paper on *the triple nature* of mathematics—Semadeni (2008)—is the content and structure of mathematics. The triple nature of mathematics consists of the following three facets, which are meant to describe "certain features of mathematics as a body of present human knowledge", not "mathematics as an activity" (op. cit., p. 4):

> The *deep idea* of a mathematical object is a well-formed abstract idea which includes the meaning of the object, its properties, its relationships with other objects [...] and its purposes. (op. cit., p. 3)

> *Surface representations* of a mathematical object are signs [...] for this object. (op. cit., p. 1)

and finally

> By a *formal model* of a mathematical object X we understand the counterpart of X in an axiomatic theory (op. cit., p. 4).

In Semadeni's perception "... deep ideas are the most important component in the triad" (op. cit., p. 17), especially because "(m)ost of mathematical reasoning is controlled by deep ideas", whereas "formalized reasoning is restricted to surface representations only" (op. cit., p. 17). Semadeni's distinctions may be seen as having some resemblance with David Tall's "three worlds of mathematics", the conceptual, the proceptual, and the axiomatic worlds (e.g. Tall 2004).

The purpose of Semadeni's paper is to counteract the "... regrettable and widening gulf between the philosophy of research mathematicians (respectively, scientists) and the philosophy of philosophers and educationalists dealing with mathematics (respectively, natural science)" (op. cit., p. 17). More specifically "... the conception of deep ideas may act as a bridge between the Platonist attitude of mathematicians and the constructivist trends among researchers in mathematics education." (op. cit., p. 18).

Speaking about the goal of mathematics education, the well-known mathematician and populariser of mathematics Keith Devlin submits (Devlin 2000, p. 17) that

> ... a major goal should be to create an awareness of the nature of mathematics and the role it plays in contemporary society. [...] An educated citizen should be able to answer the following two questions about mathematics:
>
> - What is mathematics?
> - Where and how is mathematics used?

By asking these questions, Devlin invokes most of the perspectives on mathematics listed above. His own answers to his own questions detail his overarching answer—mathematics is part of human culture—by pointing to the *four faces* mathematics shows to the world (op. cit., p. 1 and pp. 13–18):

1. *Mathematics as computation, formal reasoning, and problem solving*—the familiar face.
2. *Mathematics as a way of knowing*—mathematics is the science of patterns, unifying all the different branches of the subject.
3. *Mathematics as a creative medium*—relations between mathematics and art.
4. *Applications of mathematics*—making the invisible visible.

Devlin summarises his view of mathematics as follows:

> "All of mathematics, however, consists of variations of the same theme: the identification, abstraction, study, and application of *patterns*, using the mental tools of logical reasoning" (op. cit., p. 26).

He explicitly states that mathematics goes beyond quantitative literacy (op. cit., p. 24). Devlin's points outline an epistemology of mathematics, with particular emphasis placed on the purposes, goals, and objects of mathematics as well as on its methods and processes.

In the chapter "Mathematics in Society" (Niss 1994) in the book *Didactics of Mathematics as a Scientific Discipline*, I wrote about *the five-fold nature* of mathematics as

- A *pure science*, focused on creating knowledge and insight into matters that are entirely intra-mathematical
- An *applied science*, focused on creating knowledge and insight, by mathematical means, into matters pertaining to extra-mathematical domains

- A *system of tools for societal practice*, sometimes called "cultural techniques", which do not manifest themselves as mathematics but nevertheless depend heavily on mathematics
- A field for developing and harvesting *aesthetic experiences* and pleasures and, as a derivative of these four aspect of the nature of mathematics:
- A universal *teaching subject*, the world's largest.

The perspectives adopted here are the societal role and sociology of mathematics and the closely related purposes and goals of mathematical activity. The educational point made in my chapter is that mathematics education has to respect and reflect, in a balanced way, this four-plus-one-fold nature of mathematics, that is, not only the first aspect, or only the third aspect, as is sometimes the case. If this point of view is to be taken seriously, the teaching and learning of mathematics has to pay substantive attention to the first four aspects of the nature of mathematics while reflecting on both the commonalities and the differences in mathematics as a teaching subject across cultures and countries.

The U.C. Berkeley mathematician Hung-Hsi Wu has a long record of taking a deep and serious interest in mathematics education in the USA, where he is a prominent voice in public debates. He has given many talks and published many papers in which his views of mathematics and mathematics education have been articulated. One characteristic example—there are many others—is his plenary presentation "What is Mathematics Education?" delivered at the NCTM Annual Meeting in March 2007, with accompanying text (Wu 2007), in which he presents his views in a succinct manner. Here, he defines mathematics education as "mathematical engineering" (op. cit., p. 2), which is neither meant to be a metaphor nor an analogy but is based on his more general definition of "engineering as the customization of abstract scientific principles to satisfy human needs" (op. cit., p. 3). The job of the mathematics educator is to engineer abstract mathematics so as to "meet the needs of students and teachers in the K-12 classrooms" (op. cit., p. 5). The core of this paper is an outline of "*five basic characteristics* of mathematics" (op. cit., p. 9), alias the "unviolable scientific principles in mathematical engineering", i.e. mathematics education (op. cit., p. 8):

Precision: Mathematical statements are clear and unambiguous. At any moment, it is clear what is known and what is not known
Definitions: Bedrock of mathematical structure (no definitions, no mathematics)
Reasoning: Lifeblood of mathematics; core of problem solving
Coherence: Every concept and skill builds on previous knowledge and is part of an unfolding story
Purposefulness: Mathematics is goal-oriented. It solves specific problems

Clearly, the perspectives Wu have adopted here are the ones that deal with the content and structure, and the methods and processes of mathematics, whereas the others are not invoked. At the end of his paper, Wu deplores what he sees as the lack of mathematical engineers (in 2007) (op. cit., p. 18), resulting from the fact that mathematicians generally know (only?, my question) mathematics and educators generally know (only?, my question) education, and concludes "Let there be mathematical engineers" (op. cit., p. 20). It is worth noting that there are also mathematics educators (e.g. Wittmann 1974, 1995) who see mathematics education as an engineering (design) science. This does not imply, however, that they (in this case Wittmann) and Wu would agree on the consequences of these views.

We finish this section by contrasting two perceptions and views of the nature of mathematics. First, Frank B. Allen, emeritus professor of mathematics, a former president of the NCTM, and later national advisor for Mathematically Correct, a policy agency in the USA, gave a talk "Language and the Learning of Mathematics" (Allen 1988) to the NCTM in 1988, at a time when NCTM's well-known Curriculum and Evaluations Standards for Mathematics of 1989 (NCTM 1989) was in the making. Allen, in his paper, objects to the main tenets of the "reform movement" of the 1980's in the USA because they give too much emphasis to problem solving and applications (op. cit., pp. 2–4). The basis of Allen's objection is this: "Indeed, in a very real sense, mathematics is a language." (op. cit., p. 3) and "Mathematics is essentially a structured hierarchy of propositions forged by logic on a postulational base." (op. cit., p. 4).

Evidently, Allen's perspective on mathematics concentrates on its content and structure and the method (proof) involved in creating this structure.

Renuka Vithal, of the University of KwaZulu Natal in South Africa was one of the four panellists in the plenary panel at ICME-10, 2004, devoted to the theme "Mathematics for whom and why? The balance between mathematics education for all and for high level mathematics performance". In her contribution "A battle for the soul of the mathematics curriculum" (Vithal 2008), Vithal asks "Who decides what counts as mathematics?" (op. cit., p. 73), and states:

> ... what counts as mathematics has shifted and opened. Drawing on a broad range of disciplines, scholarship in areas such as ethnomathematics and critical mathematics education has forced a recognition of a much broader set of practices, knowledge and skills as mathematics. By holding on to a narrow definition of mathematics not only do many get excluded, pursuing a limited meaning fails to prepare the diversity of learners for life in an increasingly technological but unequal and unjust local and global world. (op. cit., p. 73)

The perspectives adopted by Vithal deal primarily with the purposes, goals, and processes of mathematics, and partly with its content and structure as well. She explicitly challenges classical perceptions of mathematics, such as Allen's, by insisting on including a notion such as ethnomathematics under the aegis of mathematics.

The examples offered above presumably suffice to allow for the conclusion that policy agents do indeed hold widely different, and sometimes conflicting views and perceptions of the nature and essence of mathematics.

Do we get a different picture if we consider *policy makers*? We have to keep in mind that many categories of policy makers remain silent about their own views, whilst often manifesting their perceptions of the nature of mathematics indirectly in terms of the nature that students are supposed to be exposed to and to experience. The policy makers we are now going to listen to are agencies or official committees which have, or have been given, decision-making power over highly influential proposals or plans on behalf of larger groups or communities, without in and of themselves being official authorities. One might argue that they could as well have been labelled policy agents, but as they act as representatives of larger collectives I found that they belong in the policy maker category.

The NCTM in the USA figures prominently here, because of the highly influential—but at the same time, in some quarters, somewhat controversial—*Curriculum and Evaluation Standards for School Mathematics*, 1989, and the more recent *Principles and Standards for School Mathematics*, 2000.

The 1989 Standards (NCTM 1989) identifies *three features* of mathematics:

> Three features of mathematics are embedded in the Standards: First, "knowing" mathematics is "doing" mathematics.
> [...]
> Second, [...] the curriculum for all students must provide opportunities to develop an understanding of mathematical models, structures and simulations applicable to many disciplines.
> [...]
> Third, [...] The new technology [...] has changed the very nature of the problems important to mathematics and the methods mathematicians use to investigate them". (op. cit., pp. 7–8)

The perspectives in play in this quotation are on the purposes and goals, as well as on the methods and processes, of mathematics, but other fields enter the game as well. In contrast, the main perspective adopted in the 2000 Standards (NCTM 2000) is that of purposes and goals of mathematics and mathematics education. The publication identifies (op. cit., p. 4) *four needs for mathematics* in a changing world:

- Mathematics for life
- Mathematics as a part of cultural heritage
- Mathematics for the workplace
- Mathematics for the scientific and technical community.

The Danish *KOM Project* (KOM: Competencies and the Learning of Mathematics), of which I had the privilege of being the director, worked on behalf of the Danish Ministry of Education to come up with analyses and proposals for a rather thorough makeover of mathematics education in Denmark, at all educational levels. The project, the official report of which was published in 2002 (Niss and Jensen 2002), and its various ramifications have proved pretty influential beyond the borders of Denmark, e.g. in the frameworks of PISA and in a number of countries. An English translation of the general parts of the original report was published in 2011 (Niss and Højgaard 2011). The perspective adopted in the KOM Project is predominantly that of the methods and processes of mathematics but also that of its purposes and goals. This is reflected in the following (updated) definition of mathematical competence and mathematical competencies:

Possessing *mathematical competence*—i.e. mastering mathematics—is an individual's capability and readiness to act appropriately, and in a knowledge-based manner, in situations and contexts in which mathematics actually plays or potentially could play a role.

A mathematical *competency* is a distinct major constituent in mathematical competence. The KOM Project identifies eight such mathematical competencies: Mathematical thinking; Mathematical problem handling; Mathematical modelling; Mathematical reasoning; Representing mathematically; Handling mathematical symbolism and formalism; Communicating mathematically; Dealing with (physical) aids and tools for mathematical activity. For details, see Niss and Højgaard (2011, pp. 52–68).

The mathematical competencies are meant to pertain to any educational level but do of course play out very differently at different levels. So, the relationship between mathematical competencies and mathematical content should be perceived as a two-dimensional one, in which the competencies are "orthogonal" to the mathematical content domains.

Whilst the mathematical competencies are activated in dealing with situations presenting mathematical challenges, the KOM Project also identifies *three kinds over overview and judgment* regarding mathematics as a discipline. These are (op. cit., pp. 73–75):

- The actual application of mathematics in other subject and practice areas
- The historical development of mathematics, both internally and from a societal point of view
- The nature of mathematics as a subject.

Evidently, these items correspond closely to the last three of the perspectives mentioned in the beginning of this section.

The *Common Core State Standards for Mathematics* initiative in the USA offers states a common platform to join in on in an attempt to create a sort of a national curriculum. In addition to being preoccupied with describing mathematical content at grade levels K-12 (Common Core State Standards Initiative 2011), the CCSSM identifies eight *Standards for Mathematical Practice* (op. cit., p. 6) which have several similarities with the mathematical competencies of the KOM Project:

1. Make sense of problems and persevere in solving them
2. Reason abstractly and quantitatively
3. Construct viable arguments and critique the reasoning of others
4. Model with mathematics
5. Use appropriate tools strategically
6. Attend to precision
7. Look for and make use of structure
8. Look for and express regularity in repeated reasoning

In this part of the CCSSM, the perspective adopted is on the methods and processes of mathematics, whereas the remaining parts have a content and structure focus.

Based on the voices present in the material presented here, we are now in a position to conclude that as far as these voices of policy agents and policy makers of today are concerned, it is possible to identify marked differences in the priorities and relative emphases expressed by these voices with regard to: the purposes and goals of mathematics and mathematical activity; the content and structure of mathematics; its methods and processes; justification of mathematical claims and results; the relationship between mathematics and other fields; and the development, societal role and sociology of mathematics. These differences, which sometimes take the form of contradictions, regarding the perceptions and views of the nature and essence of mathematics are likely to be (co-)responsible for the marked disagreement that emerges from time to time amongst policy agents and policy makers. In this paper I have not considered historical cases, such as the introduction of the so-called "new math" or "modern mathematics" in the 1960's and 1970's. However, there is little doubt that the massive controversies of those days about this movement were also rooted in very different views about the nature and essence of mathematics.

The Justification Question

There is an infinitude of responses to the question "To whom should mathematics education be provided, and why?" and to the follow-up question "What should they, hence, be taught?". Let two opposite voices of policy agents suffice to illustrate the controversies one can encounter in the spectrum of answers to these questions.

In 2011 two US mathematicians, Solomon Garfunkel, the founder and director of the Consortium for Mathematics and Its Applications (COMAP) since its establishment early in the 1980's and a protagonist in the advancement of mathematical applications and modelling in mathematics education at all levels, and David Mumford, a Fields medallist (1974) and a renowned pure and applied research mathematician of Brown University, wrote an opinion piece in the New York Times, (24th August), backed up by later unpublished elaborations (Garfunkel and Mumford 2011a; 2011b; 2011c). Their answers to the "for whom?" and "why?" questions and partly to the "how?" question are as follows:

> For some in the higher education community, mathematics education means the education of mathematicians—the replenishing of the species. (Garfunkel and Mumford 2011b, p. 7)

> ... we need a system of mathematics education that seeks first and foremost to recognize the mathematical needs of the average citizens... (Garfunkel and Mumford 2011c)

> Science and math were originally discovered together, and they are best learned together now. (Garfunkel and Mumford 2011b)

> It is through real-life applications that mathematics has emerged in the past, has flourished for centuries and connects to our culture now. (Garfunkel and Mumford 2011a)

> A math curriculum that focused on real-life problems would still expose students to the abstract tools of mathematics [...] (Garfunkel and Mumford 2011a)

> ... what we need is 'quantitative literacy' [...] and 'mathematical modeling'... (Garfunkel and Mumford 2011a)

Garfunkel and Mumford (2011b, p. 3) make a plea for the recognition of two very different student populations, the 1 % who need sophisticated, high level mathematics, and "the remaining 99 % whom we hope will use math as a tool that helps them deal with the problems of daily life". As the latter group needs a completely different diet to that of the former group, the one-size-fits all idea should be abandoned. In other words, Garfunkel and Mumford seriously question the appropriateness of a unified mathematics education in school.

The voices we've just heard are as far way as we can imagine from another US voice, when it comes to the issues of "what?" and "how?". On the homepage of the group *Mathematically Correct* the opening statements (Mathematically Correct 2012) read as follows:

> ...our children have less and less exposure to rigorous, content-rich mathematics.
> The advocates of the new, fuzzy math have practiced their rhetoric well. They speak of higher-order thinking, conceptual understanding and solving problems, but they neglect the systematic mastery of the fundamental building blocks necessary for success in any of these areas. Their focus is on things like calculators, blocks, guesswork, and group activities and they shun things like algorithms and repeated practice. The new programs are shy on fundamentals and they also lack the mathematical depth and rigor that promotes greater achievement.

Whether intended or not, when making their points regarding the "for whom?", "why?", and "what?" questions of mathematics education, the policy agents just quoted invoke certain (in fact opposite) perceptions of the nature and essence of mathematics as a subject and a discipline. Even though the perceptions of these policy agents are very explicit, it would have been easy to quote numerous other policy agents with similar views, albeit phrased less pointedly.

Mathematical Pedagogy

This section touches upon views of the best/right ways to teach and learn mathematics. In view of the fact that most of mathematics education research and development deal with these issues either directly or indirectly, it is not possible to go into any detail with them here. However, here, too, views differ along several dimensions, and, again, many of differences are closely linked to (but are not completely determined by) different perceptions of mathematics as a discipline. Below, some significant dimensions of mathematical pedagogy are depicted in bipolar terms:

- A hierarchically ordered syllabus defined solely in content terms vs. a curriculum constituted by loosely connected topics
- Examples first—theory later, or vice versa
- Meaning, sense making and understanding vs. rote learning, memorisation and skill drill
- Focus on concepts vs. focus on procedures and algorithms
- Exploration, inquiry and discovery learning vs. receiving information from teacher and textbook
- Open-ended problem solving vs. solving of closed, prototypical exercises
- Rigorous proof vs. plausible reasoning
- Collaborative activities for learning vs. individual activities for learning
- Technology as add-on tools vs. technology as a system of mediating artefacts
- Summative testing vs. formative assessment
- Textbook centred teaching and learning vs. teaching and learning by way of alternative materials

These and several other contrast pairs constitute key policy issues to all categories of policy makers and agents. In debates, these pairs are seen by many as serious dichotomies that call for a definite resolution in a pro-or-con stance. Personally I don't think they are dichotomous. But any stance taken on any of them reflects views and perceptions of the nature and essence of mathematics and mathematical activity.

Conclusion

I believe to have shown above, by way of existence proofs, that for some actually existing policy agents and policy makers, the positions they take on key issues in mathematics education policy are deeply linked to views and perceptions of mathematics as a discipline, in and of itself and in its relations to the society and to the world in general. It would be unreasonable to claim that what I have found above covers all significant issues in mathematics education policy. Likewise, it would be unreasonable—in fact wrong—to claim that views and perceptions of mathematics as a discipline solely determine the policy makers' and agents' policies. It requires comprehensive empirical research and thorough theoretical analyses to reveal the extent to which this is actually the case.

The finding that views and perceptions of mathematics are co-determinants of mathematics education policies is not surprising in itself. What may be surprising is that these views and perceptions are not only different, and based on different priorities and emphases, but that they are held, invoked, and fought for or against as if they were all deeply and fundamentally contradictory and incompatible. Some of them probably are: It is, of course, not possible at the same time to hold the view that mathematics education should be completely devoid of anything that is to do with extra-mathematical applications and modelling and the view that mathematics education should primarily deal with applications and modelling. What is possible, however, is to insist that mathematics education for all should put emphasis on serious, extra-mathematical applications and modelling and at the same time provide solid mathematical underpinning of everything that happens in the mathematics classroom. However, people (such as myself!) who hold this view are obliged to clarify how these two components should be balanced in concrete context marked by limited time and resources.

In many cases, since we cannot get everything we want in mathematics education at the same time, disagreement over mathematics education need not be based on antagonistic views and perceptions. Instead they may be explained by different emphases and different priorities concerning different valid facets of our beloved discipline. Perhaps, at the end of the day, disagreement on the allocation of time to different aspects and activities of mathematics in the presence of constraints is the key to many fights over mathematics education policy. But this, too, goes back to the nature of essence of mathematics as a discipline.

When witnessing policy debates one cannot avoid thinking that sometimes mathematics education policy is being conducted to foster and favour certain views and perceptions of the nature of mathematics as a discipline, and that sometimes the nature of mathematics as a discipline is being invoked as a means to foster and favour certain pre-determined mathematics education policies, if not politics at large. I, for one, tend to prefer the former over the latter.

References

Allen, F. B. (1988). *Language and the learning of mathematics*. A speech delivered at the NCTM annual meeting, Chicago, April 1988. http://www.mathematicallycorrect.com/allen4.htm. Accessed 31 Jan 2013.

Common Core State Standards Initiative (2011). Common Core State Standards for mathematics. http://www.corestandards.org/the-standards/mathematics. Accessed 25 Apr 2012.

Cooper, B. (1985). *Renegotiating secondary school mathematics: a study of curriculum change and stability*. London: Falmer Press.

Devlin, K. (2000). The four faces of mathematics. In *NCTM: learning mathematics for a new century. The 2000 yearbook* (pp. 16–27). Reston: National Council of Teachers of Mathematics.

Eisenberg, T., & Fried, M. (2009). Dialogue on mathematics education: two points of view on the state of the art. *ZDM. The International Journal on Mathematics Education*, *41*, 143–149.

Garfunkel, S., & Mumford, D. (2011a). How to fix our math education. *The New York Times*, August 24.

Garfunkel, S., & Mumford, D. (2011b). *What's really wrong with mathematics education—and how to fix it*. Unpublished manuscript, personal communication from S. Garfunkel.

Garfunkel, S., & Mumford, D. (2011c). Bottom line on mathematics education. *The Huffington Post*, November 30.

Hansen, V. L. (2008). The dual nature of mathematics. In M. Niss (Ed.), *Proceedings of the 10th international Congress on mathematical education*, July 4–11, 2004. Roskilde: Roskilde University Press. The CD-ROM section.

Mathematically Correct (2012). http://www.mathematically.correct. Accessed 24 Apr 2012.

NCTM (1989). *Curriculum and evaluation standards for school mathematics*. Reston: National Council of Teachers of Mathematics.

NCTM (2000). *Principles and standards for school mathematics*. Reston: National Council of Teachers of Mathematics.

Niss, M. (1994). Mathematics in society. In R. Biehler, R. W. Scholz, R. Strässer, & B. Winkelmann (Eds.), *Didactics of mathematics as a scientific discipline* (pp. 367–378). Dordrecht: Kluwer Academic.

Niss, M. & Jensen, T. H. (Eds.) (2002). *Uddannelsesstyrelsens temahæfteserie: Vol. 18. Kompetencer og Matematiklæring. Ideer og inspiration til udvikling af matematikundervisning i Danmark*. Copenhagen: The Ministry of Education.

Niss, M. & Højgaard, T. (Eds.) (2011). *Competencies and mathematical learning. Ideas and inspiration for the development of mathematics teaching and learning in Denmark*. Roskilde: Roskilde University Press. English edition. IMFUFA-tekst nr. 485.

Schoenfeld, A. (2004). The math wars. *Educational Policy*, *18*, 253–286.

Semadeni, Z. (2008). The triple nature of mathematics: deep ideas, surface representation, formal models. In M. Niss (Ed.), *Proceedings of the 10th international Congress on mathematical education*, July 4–11, 2004. Roskilde: Roskilde University Press. The CD-ROM section.

Tall, D. (2004). Building theories: the three worlds of mathematics: a comment on Inglis 23(3). *For the Learning of Mathematics*, *24*(1), 29–32.

Vithal, R. (2008). A battle for the soul of the mathematics curriculum. A section of plenary panel debate: mathematics for whom and why? The balance between mathematics education for all and for high level mathematics performance. In M. Niss (Ed.), *Proceedings of the 10th international Congress on mathematical education*, July 4–11, 2004 (pp. 56–75). Roskilde: Roskilde University Press.

Wittmann, E. (1974). Didaktik der Mathematik als Ingenieurwissenschaft. *ZDM. Zentralblatt für Didaktik der Mathematik*, *3*, 119–121.

Wittmann, W. (1995). Mathematics education as a design science. *Educational Studies in Mathematics*, *29*, 355–374.

Wu, H.-H. (2007). *What is Mathematics Education*? Text of a plenary presentation at the 2007 NCTM annual meeting, Atlanta, March 23, 2007 (pp. 1–22). http://math.berkeley.edu/~wu/C49-1.pdf. Accessed 31 January 2013.

Chapter 16
Reflections on Policy

What Should We Do, and Who Decides?

Nitsa Movshovitz-Hadar

Abstract Although mathematics is essential to mathematics-education, and mathematics-education is essential to mathematics, these claims do NOT imply that mathematics and mathematics-education are the same. Actually, they are gradually growing apart. This chapter summarizes the views of its authors on the relationship between the mathematics and mathematics-education communities with respect to policy issues believed to be important to both communities.

One argues that the professional object for mathematics teachers should be viewed as the teaching and learning of mathematics rather than mathematics in itself. Knowledge and experiences from mathematics as a discipline is necessary but not sufficient to form sustainable policy. Hence policy should benefit from being informed by mathematics-education research to a larger extend that currently.

Another view states that instructional policy is only as good as its translation to classroom practice. Without appropriate support for teachers to make the significant changes in classroom instruction being asked of them, curricular initiatives are bound to fail. Mathematicians and mathematics educators can and should collaborate to provide support to teachers in implementation of good mathematics teaching.

Yet another claim is that unlike mathematics, mathematics-education is an applied social science, and therefore research in it should be judged to a large extent, by the successful implementation of its outcome.

Last but not least is a view of mathematics and mathematics-education as two quite different areas of study, attributing many of the disputes that have arisen between mathematicians and mathematics educators with regard to what school mathematics should be, to these differences.

In conclusion, it seems necessary for the mathematics-education community and the mathematic community at large, to join forces and formulate a core of common

With contributions by
Jonas Emanuelsson, University of Gothenburg, Gothenburg, Sweden
Davida Fischman, California State University, San Bernardino, USA
Azriel Levy, Hebrew University, Jerusalem, Israel
Zalman Usiskin, Emeritus, The University of Chicago, Chicago, IL, USA

N. Movshovitz-Hadar (✉)
Dep. of Education in Science and Technology, Technion IIT, Haifa 32000, Israel
e-mail: nitsa@technion.ac.il

M.N. Fried, T. Dreyfus (eds.), *Mathematics & Mathematics Education: Searching for Common Ground*, Advances in Mathematics Education,
DOI 10.1007/978-94-007-7473-5_16,

agreements, upon which decision makers can be held accountable. Indeed, a difficult task, however without it there seem to be no hope for progress in the desired commonly agreed goal to improve the outcome of mathematics-education.

Keywords Accountability · Classroom instruction · Curriculum · Curriculum change · Experimental school teaching · Mathematics-education · Mathematics-education research · Mathematics-education practice · Pedagogical content knowledge · Policy · Policy issues in mathematics-education · Relationship between mathematics and mathematics-education · Teacher knowledge · Teacher education

Mathematics and Mathematics-Education Policy—Searching for Common Ground

Nitsa Movshovitz-Hadar

About the Relationship Between Mathematics and Mathematics-Education

The American Mathematical Society defined mathematics *implicitly* by its Mathematics Subject Classification index (MSC 2010). Mathematics-education appears as subject no. 97. But mathematics-education is not just a formal part of mathematics. Mathematics-education has been playing a central role in the development of mathematics since antiquity, and clearly, there is no future to mathematics without mathematics-education. Hence, one cannot but remain amazed at the small, almost negligible amount of time and effort the majority of contemporary research mathematicians invest in mathematics-education per se.

Not surprisingly, but nevertheless somewhat paradoxically, the MSC index (ibid.) includes mathematics sub-entries under "area 97: Mathematics-education." But Mathematics is not just 'by definition' at the heart of mathematics-education. Obviously, mathematics plays a central role in mathematics-education, and clearly, there is no future to mathematics-education without mathematics. Hence, one cannot but remain amazed at the small, almost negligible amount of time and effort the majority of contemporary research mathematics educators invest in mathematics per se.

Unlike the set theoretical implication $A \subseteq B$ and $B \subseteq A \Rightarrow A = B$, although mathematics is essential to mathematics-education, and mathematics-education is essential to mathematics, these claims do *NOT* imply that mathematics and math education are the same. Actually, they are gradually growing apart!

A similar paradoxical relationship between education and mathematics-education can easily be formed. Is it because as yet we have not defined clearly what mathematics-education is all about? Or, at least its policy?

Let us examine the relationship through a few analogies, which should be considered 'with a grain of salt'.

In a way—

- Mathematics-education (abbr. ME) is to Mathematics, as Sentential Logic (abbr. SL) is to Mathematics. Both are Meta-mathematics. SL is about the language of, and about the nature of deduction in Mathematics. ME is about the comprehension of, and about the nature of didactics in Mathematics.
- Mathematics-education is to Mathematics, as conducting a concert is to composing the music. Both ME and conducting a concert are not a free creation. Both are subject to human and real-life constraints.
- Mathematics-education is to Mathematics and behavioral sciences, as Architecture is to Mathematics and natural sciences. Their designs and constructs are interdisciplinary.
- Lastly, Mathematics-education research is to Mathematics-education practice, as Medical research is to Medical practice. Both of these research domains identify basic symptoms, and run empirical studies to develop and test innovative remedies for common cognitive/bodily diseases. Their results are implemented by practitioners for the benefit of their target populations.

Anyhow, mathematics and mathematics-education are definitely *NOT* mutually exclusive disciplines. They are inseparable, and in many ways are also complementary.

About Policy Issues in Mathematics Education

As commonly understood, policy refers to the collection of governing principles and plans instituted in order to operate some system, or a group of individuals, or sometime even one person.[1] Unlike a law, which prohibits or enforces certain behavior, some policies merely guide and help decision making about actions that are most likely to achieve a desired outcome (e.g. to increase the visibility of mathematics as a field of study[2]) or avoid some negative effect that has been noticed (e.g. declining enrollment in mathematics programs (see footnote 2)). A policy can be considered as a "Statement of Intent", not necessarily including implementation procedures.[3]

How likely is it to reach a consensus about policy among mathematicians and mathematics-educators, beyond the obvious intent: 'to improve the outcome of mathematics-education'? A major concensus-seeking process was carried out by NCTM in preparing its policy book on mathematics-education (2000): *Principles*

[1]Note that Mogens Niss, in his Chap. 15 in this book, defines policy as something much more practical: decisions and actions, rather than the principles on which these decisions and actions are based.

[2]Example taken from The Joint Policy Board for Mathematics (JPBM). http://www.mathaware.org/about.jpbm.html. Accessed 13 June 2013.

[3]Note, once again that this is less in accord with Mogens Niss' definition that appears in Chap. 15 of this book.

and Standards for School Mathematics. Regretfully, one policy principle was left out: a commitment to periodic review and continuous discussion. (This is not saying that NCTM has no intent to do it. Rather the opposite is implied by the fact that NCTM published in the past three related important documents[4]).

Clearly, mathematics itself is growing very rapidly and it permeates almost all walks of life. Which policies could/should be adopted in order to adequately prepare future mathematicians and scientists for the explosion of mathematics knowledge?

And not less important—What policies are appropriate for the education of the majority whose career may not be directly related to mathematics, but will be greatly affected by it?

Mathematics has become multifaceted to the extent that two mathematicians specializing in two different areas find it difficult to communicate. Who, then, can be approached for a comprehensive view of contemporary mathematics? Or, for an overview of its essence? Who, then, is there to be able to tell: What is important to teach at the pre-university level? And why is it important to teach it? Obviously, only after the policy is clear, can one deal with the issues of how to do it, and who can do it. How else can we relate to questions such as: Why and what for are we teaching the Pythagorean Theorem? Why are we not teaching at least 1 % of its proofs? Nor its generalization obtained by replacing the squares on each side by any similar polygons or semi circles? Nor do we clearly explicate that the cosine theorem is actually its extension to triangles that have no right angle?

And why don't we expose school students to the fact that 2 is the only power for which there is a solution (in fact infinitely many ones) to the Diophantine equation $a^n + b^n = c^n$, which makes the Pythagorean Theorem unique indeed?

How else can we justify why Euler's Polyhedron Theorem $V + F = E + 2$ is absent from most school curricula? And the infinitude of primes, or the theorem about 2-coloring of every map that is formed by intersecting circles?

Are these less accessible? Less important than the solution of quadratic equations? Or are they not sufficiently beautiful? Less powerful in terms of applications???

Setting criteria for sifting priorities cannot be an informed process without commonly agreed upon 'foundations' of mathematics-education, and without a periodical consensus-seeking discussion.

We witness long and hot debates about including this topic or omitting another one from the curriculum, employing this teaching strategy or avoiding another one, but they seem to lead nowhere, as there has not been enough investment in laying out the basic assumptions, the 'axiomatic system' of mathematics-education.

---

[4]The three prior publications by NCTM are:

- Curriculum and Evaluation Standards for School Mathematics (1989), which outlined what students should learn and how to measure the outcomes.
- Professional Standards for Teaching Mathematics (1991), which includes best practices for teaching mathematics.
- Assessment Standards for School Mathematics (1995), which focused on employing accurate assessment methods.

Isn't this the only area in the MSC2010 index (ibid.) that so far no attempt has been made to axiomatize? Nevertheless, there have been attempts at forming some mathematics-education statements as meta-mathematical theorems (e.g.: Every mathematics theorem is a boxful of surprises, Movshovitz-Hadar 1988, p. 39; 1993, p. 267).

It seems necessary for the Mathematics-education community and the mathematic community at large, to join forces and formulate a core of common agreements, upon which decision makers can be held accountable. Indeed, a difficult task, however without it there seem to be no hope for progress in the desired commonly agreed goal to improve the outcome of mathematics-education.

Let's not take too seriously the analogy between a discussion of mathematics-education policy and the idea of axiomatizing it as a mathematics sub-area. Instituting a policy is usually based upon accumulated experience, beliefs and research-based evidence. A concensus-seeking process takes compromise. We ought to be ready to stick to it for a while, but we also have to be prepared to reconsider it in view of recent developments—not only in mathematics itself, but also in the technology that becomes available, and in developments in other related areas that mathematics-education is leaning upon.

Let me conclude this part by two open questions:

Q1. Narrowing the gap between school mathematics and contemporary mathematics—Is it a non-realizable dream, or can some curriculum-policy encourage it?
Q2. Alongside the teaching and learning of mathematics, should mathematics serve also as a vehicle for human-values education? Or are these two 'orthogonal'?

On the Professional Object of Teachers

Jonas Emanuelsson

I have participated in educational research with a special interest in the teaching and learning in and about mathematics for a more than the past decade. During last years I have mainly done classroom studies of learning in an international context. I am the head of a department specialized in education rather than mathematics. This is an environment where education and training of becoming teachers, many of them mathematics teachers, heavily influences the daily life at the department. How to bridge the gap between mathematicians and mathematics educators in the context of teacher education is on the agenda almost every day.

One general point I want to emphasize here is that the professional object of mathematics teacher is better viewed as the *teaching and learning of mathematics* rather than viewed as mathematics in itself. In their teaching teachers should be oriented towards how learners respond to the mathematics taught instead of towards the mathematical content in itself. The content in an educator's mind while teaching should be mathematics as understood and handled by their students. Both working teachers and pre-service teachers should, in my view, acquire sensitivity to discern the learners' understanding of mathematics. An integral part of policy in relation to mathematics education should consequently be directed towards the support of building knowledge in and about the learners' perspective on mathematics. Formulated in this vain policy have a possibility to become more helpful in answering questions on what teachers should know and hence what should be taught in teacher education programs.

At Gothenburg University prospective teachers study mathematics in one part of the organization (faculty of Science) and mathematics-education in another (faculty of Education). Chalmers University of Technology is the host for another teacher education program. At both universities, the mathematicians teaching mathematics in the teacher education programs and the mathematics educators teaching in the same programs do not always have constructive dialogues. There are elements of struggle for students and resources rather than a constructive collaboration aiming at catering for the best quality in teacher education. Many mathematicians tend to see the teaching of mathematics as something you learn in the course of teaching rather than something that can be informed by mathematics-education research or by extensive and documented experience from teaching in schools. On the other hand mathematics educators often see mathematicians as only interested in those students who are of "the right stuff" to become research students in mathematics and hence, in their view, tend to neglect a major part of the students.

Teacher education at the two universities of Gothenburg broadly follows three different paradigms or approaches with respect to teaching. These approaches are not mutually exclusive. Instead they usually exist side-by-side often within the same program and sometimes even in the same course. These paradigms can be summarized as follows:

- Teaching through knowing mathematics more solidly. The capability to teach improves with mere practice
- Learning to teach by imitating exemplary mathematics teaching
- Learning to teach by drawing upon research in mathematics-education

These paradigms place mathematics in very different roles. In the first paradigm mathematical knowledge is placed at the centre. Knowledge about teaching and the students learning falls to the background and is diminished to something that is personal and learned through practice. The second approach tends to underemphasize the systematic knowledge of both mathematics and mathematics-education research. If not balanced with solid knowledge in mathematics as a discipline the third approach runs the risk of being mathematically shallow. One could ask if this oversimplified description of different types of approaches illustrates the often argued gaps between mathematicians, mathematics-education researchers and mathematics educators?

There is considerable agreement that an effort is needed to bridge the gap between mathematicians, mathematics educators and researchers in mathematics-education as to how the nature of mathematics and mathematics-education are perceived. The bringing of people together in a joint knowledge building process is foolish to argue against, and we should strive to bring it about in our respective communities. However I would like the different communities scrutinize what we possibly can gain from such bridging and how we should proceed to accomplish that. These are important and I think fair questions.

Mathematics-education, in my understanding, is best viewed as a hybrid science with a strong foundation in social science but with connections to mathematics both as a discipline and a school topic. The objects of study are human behavior, reasoning, problem solving while keeping a sharp focus on mathematical content as perceived, handled and treated by humans and organizations. We use methods and theories, sometimes fine-tuned to fit our specific purposes but borrowed from other social sciences rather than from mathematics in our efforts in understanding education, teaching, instruction, learning in and about mathematics. Our objects of research hence come from the human, cultural, social and psychological world. They do not belong to the world of mathematics. Mathematics-education is hence hard to conceptualize as an applied form of mathematics. Deep knowledge of mathematics is of course a necessary but not sufficient condition to practice mathematics-education research (or to teach it).

When forming policy about educational issues such as curriculum development, teacher education (both content and form), I argue that policy should be informed by research and practice in mathematics-education to a larger extent than presently.

To sum up I want to pose a series of questions that address issues raised above and in the paper by Mogens Niss in Chap. 15 this volume (the enumeration of open questions is continual throughout this chapter).

Q3. In the discussion on policy e.g. teacher education policy we often state that results and experiences from both research in mathematics and research in mathematics-education are much needed. Furthermore we often argue that policy can benefit from bridging the gap between these two fields. What more precisely can we expect, or hope to, gain from such bridging?

Q4. On what basis can we make well informed decisions on what to include in school curriculum and curriculum for (mathematics) teacher education? At the lowest level and at a minimum I argue that we would like policy makers, curriculum developers, teachers and becoming teachers to know something stable and systematic about:

(a) Different ways of teaching mathematics and the corresponding learning that might be occurring (in relation to different aspects of mathematics, in different settings, during different contextualization, with kids of different backgrounds and experiences and so on).
(b) The targeted age groups ways of using mathematics in everyday situations (also outside schools and other institutions).
(c) Adults ways of using mathematics in future life (everyday life, in further academic studies and in different professions).
(d) Ways of using mathematics in other school subjects.
(e) Ways of using mathematics in other scientific disciplines.

(f) We also need to know mathematics in terms of its historical development, use, nature, structure, affordances and constrains, as well as facts and procedures.

I am confident that mathematicians can contribute in answering these questions and that discussion on a shared arena can improve the answers.

I want to know, hear, and see more contributions from mathematicians. The idea and effort of building relations between education (as a practice and as a university discipline) and the mathematics discipline gave rise to mathematics-education. Research in mathematics-education is now developing as a discipline on its own with its own journals, conferences and organizations. I want to defend this as a specific area of expertise and invite mathematicians to take part. As other authors in this section argue I agree that mathematicians should become more active in the field of mathematics-education and participate in conferences and publish in mathematics-education research journals.

From Policy to Practice

Davida Fischman

Instructional policy is only as good as is its translation to classroom practice; it is useless to have stellar curriculum and instructional approaches determined at the national level, which then go through multiple interpretations and simplifications until they reach the classrooms of most teachers as a set of sterile packages of information and rules, which are then implemented as a laundry list of skills and algorithms. A typical chain of interpretations in the United States involves federal policy makers (whose policies determine allocation of funds to states as well as instructional standards), state policy makers (who refine instructional standards and determine state assessments of students), district mandates, school imperatives, and finally—finally!—the teacher, who puts all of this into practice in the classroom. In every consideration of mathematics and mathematics education policy, it is imperative that we consider the practical consequences for teachers and their students. Ongoing policies with an emphasis on multiple choice tests, along with the common view of school mathematics as computational, have continued to support teaching that is focused almost exclusively on computation and students who are afraid of mathematics, or if not, they love it because school mathematics primarily (in their experience) is algorithmic and computational, with few opportunities for creative thinking and little demand for real understanding.

On a recent plane trip, my neighbor asked what I do. When I responded: "I teach mathematics", he replied: "Oh, I'm *crazy* about math!" He went on to tell me that he liked to play with numbers and find out all kinds of things about them. While his level of knowledge of mathematics might not have been very advanced, his teacher(s) had seemingly succeeded in the goal of "... lead[ing] students to appreciate the power and beauty of mathematical thought" (Dreyfus and Eisenberg 1986).

Some of us (mathematicians and mathematics educators) love mathematics for its power to help us understand and tame the world around us, others for the beauty of the structures of mathematics. Is it not incumbent upon us to support the creation and implementation of policies and curricula that lead the majority of students to view mathematics as powerful and beautiful?

Unfortunately, these goals are often at odds with the perceived need for accountability in the education system, which generally translates to measuring teachers' effectiveness by their students' scores on standardized, state-sponsored, tests of skill in applying mathematical algorithms and accurate computation. This behaviorist approach to education was lamented by Ted many years ago (Eisenberg 1975), and its use has only increased since then. He notes: "Behaviorists claim that education is an observable change in behavior, which is measurable, and hopefully permanent.... The student must be able to 'do something' as a result of instruction." And he goes on to say "It is incredulous that the State Departments of Instruction confuse education with training." Where is the "beauty and power" of mathematics in this approach? It is lost in the avalanche of paperwork, skills testing, and fear generated by teacher and administrator evaluation based almost exclusively on student scores in such tests.

With the advent of the Common Core State Standards, there is hope that student assessment in the United States will shift its focus to assessing understanding of mathematics, the ability to synthesize and apply mathematical ideas, and the ability to engage in and articulate mathematical thinking. How will school districts, teachers, and university mathematics educators respond to these changes and to the challenges that come along with them? It seems self-evident that in order to teach mathematics well, a teacher must have a deep understanding and appreciation of mathematics; mathematicians as well as mathematics educators will not, in general, argue with this statement. However, this statement is unfortunately vague—what constitutes "deep understanding"? Is such understanding developed by an undergraduate degree in mathematics? An advanced degree in mathematics? Special types of courses in mathematics education? A great deal of work has been done to answer these questions and to define appropriate concepts (for example, a synthesis of work involving pedagogical content knowledge can be found in Graeber and Tirosh 2008), but while there has been progress in understanding what is an appropriate knowledge base for a good mathematics teacher, there is still no consensus nor a generally accepted definition of the required concepts. If we are to arrive at policies that generally lead to high quality teaching, we would do well to include both mathematicians and mathematics educators in these discussions; this seems to be an excellent area for productive collaboration of mathematicians and mathematics educators to enrich the field in partnership.

Currently, mathematicians seldom view themselves—or are viewed by mathematics educators—as able to contribute to these discussions. Pre-service teachers typically are taught mathematics content by mathematicians, and teaching methods by mathematics educators; graduate programs all too often continue this separation of areas. Teachers, having been educated in a culture that separates content from pedagogy, are sent out into the professional world inadequately prepared to merge the two into a productive instructional program.

Thus teachers are asked to teach content for which they have been insufficiently prepared and students continue to learn mathematics primarily as a collection of procedures rather than a vital, deep, and beautiful discipline, and many very quickly develop a poor attitude toward what they believe to be mathematics.

At the other end of the spectrum are the mandated content and practice standards for students. In the US, state content standards are frequently far higher than the mathematics in actual classroom instruction, although not necessarily coherent or designed to lead to a good understanding of the fundamentals of mathematics. In recent years, there has been a well-orchestrated effort to construct common core content standards that are mathematically coherent, and standards for mathematical practice as a framework to support high quality mathematics work in the classroom. In addition to the standards document itself, the writing team is involved with writing supporting documents to clarify the content, and in designing assessments that reflect this approach and content.

What support will be provided to teachers as they are asked to change their approach to mathematics and to teaching mathematics, and to improve their content knowledge in order to bring their instruction to the level expected by the new standards? Over the years, various reforms have been designed and introduced with great fanfare and high hopes—only to crash on the rocks of classroom reality.

Teachers are the bridge between policy and practice. Ultimately, it is classroom instruction that makes or breaks a student's education—but classroom practice cannot be legislated. There is a human process that must take place in order to enhance instruction, and policy makers must consider this process at least as important as creating standards documents and determining textbooks. Without adequate support for teachers to make this transition, and time to practice new approaches without fear of immediate criticism, nothing substantive can change in classrooms throughout the country.

What is the role of mathematicians in this story? Typically mathematicians engage in the doing of mathematics, and leave the education to those who prefer to focus on education. As a research mathematician, that was certainly my approach to mathematics education for years. But then—if we mathematicians are unwilling to contribute to the world of education, what right have we to complain that it is done poorly? If we wish to see K-12 students learn mathematics as we believe it should be learned, it is up to us to participate in the design and implementation of good curricula and support what we understand to be good mathematical practice. In order to do this effectively, we need to learn more, and become more reflective, about mathematics education.

Clearly, research in mathematics and research in mathematics education are very different animals. Having "grown up" professionally as a mathematician, and currently learning to engage in high-quality mathematics education research, I have experienced the enormous differences in these two types of research; some of these differences are described in other papers in this monograph (e.g. other contributions to this policy discussion).

And yet... both revolve around mathematics and the doing of mathematics, albeit from different perspectives and in different contexts. Without a serious exchange of ideas between mathematicians and mathematics educators, both disciplines are less than they might be. There are mathematics faculty who have two distinct standards for mathematics: they themselves carry out first-rate mathematics research, but for the majority of their students they hold little hope of a deep understanding of mathematics, and the kind of teaching in which they engage does not lead to students stretching their thinking and developing good mathematical practice. We also know mathematics educators whose teaching revolves to a great extent around process and pedagogical strategies at the expense of content. Would we not be better off if there were more education in the world of the mathematician, and more mathematics in the world of the educator?

Some examples of university policies that support such collaboration include funding travel of mathematics faculty to meetings and conferences focusing on teaching, support for education and content faculty to collaborate on teaching, research, and grant activities, mentoring of junior mathematics faculty by senior mathematics and education faculty in learning about education and grant opportunities, and last but not least: recognition in the promotion and tenure process for mathematics faculty who engage in research and professional development activities related to mathematics education. In order that involvement in the work of education not be relegated only to veteran, tenured, mathematicians or to adjunct faculty or lecturers who teach mathematics education courses because they have no choice, university policy would need to change in many universities to include at least some of the items listed here. In order for young tenure-track mathematicians to become involved in education, their departments/colleges/universities will need to create policies (particularly regarding promotion and tenure) that support their participation in such activities. This would in turn require a major change in perspective, and would be potentially quite contentious—but without this sort of change, there is little hope that large numbers of mathematicians will become involved in education.

Some questions for consideration (the enumeration of open questions is continual throughout this chapter):

Q5. What makes "good" mathematics education research?

Q6. Would greater emphasis on mathematics as a discipline in policy decision-making lead to better, more effective, education policies?

Q7. How can the education system reconcile its perceived need for accountability with its stated goal of teaching students to think mathematically and appreciate the power and beauty of mathematics? Can accountability systems be used towards this end?

Q8. In what ways and to what extent should mathematicians be involved in mathematics education, and what policies should universities enact to provide recognition and encouragement to young mathematicians who participate in this work?

On the (Almost) Separate Roles of Mathematicians and Mathematics Educators

Azriel Levy

When one looks at the areas of mathematics and mathematics-education, almost everything mathematicians do has nothing to do with Math Education. And almost everything mathematics educators do has very little to do with what mathematicians do best, which is discover new mathematics. We can rely on mathematicians to tell us what parts of mathematics are more suitable for youngsters which intend to have a career in mathematics or science, but they have no advantage over mathematics educators when it comes to decide what mathematics to teach to youngsters who intend to become cooks or taxi drivers. Therefore many mathematicians are involved in mathematics-education, mostly in advanced high school education and the involvement goes down as you go down the ladders of the depth of the mathematics and of the school years. Of course, a mathematician can contribute much even to nursery school mathematics but this is a result of his personal qualities not of his expertise as a mathematician. The reasons for the higher involvement of mathematicians in advanced level high school mathematics are not only because this is the part of school mathematics which is closest to the mathematics done at the university but also because the graduates of high school mathematics are the beginning students of the university mathematics.

In Israel I did not evidence any rift or disputes between mathematicians and math educators because of the tacit agreement that the contents of the advanced level high school mathematics is determined mostly by the mathematicians, and as you go down the ladder of depth and school years the weight shifts more to the mathematics educators. Still, in the past some disputes occurred because the mathematicians and the mathematics educators belong to separate social groups, and a fad which carries over one group does not necessarily carry over the other. This was the case with the Cuisenaire rods.

A practical advantage which mathematics educators have over the mathematicians is that they control directly experimental school teaching, mostly by means of their graduate students, and as a result they can come up with realistic changes to the curriculum. Now we come to a problematic point, where the difference between mathematics and mathematics education comes to play. When a mathematician proves a good new theorem he writes up the proof and publishes it, and this is the end of the story of that theorem, and this is his contribution to mankind. Personally, this publication can mean to him a Ph.D. degree or a promotion at the university. When a mathematics educator carries out some experimental teaching and it turns out successfully he also writes up the description of what he did, publishes it, and gets promoted or gets his Ph.D. degree. The difference between the mathematician and the mathematics educator is that the theorem that the mathematician proved becomes a part of human knowledge and stays there forever, but the results of the experimental teaching are usually valid only for the present time and place, and are not a lasting contribution to mankind.[5] Therefore, the mere publication of the mathematics educator's innovation should not be the end of the story, and the mathematics educators should invest a part of their time and energy to make what they create, part of the current curriculum, or the prevalent teaching method. This is not always a very enjoyable activity in an educational system which is, by its very nature, conservative, but since mathematics-education is an applied social science its practitioners should not avoid that part of their work.

It is therefore up to the universities to judge the mathematics educators not only on their published work but also on the implementation of their results. This is not a problem of mathematics-education only, but of many applied sciences at the university. I understand that one can be a great surgeon but one will not be promoted without writing some mediocre papers. My feeling is that in Israel many good ideas in the area of mathematics-education ended with a Ph.D. thesis or a university promotion.

The same conflict between theorizing and applying is also evident in mathematics-education conferences. I attended several conferences on technology in mathematics-education. The main problem now is not the invention of new software and hardware but the mass implementation and the efficient use of those which became well known. Yet most of the talks were about new software and hardware or about novel uses of existing software rather than the efficient wide scale use of the bread and butter software. It is always more enjoyable to expose your bright new ideas than to describe the uphill fight of getting more and more students to use efficiently the available resources.

Mathematics and Mathematics Education: Two Quite Different Perspectives on the Same Subject

Zalman Usiskin

Allow me to begin with a comment on Mogens Niss view expressed in Chap. 15 of this volume. Unlike Mogens Niss, I distinguish a *policy* from a *practice*. A policy is something that is written down and is usually decided by a committee. Because a committee is involved, it is difficult to change a policy, whereas an individual can change a practice. Also, because it is decided by a committee, there are people from a variety of opinions on the committee; otherwise there is no need for a committee. Because of that variety, there are bound to be disagreements on the setting of policies, the beliefs of the committee members come into play, and beliefs trump data.

As an example of disagreement, many people in the United States think that students are worse today than students a generation or two ago even though widely

[5]Nevertheless, the reader may note the series of short publications by the Education Committee of the European Mathematical Society (EMS) on solid findings in mathematics education; One article of this series has been published in every Newsletters of the EMS since September 2011.

available data—from the National Assessment of Educational Progress—indicate that students in the United States—particularly elementary school students—are performing better than they ever have, an estimated one to two years ahead of where they used to be. Students are also taking more mathematics than ever before, so much calculus in high school that more students in the US take first-year calculus in high school than in college, and the best students fulfill their college mathematics requirements in high school.

So why do people not believe the data? I think it is because, in the past, college mathematics departments recruited their mathematics majors from their best calculus students, but today many of these students are never seen by those departments. In their place are students who in prior years would not have gone to college or, if they had gone, would not have had to take as much mathematics. So it is the case that the students college faculty see, are not as good as they used to be, but not because students in general are worse now than before.

When mathematics in upper secondary school was taken only by those who were going on to study in the physical sciences or engineering, there was not much dispute between mathematicians and mathematics educators because these students would all take calculus and so there was obvious preparation in school for that. It is with the notion that *all* students should study mathematics through secondary school that tensions have come to the forefront. The battle is of a common type in society when decisions have to be made that affect different groups: it is a turf war.

Everyone in Israel knows how bitter turf wars can be and how difficult they are to settle. But they must ultimately be settled and the directions for settling them are rather well-known even if difficult to establish. That is what we are trying to do here: discussions of our common goals and searching for common ground; frank discussions of our differences in a civil manner and in a way that clarifies the reasons for these differences and tries to erase misconceptions; all of this aiming at mutual respect and tolerance.

Our fields are not alike. Even a cursory look at journals in mathematics and mathematics-education shows these fields to be fundamentally different. For the most part, the ultimate objects of mathematics are concepts and problems; the ultimate objects of mathematics-education are students. The objects of mathematics are inanimate and eternal; the objects of mathematics-education are animate—indeed, often quite animated—and constantly changing. Truth in mathematics is established by a logical proof, while truth in mathematics-education is dependent on data, and data fluctuate, so a result in mathematics-education in one place might not apply to another.

Póya, an acknowledged expert in both fields, described the first step in solving a difficult problem as understanding the problem. The underlying problem that fuels mathematics-education and brings mathematicians into mathematics-education is the perception that students do not know as much mathematics as we would like them to know. There has never been a time or a place where this problem is not perceived. Even in Singapore and Shanghai people have these beliefs. Then the question is: Who can best address this problem?

Ted and Michael Fried pointed out in their paper (Eisenberg and Fried 2009) that mathematics educators do not agree on many fundamental questions, such as the best ways to do research and what are important questions to ask, which is true, but the reality is even more frustrating because even on those things in which there might be agreement today, conditions could change tomorrow. There are disagreements in mathematics, too, such as whether computer proofs should be considered as valid or, if we want to go back into history, whether we want to allow the axiom of choice or an equivalent, but there is more agreement in mathematics than in mathematics-education, as one would expect of a natural science over a social science.

Unlike Jonas Emanuelsson and some others represented in the book, I find it very appealing to view mathematics-education as one type of applied mathematics. And like other fields that apply mathematics, such as statistics, computer science, physics, or operations research, although the field is grounded in mathematics, the problems that fuel mathematics-education emerge from the world outside of mathematics. Additionally, there are aspects to the field of mathematics-education that are not mathematical at all, and other aspects that are border-line.

The policy-maker dealing with mathematics curriculum, the area of my major work, deals with the selection of content to be covered in school, who should encounter that content, in what sequence, and at what age. Concerning the selection of content, is statistics mathematics? Is formal logic a part of mathematics? Is physics mathematics? In general, when if ever does applied mathematics cease to be mathematics? Should telling time be a part of the mathematics curriculum? What about reading tables of data or locating one's home town on a map? What about doing a logic puzzle such as a Sudoku puzzle? What about a discussion of lucky numbers and favorite numbers and unlucky numbers? Is computing using a calculator *doing* mathematics or *avoiding* it? Is conjecturing *mathematics* or is it *proto-mathematics*, that is, not the real thing but leading up to the real thing. These questions bring out differences both between and within our groups in what we think mathematics is, and differences in what we think is *real* or *good* mathematics.

Pólya's second step in problem solving he called "devising a plan", and one bit of his advice in this regard is to look at a related problem. A discussion like this one could involve statisticians and mathematicians rather than mathematics educators and mathematicians. We all know that while in some universities statisticians reside in a department of mathematics, in other universities they have their own department. The statisticians on my campus are all very knowledgeable about mathematics, but they do not view their discipline as a sub-branch of mathematics. My own background reflects the difference—I minored in statistics as an undergraduate in a department of mathematics and at no time in any of my statistics courses was I ever asked to examine a data set. Our distributions were all theoretical. In contrast, the statisticians on my campus feel that data is the starting point for all statistics, just as we in mathematics-education have to begin with the learner, the teacher, or the school situation, not with mathematics.

An effect of this difference in view is seen in the Common Core State Standards in the US, where statisticians had very little influence. Only a few years ago a committee of the American Statistical Association (ASA) published a report with detailed guidelines and examples for a curriculum in statistics K-12 (Franklin et al.

2007). The Common Core document references this report but it seems that is as far as the Common Core authors went. There is no statistics in grades K-5 and the statistics that is in grades 6–12 is not developed with anything near the care that is in the ASA document.

The difference in views of statistics reflects a fundamental difference between school mathematics and the mathematics as done even in the most inclusive college mathematics departments. School mathematics covers a far broader agenda than mathematics. Just as statistics involves such things as the design of experiments, a topic that is not on the radar screen of pure mathematics, the school mathematics teacher, and thus the school mathematics curriculum, is obligated to cover all the basics of the mathematical *sciences* that the public needs, including quantitative literacy and other topics that do not fit any sort of logical mathematical system. This might explain circle graphs or the metric system, both topics hard to fit into a discussion of the properties of numbers, are not mentioned in the recent Common Core standards in the United States and why much more attention is given to fractions, easily associated with division, than to decimals, a numeration system.

The natural sequence in mathematical research is logical. Accordingly, when I first wrote curriculum, I thought that all children would learn easily from a carefully-written well-explained logical mathematical argument. I was mistaken. The learner brings elements into learning mathematics that do not fit mathematical logic. Young children are not typically convinced by a logical argument. Even adults are not always convinced by logical arguments; they tend to view an argument as valid if they believe the conclusion, and invalid if they do not believe the conclusion. It's not too strange to feel that way. We have to teach students to have faith in mathematical logic and in mathematical systems; we have to teach students that one can proceed logically from a false statement to another false statement. We cannot assume that such thinking is innate.

The belief in the primality of a logical sequence in mathematics curriculum is related to what is meant by understanding of a mathematical concept. Does a person fully understand the division of fractions because they can derive the general rule from other properties? I would say they do not. The full understanding involves knowing alternate algorithms for finding the quotient, recognizing and being able to apply the division of fractions in problem situations, and being aware of the history of the idea. Some mathematics educators and psychologists might add to this the ability to represent division of fractions in some sort of iconic way. Mathematics educators tend to harbor a broader view of understanding than mathematicians.

Sticking to the notion that a logical sequence is needed in order to understand a mathematical concept can narrow what students encounter in their mathematics experience. There is no mention of infinite decimals in the Common Core, perhaps because some on the writing committee felt that an understanding of limits is needed to understand infinite decimals. Rarely does one find theorems such as the Isoperimetric Inequalities in the plane and 3-space before college even though they have many applications, probably because the proofs of these statements require college-level mathematics.

Agreeing that a logical sequence is not appropriate for all topics in school mathematics, some people hold the notion that the optimal sequence through elementary mathematics should follow the sequence of their invention by mathematicians. How else can we explain that negative numbers do not appear in the Common Core curriculum until grade 6, after fractions and decimals? I remember well a conference in the early 1970s at Southern Illinois University where Frederique Papy spoke about how negative numbers were introduced in their elementary school mathematics curriculum in Belgium. In first grade, from the beginning of school, the teacher and students graph the high and low temperatures for each day. And around November the low temperatures first go below freezing, so voila (but not in Beer Sheva!)! Pure mathematics avoids everyday experiences, but mathematics-education not only cannot avoid them, it is well-advised to use them.

Although the fields of mathematics and mathematics-education are so different, in both fields we teach mathematics, and in both fields we think deeply about mathematical concepts, and we think that our perspective gives us special insights into that mathematics. Some of us—perhaps most of us here—would like to think that we straddle the fields, but for most people these differences constitute the way it is.

Thus we should not be surprised that there are conflicting views towards a host of curricular issues. In some places, the diversity of views is welcomed and there is mutual respect between mathematicians and mathematics educators. However, in the US, these conflicts have not been resolved; whoever is in power locally or nationally rules the day, and the conflicts are being played out as we speak in the implementation of the Common Core, where the stakes for agreement are higher than they have ever been. We should take advantage of the fact that mathematics is an international language and work together for the common good.

With the entry of mathematicians into the mathematics-education arena, one of the nicest things that has happened is that mathematicians have been outspoken in the view that elementary mathematical concepts can be quite complex, and that understanding these concepts is not trivial and requires deep thought.

But other mathematicians have not been so thoughtful. They know a little about mathematics-education but they think they know more than the people who have spent their lives in the field. Their writing is a combination of accurate statements and silly pronouncements, hidden behind mathematical arguments to exhibit their knowledge of theory.

It is also the case that there is nonsense research in mathematics-education. We need some of our mathematicians to speak out against those mathematicians who are preaching nonsense. And, at the same time, more of us mathematics educators need to speak out against nonsense in our field.

Reflective Summary

Nitsa Movshovitz-Hadar

The focus of this chapter, mathematics and mathematics-education policy, is multifaceted. The multitude of ideas and connections embedded in mathematics, and the characteristics of research in mathematics as well as in mathematics-education make it almost impossible to cover many policy issues under the constraints of one chapter in a book or one panel in a symposium. We have touched upon several issues but more are left implicit, untouched or as open questions specified explicitly.

Nevertheless, mathematics teachers' preparation and the school curriculum are the two pillars of mathematics-education. Teacher education policy and curriculum policy are therefore at the heart of mathematics-education policy. Striving for vision may yield an overview of the essence of contemporary mathematics. In collaboration between mathematicians and mathematics-educators, such an overview may yield criteria to examine pre-university mathematics subjects for their educational potential. This would be an invaluable contribution towards a curriculum policy.

A collaborative process of investigation about the requisite mathematical knowledge of mathematics teachers, its acquisition and lifelong development (such as Gutfreund and Rosenberg 2012) will hopefully yield some new policy concerning teacher preparation.

To the extent that policy concerns curriculum and standards, it demands informed opinions on the kinds of fundamental issues we spoke about.

To the extent that it should make decisions about such things, it must face the difficulty of reconciling seemingly conflicting ends—for example, between being mathematically precise and rigorous and being intuitive or heuristic for the sake of creating steps towards further learning.

A forum where policy is discussed and decided becomes thus a natural occasion for discussing matters of common interest to mathematics-education as well as a context requiring mathematicians and mathematics educators to come together and bring with them their own special perspectives in order to sift priorities.

Pre-university mathematics-education is facing many challenges: e.g. creating motivation, and maintaining it; mathematics-education should go hand in hand with education for human values. Mathematics-education must go hand in hand with mathematics, namely exposing students of all ages to the broad spectrum of contemporary mathematics that permeates almost all walks of life, to the true nature of mathematics as an ever-growing body of knowledge, its applications, its beauty and its rich intellectual challenge.

Math education policy should be adopted to help in coping with these challenges. The "how to" will follow if the policy statements are clear.

References

American Mathematical Society (2010). Mathematics Subject Classification index (MSC2010). http://www.ams.org/mathscinet/msc/msc2010.html. http://msc2010.org. Accessed 12 June 2013.

ASSM (1995). *Assessment standards for school mathematics*. NCTM—National Council of Teachers of Mathematics, Reston, VA.

CESSM (1989). *Curriculum and evaluation standards for school mathematics*. NCTM—National Council of Teachers of Mathematics, Reston, VA.

Dreyfus, T., & Eisenberg, T. (1986). On the aesthetics of mathematical thought. *For the Learning of Mathematics*, *6*(1), 2–10.

Eisenberg, T. A. (1975). Behaviorism: the bane of school mathematics. *International Journal of Mathematical Education in Science and Technology*, *6*, 163–171.

Eisenberg, T., & Fried, M. N. (2009). Dialogue on mathematics education: two points of view on the state of the art. *ZDM. The International Journal on Mathematics Education*, *41*, 143–149.

Franklin, C., Kader, C., Mewborn, D., Moreno, J., Peck, R., Perry, M., & Scheaffer, R. (2007). *Guidelines for assessment and instruction in statistics education (GAISE) report: a pre-K-12 curriculum framework*. Alexandria: Am. Statist. Assoc.

Graeber, A., & Tirosh, D. (2008). Pedagogical content knowledge: useful concept or elusive notion. In P. Sullivan & T. Wood (Eds.), *The international handbook of mathematics teacher education* (Vol. 1, pp. 117–132). Rotterdam: Sense Publishers.

Gutfreund, H., & Rosenberg, Y. (Eds.) (2012). *Knowledge and training of secondary school mathematics teachers—a report of the committee on the knowledge-base for teaching mathematics*. Jerusalem: Israel Academy of Sciences and Humanities. http://education.academy.ac.il/English/PublicationDetails.aspx?FromHomePage=true&PublicationID=27. Accessed 12 June 2013.

Movshovitz-Hadar, N. (1988). School mathematics theorems—an endless source of surprise. *For the Learning of Mathematics*, *8*(3), 34–40.

Movshovitz-Hadar, N. (1993). The false coin problem, mathematical induction and knowledge fragility. *The Journal of Mathematical Behavior*, *12*, 253–268.

PSSM (2000). *Principles and standards for school mathematics*. NCTM—National Council of Teachers of Mathematics, Reston, VA.

PSTM (1991). *Professional standards for teaching mathematics*. NCTM—National Council of Teachers of Mathematics, Reston, VA.

Collaboration Between Mathematics and Mathematics Education

Chapter 17
Mathematics and Education: Collaboration in Practice

Hyman Bass and Deborah Loewenberg Ball

Abstract This chapter describes the work of a longstanding collaboration between a research mathematician and a practicing teacher and education researcher. In addition to showcasing specific aspects of the joint work, and the ideas that have been produced, the chapter also examines the nature, challenges, and opportunities of this unusual cross-disciplinary collaboration.

Keywords Collaboration · Mathematics instruction · Mathematical knowledge

Over the last 15 years, we have been studying the work of teaching mathematics in primary school classrooms (Ball 1999; Ball and Bass 2000a, 2000b, 2003a, 2003b, 2008, 2009; Bass 2005; Ball et al. 2005a, 2005b). Our goal, as research mathematician and education researcher, has been to understand what it takes to teach mathematics with integrity, and to contribute to the improvement of teachers' training. We have asked: What is mathematical about mathematics teaching, and what are the mathematical demands of that work?

We brought to this research complementary training, skills, knowledge, and perspectives. We studied primary records of practice so as to focus on common artifacts that we could examine, analyze, discuss, and unpack. These records—most often videotapes of lessons—also enabled us to hold teaching still, and to study and re-study the same moments, interactions, explanations, questions, and tasks.

The perspective that we have developed is that mathematics teaching is a special form of mathematical practice—a form of applied mathematics (Bass 2005)—and it has been this frame that has both required our collaboration and supported its development. We will illustrate this with examples, showing both the affordances of our work together, as research mathematician and education researcher, and also its challenges and the problems we have had to solve.

H. Bass (✉) · D.L. Ball
University of Michigan, Ann Arbor, MI, USA
e-mail: hybass@umich.edu

M.N. Fried, T. Dreyfus (eds.), *Mathematics & Mathematics Education: Searching for Common Ground*, Advances in Mathematics Education,
DOI 10.1007/978-94-007-7473-5_17,

Two premises guide our work:

- Children are not learning mathematics nearly as well as they could, and should
- Improving mathematics instruction and student learning depends on expertise both in disciplinary mathematics and in instruction

Addressing these issues has been our driving purpose; what we have learned about collaboration is a by-product of our actual work together. We did not set out to study collaboration; our work demanded it. But that work taught us much about the entailments of collaboration.

Our work together started in 1994, with a question from Ball to Bass: "What mathematics do you see in this episode of elementary teaching?" This exchange illustrates several characteristic features of our ongoing work. First, its motivation was to better understand the mathematical resources needed for the work of teaching mathematics. Second, to seek answers to this problem, an education researcher was enlisting the views of a research mathematician, a not altogether commonplace impulse, particularly in the case of primary instruction. Third, rather than seek an answer based on mathematical reflections of a disciplinary practitioner, or on an analysis of the school curriculum, the query was situated in examination of a primary record of practice, not in some imagined or presumed reality of elementary mathematics instruction. In other words, whatever conclusions we drew about the mathematical demands of mathematics teaching were "practice-based," i.e. grounded in a close study and analysis of teaching practice itself. It was this direct link to practice that provided some assurance that the theoretical ideas that we developed would be closely enough tied to practice to inform the improvement of mathematics instruction.

Herein rests the first lesson of our collaboration—namely, that using artifacts of practice was fundamentally important to grounding our joint inquiry. As different as our perspectives were, situating their use in a shared example of learning and teaching enabled us to bring our different expertise and knowledge, as well as dispositions and curiosities, to bear. The fact that we were looking at the same classroom lesson, example of student work, or episode of children talk helped us to focus our inquiry in common ground. This is a very different approach than, for example, writing a curriculum, or discussing standards for student learning, neither of which would be similarly disciplined by actual classroom interactions and mathematical discourse.

The Elementary Mathematics Laboratory (EML)

We focus here on one way that we have institutionalized this practice-based approach to our work, through establishment of the Elementary Mathematics Laboratory (EML). The EML is now conducted under the auspices of TeachingWorks (http://www.teachingworks.org/), a new organization at the University of Michigan whose mission is to raise the standard for classroom teaching practice by transforming how teachers are prepared and professionally supported.

The EML is an intensive two-week summer program for upper primary students who have not been successful in school mathematics. It is based on "turn-around" (in contrast with remedial) instruction, which advances students' mathematical development (knowledge, skills, and disposition) at the same time that it strengthens sometimes underdeveloped basic skills. It develops mathematical proficiencies around fractions, basic arithmetic skills, mathematical explanation and reasoning, and the use of language and representations in mathematics. It also attends to the development of study and learning skills, such as participating productively in class, doing homework, preparing for a test, keeping notebook records, etc.

What makes the EML a "laboratory" is that it is a setting for the real-time study of the interplay of instructional design, teaching, and learning. The features that enable this include these:

- The teaching is public and designed to be open for collective study, by a diverse professional community of observers (Ball et al. 2013)
- There is careful documentation (high quality audio and video, detailed lesson plans, student notebooks, etc.) (Suzuka 2013)
- The teaching and learning are deliberately made as public and visible as possible (Ball et al. 2013; Suzuka 2013; Mann and Thames 2013)
- There are planned instructional experiments
- The physical environment borrows some features from the "surgical theatres" commonly used in the training of medical professionals (Suzuka 2013)

The research questions whose investigation has been supported by the EML include the following:

- What are some of the key sites of using mathematical knowledge in and for teaching (MKT)?
- What constitutes "knowledge at the mathematical horizon" and how is it entailed by and used in practice?
- What mathematical problems are central to beginning teaching, and how might this inform high-leverage initial professional training in mathematics instruction?
- What are some of the challenges and affordances of using the number line as a central mathematical object in the teaching and learning of rational numbers in elementary school?
- What dilemmas are inherent in seeking to help students develop basic skills in active mathematical practice?
- What are the mathematically intensive instructional resources that are highest leverage to advancing children's mathematics learning and engagement, and how can they be more effectively used?
- What is involved in "turnaround" instruction that seeks to fill in gaps that students have accumulated while also providing challenging work that accelerates their mathematical opportunities and progress?
- What are the problems and challenges of designing and using homework in ways that advance children's academic development and that are equitable and sensitive to home-school connections?

Our research work is situated in other sites as well as the EML. For example we use records of practice from other settings and classrooms. Also we have begun a practice-based study of the resources most immediately and critically needed by beginning mathematics teachers (both primary and secondary).

To illustrate what we have learned *through*—and *about*—collaboration, we turn next to make two claims that are products of our joint inquiry. Our discussion of each claim is grounded in episodes of instruction.

Claim 1 *Mathematics in instruction is importantly different from mathematics considered in the abstract, or in the curriculum.*

Discourse around mathematics education in the US is currently influenced by the new Common Core State Standards for mathematics (http://www.corestandards.org/), an initiative to build a common national curriculum. In addition to a focusing and ordering of the mathematical topics of the curriculum, the Common Core includes standards for *mathematical practice*:

MP1. Make sense of problems and persevere in solving them.
MP2. Reason abstractly and quantitatively.
MP3. Construct viable arguments and critique the reasoning of others.
MP4. Model with mathematics.
MP5. Use appropriate tools strategically.
MP6. Attend to precision.
MP7. Look for and make use of structure.
MP8. Look for and express regularity in repeated reasoning.

It is a challenge to teachers to assign tangible meaning to these practices, and to acquire some vivid images of what it could mean to teach, and to learn them. Mathematical practice provides one helpful frame for understanding the mathematics we might see in the practice of teaching. We use this in an analysis of an episode of teaching in an EML class.

It is the beginning of the third day of class, with the following "warm-up" problem on a poster in front of the class as the students enter:

> *How many different three-digit numbers can you make using the digits 1, 2, and 3, and using each digit only once?*
> *Show all the three-digit numbers that you found.*
> *How do you know that you found them all?*

On the face of it, the mathematics involved in this problem consists of finding all the permutations of three objects, and proving the completeness of the solution set. What, further, is the mathematics involved in an instructional enactment of this problem with upper primary students who have been unsuccessful with school mathematics, and who may not have been regularly presented with mathematically challenging work? The problem appears to be sensible and accessible, and it is clearly stated. So it would seem that the children are well situated to set off individually working on the problem. But the instruction we shall now see presumes none of this.

Teacher: . . . Jamal, could you read it nice and loud?

Jamal reads the problem.

Teacher: Can you tell us what you think the problem is telling you to do?
Jamal: It's asking to make as many three-digit numbers as you can with one, two, and three. And—
Teacher: Do other people agree with what Jamal is saying?
Students: Yes.
Teacher: Can somebody give an example of a number that would *not* be an answer to this question? What's a number that would not be an answer to this? Sean?
Sean: One two three four?
Teacher: And tell us why that wouldn't be one?
Sean: Because it's a four-digit number.
Teacher: Excellent. What else does it do that doesn't fit? There's one other thing that's not good about it. What is it, Eli?
Eli: It's a thousand.
Teacher: It's a thousand, but there's something else that doesn't fit the conditions of the problem. So he put two wrong things into it, in a way. What was the other thing? Susanna?
Susanna: He's using the number four.
Teacher: He used the number four. Is that allowed?
Student: No.

The instruction begins with a public student reading, and interpretation, of the problem. Giving one solution of this problem means providing a number that meets these conditions: (1) It is a three-digit number; (2) It uses only the digits 1, 2, and 3; and (3) It uses each of these digits only once. This apparently straightforward, but crucially important, understanding of the problem is not a part of the habitual thinking or experience of many of these children; it is a habit that can, and must, be taught, if these children are to be successful mathematically.

Interestingly, the teacher first directs their attention not to what a solution looks like, but rather to non-solutions, for which the children are asked to identify the conditions whose violation makes an example a non-solution. We argue that recognition of this explicitly needed resource for student work on the problem, as well as the approach taken to this by the teacher, represents a kind of thinking and problem solving that is as much mathematical as pedagogical. The teacher goes on to elicit other student nominated non-solutions (like "two billion" and "zero"), though none of these violate condition (3), a case she wanted in order to complete the account of ways that a proposed solution might fail. Rather than address this directly, she next asks for possible solutions. But then she interjects an example violating condition (3).

Barak: Three—Three one two.
Teacher: Okay. Can you explain why you think that is one of the answers?
Barak: It's using three-digit numbers and it's using the numbers.
Teacher: It's used three-digit number—It uses the—Say it again?
Barak: The numbers one and two and three.
Teacher: And one more thing. What does it say?
Barak: And using each name—digit only once.
Teacher: Did you use each digit only once?
Barak: Yes.
Teacher: Okay, so would this be an answer? (*Writes "221" on the whiteboard.*)
Student: Yep.
Students: No.
Teacher: Can someone explain why that one is not an answer to this? Why is this one not an answer? Lucas?
Lucas: You used the two twice.
Teacher: It uses the two twice. Is it a three-digit number?
Lucas: Yes.
Teacher: Does it use only the numbers one, two, or three?
Students: Yes.
Teacher: But? What's wrong with it again, Lucas?
Lucas: It used the two twice.

At this point the teacher gathers more solutions from the students, in each case recording the solution on the board, and asking a student other than the contributor to explain why it is a solution, explicitly verifying the three conditions. When the students cease to produce new solutions, the teacher asks if they think that they have them all, and to give reasons for their belief, and finally she asks if they could *prove* that they have them all.

What is the mathematics involved in this instruction? In terms of mathematical content it is clearly about finding all three-digit numbers using the digits 1, 2, and 3, each only once, and proving that one has all (six) of them. In the course of this students encounter the notions of digit, three-digit number, the concept of zero as a number, etc. But there is more. In fact the teaching is also focused on developing mathematical practices and habits of mind that are foundational to doing mathematics, but that are often left implicit. Among the practices in which the teacher engages the students are the following, especially MP1 and MP3:

MP1. Make sense of problems and persevere in solving them.
MP3. Construct viable arguments and critique the reasoning of others.
MP6. Attend to precision.
MP7. Look for and make use of structure.

What kinds of mathematical work and knowledge are being deployed by the teacher in this instruction? One is the design of the mathematical task, to accessibly engage the students in a rich problem with which the above mathematical practices could be mobilized and made an explicit goal of the instruction, enacted in a way that enlists the interactive participation of an entire class. A pre-condition for this is the teacher's awareness of the importance and significance of these practices, and knowing the kinds of questioning and scaffolding that could induct the students into them.

What does this examination of an episode of instruction illustrate about Claim 1? First, there is far more mathematics visible in watching it in action than one can see from the curriculum alone. Second, the students' opportunities to learn mathematics depend on the mathematical transactions that occur during instruction. And, finally, the more we examined mathematics instruction, the more we uncovered about the content, the learning, and the work of building bridges between the mathematics and the children.

A second lesson from our collaboration is that the interplay of mathematical and pedagogical perspectives in our analyses permits a method of study that is neither solely a mathematical analysis nor a pedagogical one (Thames 2009). When we first began to work, Ball asked Bass to provide a "mathematical commentary" on some episodes of instruction (Ball 1999) and provided specific questions to guide his viewing and annotation. Quickly, however, his comments were the object of disagreement as we discussed and compared our interpretation of events and interactions. For example, Bass would ascribe to students significant mathematical insights for which there was little evidence. He often interpreted and explained teacher reasoning from his perspective instead of considering the mathematical and pedagogical dilemmas in which instructional practice is managed. As we worked through our examination of specific data and evidence in video records of practice, and argued, often vigorously, about the mathematics in instruction and the instruction of the mathematics, about student thinking and about teacher moves, we began to meld an approach to analysis of mathematics instruction that was sensitive to the multiple considerations that are the foundation of practice. As we grew more skilled, the arguments did not recede, but our ability to examine the mathematics in practice improved. It was from this development that our ability to consider a new perspective on teacher knowledge emerged: a concept that we and our research group came to label "mathematical knowledge of teaching" (Ball and Bass 2000a, 2000b, 2003a, 2003b; Ball et al. 2005a).

Claim 2 *Mathematics instruction is fundamentally a special kind of mathematical practice.*

First of all, what do we mean by "mathematics instruction?" It is *not*:

- What teachers do
- The cause of student learning
- Equivalent to the curriculum

Rather, in our view, teaching is what is *co-produced* by students and teachers in contexts, around specific mathematics and curriculum (Cohen et al. 2003).

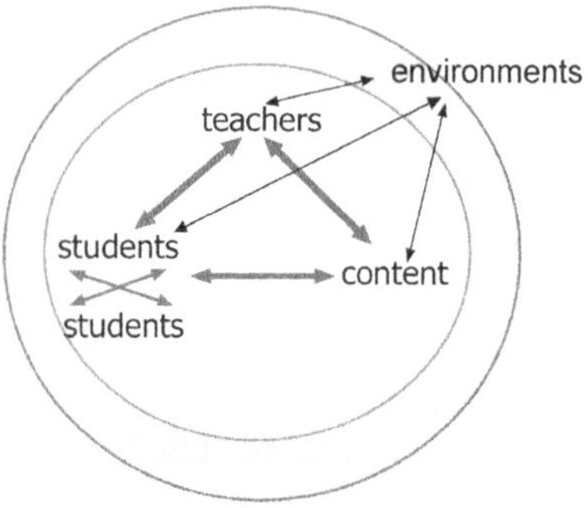

To illustrate, and support this claim we situate our argument in the examination of an episode of practice. The instruction in this example strove to achieve three commitments: (1) to work on substantial mathematics and treat the mathematics with integrity; (2) to take students' thinking seriously and make it an integral part of the instruction; and (3) to treat the construction of mathematical knowledge as the work of an intellectual collective, with mathematical justification and critical evaluation of solutions and claims being a central demand of the student work. Our interest here is to understand the *mathematical* demands of teaching to fulfill those commitments.

The third grade class (eight-year olds) analyzed here was culturally and linguistically diverse (with many children speaking English as a second language, and some only recently arrived in the United States). It is mid-year, and the children were working on even and odd numbers. They came to third grade "knowing" which (small numbers) were even and which were odd, but without any formal definition of these notions. The topic was introduced through investigation of problems such as this one: *Mick has 30 cents in his pocket and he wants to spend it all (on gum, 2 cents, and pretzels, 7 cents) and not have any change left in his pocket. What can he buy for 30 cents? What are his choices?* The solution to this problem pushed the children into encounters with notions of even and odd numbers, and eventually to make conjectures about their arithmetic properties (e.g., even + odd = odd, odd + odd = even, etc.).

On one particular day, the students were asked about the meeting they had with the fourth graders to discuss whether zero was even or odd or neither. One of the boys, Sean, reflecting on a discussion they had the previous day, raises his hand and says,

> I don't have anything about the meeting, but I was just thinking about six, that it's a... I'm just thinking it can be an odd number, too, cause there could be two, four, six, and two, three twos, that'd make six.... And two threes, that it could be an odd and an even number. Both! Three things to make it, and there could be two things to make it.

The teacher does not challenge or correct Sean. She first re-voices and tries to publicly clarify what he is saying, and then she invites comments from the class. His classmates quickly disagree. They already knew from second grade that six is even. We watch as this mathematical debate unfolds, attending to how the children are processing mathematical ideas and claims, and to the mathematical moves of the teacher to shepherd this discussion. Cassandra, the first to object, points to the number line above the blackboard, saying,

> Six can't be an odd number because this is (she points to the number line, starting with zero) even, odd, even, odd, even, odd, even, ... Because zero's not an odd number.

Sean persists,

> . . . because there can be three of something to make six, and three of something is like odd. . .

Keith protests, "That doesn't necessarily mean that six is odd." Several students chime in, "Yeah." When the teacher asks Keith, "Why not?" he responds,

> Just because two odd numbers add up to an even number doesn't mean it has to be odd.

We note that the reasoning of both Cassandra and Keith is sound, though different. Cassandra argues from a different definition (of even and odd) to a contrary conclusion. Keith, on the other hand, directly challenges Sean's argument, not simply his conclusion. Meanwhile, the teacher leaves it to the students to reconcile the disagreement, while carefully moderating the discourse. Thinking that Sean may be just confused about the meaning of "even", she makes an important mathematical move, and asks Sean,

> What's our working definition of an even number? Do you remember from the other day the working definition we're using?

When Sean does not seem to recall, she asks several other students, until Jillian offers,

> It is, um, if you have a number that you can split up fairly without having to make (long pause) to split one in half, then, um, it's an even number.

When the teacher then asks Sean if he can do that with six, he affirms, and so she says, "So then it would fit our working definition; then it would be even, okay?" To which Sean comfortably concurs, adding, "And it could be odd. Three twos could make it." Sean, defying the tacit understanding of the class, seems to allow that a number can be both even and odd. The teacher then realizes that to mediate this discussion requires a definition of odd numbers as well as one for evens, something she had not before thought necessary. After some discussion, the class agreed that odd numbers were those you could not split up fairly into two groups, or that, when you group them in twos, there is one left over. But Sean is tenacious, saying that, "You could split six fairly (two threes) and not fairly (three twos)."

The teacher pursues a new line of questioning and asks Sean if he thinks *all* numbers are odd then. When he says no, she asks him which numbers are not odd. He says that 2, 4 and 8 are not odd, but that 6 can be odd or even. Several students then shout, "No!" And Tembe challenges him: "Show us!" Sean only repeats, "There are three twos; one, two; three, four; and five, six." Unconvinced, Cassandra and Tembe insist, "Prove it to us that it can be odd." The teacher then invites Sean to prove it to the class and asks everyone to pay close attention. Sean goes to the board, where there is a drawing of six circles, which he then proceeds to separate into groups of two,

saying, "There's two, two, and two. And that would make six." To which Cassandra rejoins, "I know, which is even!" And Tembe backs her up.

Then Mei raises her hand to say, "I think I know what he is saying." The teacher asks Sean to remain at the board while Mei explains,

> ...I think what he's saying is that you have three groups of two. And three is a odd number so six can be an odd number and an even number.

It seems that the question is no longer whether Sean is right or wrong, but whether Mei has correctly interpreted Sean's idea and argument. The teacher first gets Sean's confirmation of this, and then she asks if others agree with Sean. After having clearly articulated Sean's argument, Mei herself then says,

> I disagree with that because it's not according to like... here, can I show it on the board?

At the board, facing Sean, Mei continues,

> It's not according to like... how many groups it is. Let's say that I have (long pause while she thinks).... Let's see. If you call six an odd number, why don't (pause)... let's see (pause)... let's see—ten. One, two,... (draws circles on board) and here are ten circles. And then you would split them; let's say I wanted to split, split them, split them by twos... (she draws the dividing lines and counts the groups of two). One, two, three, four, five,... Then why do you not call ten a, like... an odd number and an even number, or why don't you call other numbers an odd number and an even number?

What is Mei doing here? First she has understood and explained Sean's idea, one with which she in fact disagrees, and she has pinpointed the fault in Sean's argument. ("It's not according to how many groups (of two).") But she goes well beyond the mere statement of that critique. She embraces Sean's own reasoning, and cleverly constructs an argument that she is persuaded will make Sean, in his own terms, see the error of his ways. She generalizes the principle of Sean's reasoning—that six is made of an odd number of groups of two—and so sees that this same criterion would usher in an unlimited supply of new odd-and-even numbers, to her a menacingly chaotic predicament that she fully expected Sean to back away from. Her reflective pauses were needed to search mentally, while the class waited quietly, for the next example—10—of an odd number of groups of two. To Mei's surprise, and then dismay, Sean responds,

> I disagree with myself... I didn't think of it that way. Thank you for bringing it up; so, I say it's... ten can be an odd and an even.

In this exchange, Mei, who seems to seek to persuade Sean with the implications of his argument, in fact succeeds instead in giving Sean an expanded understanding and appreciation of his own idea, which he embraces with thanks. Mei's argument is mathematically solid, well expressed, and well understood by Sean (and the class, as we later see). Mei and Sean differ in the significance that they each attach to it. Exasperated, Mei then proclaims,

> Yeah, but what about other numbers?! Like, if you keep on going on like that and you say that other numbers are odd and even, maybe we'll end it up with all numbers are odd and even. Then it won't make sense that all numbers should be odd and even, because if all numbers were odd and even, we wouldn't even be having this discussion!

Noteworthy here is Mei's mathematical sensibility about definitions, noting that they fail in their purpose if they lose the capacity to make significant distinctions, to give concepts appropriately sharp boundaries.

In these few moments of instruction, what can we observe about the mathematical work going on, by the students and by the teacher? On one level the children are exploring aspects of even and odd numbers. But, perhaps more significantly, they are deeply engaged in several mathematical practices:

MP1. Make sense of problems and persevere in solving them.
MP2. Reason abstractly and quantitatively.
MP3. Construct viable arguments and critique the reasoning of others.
MP5. Use appropriate tools strategically.
MP6. Attend to precision.
MP7. Look for and make use of structure.
MP8. Look for and express regularity in repeated reasoning.

They are making mathematical claims and counterclaims, and critically examining each other's ideas. There is an imperative for justification of claims that the children seem to be used to doing and to which they hold each other accountable. They are developing and using mathematical language and representations. They are making mathematical generalizations. Such mathematical practices, much as we rhetorically encourage them, are not learned if they are not taught and practiced. That entails an instructional investment that we have not fully seen, but whose benefits we can see manifested in this episode.

To reconcile mathematical disagreement, the teacher recognizes the need for definitions of the mathematical terms in play. She asks the class to make explicit the "working definition" of even number. In fact three definitions of even (and odd) numbers are implicitly in use: fair share (a number is even if it can be split into two equal groups), pair (a number is even if it is composed of groups of two), and alternating (the even and odd numbers alternate on the number line, with zero being even). These are not all explicitly stated or shown to be mathematically equivalent, but they are tacitly assumed to be so. Most students (not Sean) assume the "even" implies "not odd". Noticing these different definitions in the children's reasoning, realizing the need to reconcile them, and considering what is entailed in establishing their equivalence are all crucial aspects of the teacher's mathematical thinking. It is also important that the teacher to knows what are mathematically appropriate and usable definitions of even and odd numbers for third graders. Mathematical reasoning is not feasible without some careful attention to commonly understood mathematical definitions. For example, the children's later proofs of conjectures (e.g., odd + odd = even) depend crucially on the use of such definitions.

Though Sean misuses the mathematical term "odd", he nonetheless has a clear mathematical idea about six: he notices that it has "an odd way of being even." But, lacking vocabulary to name this (well specified) feature, he disconcertingly appropriates the name "odd-and-even" for it. Sean is thinking only about six. But Mei recognizes that Sean's argument about six is generalizable and opens the door to far-reaching possibilities that she assumes would cause Sean to retreat from his claim, but he does not do so.

What are these "Sean numbers" (as the teacher came later to call them) introduced by Sean and Mei? They are the odd multiples of two. Is this a topic worthy of instructional time? The concept of even and odd is about mod 2 arithmetic. Sean and Mei have cracked the door open on mod 4 arithmetic, identifying numbers congruent to 2 mod 4. These numbers turn out also to be exactly those natural numbers that are not a difference of two squares, and were studied by the ancient Greeks. So, the idea surfaced by Sean's natural curiosity about numbers in fact has some interesting mathematical significance that he could not have anticipated, but that might figure in the teacher's evaluation of how much instructional play to give it. Indeed, once Mei had essentially defined these Sean numbers, the students eventually began an exploration of their properties—finding patterns (every fourth number, starting with two, is one); making and proving conjectures (a sum of two Sean numbers is not one); etc.

But, more important, what the children are learning goes well beyond the properties of Sean numbers. It includes the skills of mathematical exploration and reasoning, hearing, using, and critiquing the reasoning of others, generalizing, using mathematical definitions and representations, etc. For those who wonder in frustration over our students' failure to gain proficiency with or appreciation of such mathematical practices, you might consider that this episode provides one image of what it might look like for young children to begin to develop such skills.

But these practices do not develop naturally in a classroom. They must be cultivated and scaffolded through deliberate instruction, itself informed by such mathematical practices as: posing productive and accessible tasks; asking strategically purposeful mathematical questions; using appropriate mathematical models and representations; prompting promising mathematical explorations; encouraging speculative thinking and conjecture; asking for mathematical justifications of solutions and claims; evaluating mathematical arguments; developing mathematical language and its careful use; and being attentive to mathematical structure. This illustrates some of what makes teaching a special kind of mathematical practice.

The third lesson from our almost twenty years of collaborative work, by a research mathematician and an educational researcher, is that the norms of evidence and argument needed to make claims about instruction in the analyses we do have little precedent in our respective fields and practices of research. Mathematical arguments and reasoning rest on principles contained within the discipline, although claims made in qualitative approaches to the study of education most often rest on theoretical frameworks that guide the sorts of questions asked and the kinds of evidence marshaled to answer them. Instead, in our work, we have had to develop standards of argument and evidence to support our efforts to use mathematics as a lens to investigate instruction and the resources (such as teacher knowledge) involved in its practice, and our simultaneous efforts to develop an instructional and pedagogical set of tools for an applied mathematics of teaching and learning. Our interactions have often been heated and determined as we hammer out our different perspectives, and the work is intellectually challenging. It is challenging because there is no road map for what we have tried to do, and because we have needed the strengths of our own respective disciplinary perspectives and training to do the work. Our commitment to find the interplay of perspectives and in so doing to uncover new ways to understand, explain, and develop mathematics instruction has supported our efforts, but it is work not for the timid of intellectual spirit.

What Have We Learned?

Improving mathematics learning depends on intertwining deep expertise in the practice of both mathematics and instruction. These connections can be built strongly when collaborations engage in practice—through direct engagement in instruction, through artifacts that can be discussed, studied, and re-examined over and over. This involves cross-disciplinary and new interdisciplinary work—about what are the key questions, what counts as a claim, and what counts as evidence and warrants. A crucial foundation for such collaboration is mutual respect; another is solid grounding in the domains of mathematics as a discipline and in the actual practice of instruction as well as its close and disciplined study. With almost twenty years of experience with this work, we can see our progress as well as the hard knocks of the arguments it has taken to get here. We are encouraged by the results and interested in articulating more fully the methods involved so that others can also engaged in such partnerships.

References

Ball, D. L. (1999). Crossing boundaries to examine the mathematics entailed in elementary teaching. In T.-Y. Lam & A. R. Magid (Eds.), *Contemporary mathematics. Algebraic K-theory, groups, and education: on the occasion of Hyman Bass' 65th birthday* (pp. 15–36). Providence: Am. Math. Soc.

Ball, D. L., & Bass, H. (2000a). Interweaving content and pedagogy in teaching and learning to teach: knowing and using mathematics. In J. Boaler (Ed.), *Multiple perspectives on the teaching and learning of mathematics* (pp. 83–104). Westport: Ablex.

Ball, D. L., & Bass, H. (2000b). Making believe: the collective construction of public mathematical knowledge in the elementary classroom. In D. C. Phillips (Ed.), *Yearbook of the national society for the study of education, constructivism in education* (pp. 193–224). Chicago: University of Chicago Press.

Ball, D. L., & Bass, H. (2003a). Making mathematics reasonable in school. In G. Martin (Ed.), *Research compendium for the principles and standards for school mathematics* (pp. 1–39). Reston: National Council of Teachers of Mathematics.

Ball, D. L., & Bass, H. (2003b). Toward a practice-based theory of mathematical knowledge for teaching. In B. Davis & E. Simmt (Eds.), *Proceedings of the 2002 annual meeting of the Canadian mathematics education study group* (pp. 3–14). Edmonton: CMESG/GCEDM.

Ball, D. L., & Bass, H. (2008). The role of mathematics in education for democracy. In G. Fenstermacher, D. Coulter, & J. Wiens (Eds.), *Yearbook of the national society for the study of education: Vol. 107(1). Why do we educate in a democratic society?* (pp. 171–184). Malden: Blackwell.

Ball, D. L., & Bass, H. (2009). *With an eye on the mathematical horizon: knowing mathematics for teaching to learners' mathematical futures*. Paper prepared based on keynote address at the 43rd Jahrestagung für Didaktik der Mathematik held in Oldenburg, Germany, March 1–4, 2009.

Ball, D. L., Goffney, I. M., & Bass, H. (2005b). The role of mathematics instruction in building a socially just and diverse democracy. *The Mathematics Educator, 15*, 2–6.

Ball, D. L., Hill, H. C., & Bass, H. (2005a). Who knows mathematics well enough to teach third grade, and how can we decide? *American Educator, 30*, 14–17, 20–22, 43–46.

Ball, D. L., Sleep, L., Boerst, T., & Bass, H. (2009). Combining the development of practice and the practice of development in teacher education. *The Elementary School Journal, 109*, 458–476.

Ball, D. L., Shaughnessy, M., & Sleep, L. (2013). *Public teaching: the work of making one's own teaching practice studyable by observers*. Paper presented at the annual meeting of the American educational research association, San Francisco, CA.

Bass, H. (2005). Mathematics, mathematicians, and mathematics education. *Bulletin of the American Mathematical Society, 42*, 417–430.

Cohen, D., Raudenbush, S., & Ball, D. L. (2003). Resources, instruction, and research. *Educational Evaluation and Policy Analysis, 25*, 1–24.

Mann, L., & Thames, M. (2013). *The role of "lesson plans" in collective work on teaching*. Paper presented at the annual meeting of the American educational research association, San Francisco, CA.

Shaughnessy, M. M. (2013). *Designing for the interactive mathematical work of teaching*. Paper presented at the annual meeting of the American educational research association, San Francisco, CA.

Suzuka, K. (2013). *Designing laboratory space and documentation to support the examination of teaching and learning*. Paper presented at the annual meeting of the American educational research association, San Francisco, CA.

Thames, M. (2009). *Coordinating mathematical and pedagogical perspectives in practice-based and discipline-grounded approaches to studying mathematical knowledge for teaching*. Unpublished dissertation, University of Michigan, Ann Arbor, MI.

Chapter 18
Reflections on Collaboration Between Mathematics and Mathematics Education

Patrick W. Thompson

Abstract Our chapter is in four sections. Michèle Artigue tells the story of her transition from mathematical logic to mathematics education and of collaborations at a wide variety of institutional levels. Günter Törner gives a history of collaboration between mathematics and mathematics education in Germany along with a list of recommendations to foster collaboration. Ehud de Shalit shares lessons learned from personal experiences collaborating in the production of a math fair and in the design of a mathematics education major. Pat Thompson tells of several collaborative efforts at his home institution and examines ways that mathematics education contributed mathematically to them. A concluding section provides a reflection on our charge—structural and cultural issues involved in collaborations between mathematics and mathematics education.

Keywords Collaboration · Constraints · Affordances · Examples · Mathematics · Mathematics education

Introduction

The editors of this book asked our group to address the matter of collaboration between mathematics and mathematics education. For some time we debated whether to change our charge so that it referred to people rather than to disciplines—collaboration between *mathematicians* and mathematics *educators*. We finally decided there was much wisdom in the editors' original charge. We therefore attempted to focus on aspects of the disciplines and their organizations that might lend to collaboration among the people populating them.

With contributions by
Michèle Artigue, Université Paris 7 Diderot, Paris, France
Günter Törner, Universität Duisburg-Essen, Duisburg, Germany
Ehud de Shalit, The Hebrew University of Jerusalem, Jerusalem, Israel

P.W. Thompson (✉)
Arizona State University, Tempe AZ, USA
e-mail: pat@pat-thompson.net

M.N. Fried, T. Dreyfus (eds.), *Mathematics & Mathematics Education: Searching for Common Ground*, Advances in Mathematics Education,
DOI 10.1007/978-94-007-7473-5_18,

Collaboration Between Mathematics and Mathematics Education: Personal Experiences in France and Abroad

Michèle Artigue

Collaboration between mathematics and mathematics education is first collaboration between individuals who belong to the corresponding communities or navigate at their interface. Preparing my contribution on this theme has given me an opportunity for reflecting on my personal experience and for trying to draw some lessons from it. In this contribution, I summarize this reflection and its outcomes.

Such a reflection is necessarily subjective. For that reason, it is important that I start by pointing out some characteristics of my professional life that necessarily influence my perception. I was trained as a mathematician at the Ecole Normale Supérieure in Paris, and logic was my first research area. My Ph.D. was on recursivity issues and then I got a position at the mathematics department of the University Paris 7 and entered a research group working on non-standard models of arithmetics and bicommutability between theories. One of my professors at the Ecole Normale Supérieure, André Revuz, had been recently recruited there and he was in charge of a new and original institution, called IREM (Institute of Research on Mathematics Teaching). IREMs are specific structures attached to universities with close links to mathematics departments. The first three IREMs were created in 1969, but there are now 28 that form a network covering the whole country, and there are even some IREMs abroad (http://www.univ-irem.fr). Their mission is to contribute to teacher professional development, to develop innovation and research, and to produce resources both for teaching and for teacher education. For fulfilling these missions, the IREMs create mixed thematic groups including university mathematicians, teachers and teacher educators working part time collaboratively. For instance, at the creation of the IREM of Paris at the University Paris 7 in 1969, the mathematics department was allocated six specific positions by the Ministry of Education, and mathematicians from the department were invited to spend part of their academic duty contributing to IREM activities, in collaboration with the 20 secondary teachers delegated half time to the IREM by the academic authorities.[1]

I was soon invited by André Revuz to join the IREM team, and this was the origin of my engagement in educational issues. In the mid 1970s, Revuz proposed that two colleagues and I take charge of mathematics teaching in an experimental elementary school that had been recently attached to the IREM. We had a lot of freedom for organizing mathematics teaching and learning in that school, and in this capacity I collaborated with Guy Brousseau. Brousseau had obtained the creation of a similar school attached to the IREM of Bordeaux, a laboratory where the main constructs of the theory of didactic situations were being developed and put to the test (Brousseau 1997).

[1]The current means of the IREM are far from this idyllic state, for instance the many secondary teachers contributing to its activities only receive some extra salary.

I thus entered the emerging community that would be later known as the French didactic community, and had the possibility to contribute to its development while pursuing my research in logic. This context also led to the fact that, even when years later I stopped doing research in mathematics, I could continue to collaborate with mathematicians. I taught with them at the undergraduate level or in teacher education programs; I worked with them at the IREM as well as in different academic institutions and commissions, such as the CNU (National Council of Universities) which is in charge of the qualification and promotion of university academics[2] and, later on, worked with mathematicians in the CREM (Commission of Reflection on Mathematics Teaching) presided by the mathematician Jean-Pierre Kahane and in ICMI, the International Commission on Mathematical Instruction. These characteristics of my professional life made me move regularly at the interface between communities, and they certainly influence my vision of collaboration between mathematicians and didacticians.

Collaboration between mathematicians and didacticians is necessary, and I am personally convinced that no substantial and sustainable improvement of mathematics education can be obtained without building on the complementarity of their respective expertise, without their common engagement and coordinated efforts. However, I am perfectly aware that productive collaboration is not easy to create and that maintaining it, once established, requires continued effort.

The situation was certainly different in the sixties and even the seventies, a time when didactic research was just emerging. The proceedings of the first ICME congresses, for instance, attest to the existence of such collaborations, as well as to the existence of many individuals combining research activity both in mathematics and mathematics education. But, as I explained in the closing lecture of the symposium organized for celebrating the centennial of ICMI in Rome in 2008 (Artigue 2009), the development and professionalization of research in mathematics education and the increasing pressure put on researchers—whatever their field of expertise—inexorably increases the distance between the communities and their respective agendas. This distance makes individuals who can maintain a substantial and recognized research activity both in mathematics and in mathematics education more and more an exception. Of course, there are still some people who span the boundary between mathematics and mathematics education, and they play a particularly important role in maintaining and even strengthening the connections between the communities. However, the quality of relationships between mathematicians and didacticians depends increasingly on the establishment of productive collaboration between individuals or groups who do not have full expertise in both domains, but who think that collaboration is needed for the improvement of mathematics education and are ready to invest part of their time and energy for making this possible and productive. As I also pointed out in my Rome lecture, the development of mathematics education as a genuine field of research has led to the building of theoretical

---

[2]In France, most didacticians are attached to the section of CNU in charge of applied mathematics and applications of mathematics.

frameworks, constructs, of technical terms that make communication more problematic. Mathematics education has progressively built a specific form of discourse in which research articles are written and results expressed. Making these results and ideas accessible outside the research community needs appropriate transposition of discourse. I am not sure that the didactic community does this so well. The difficulty of the work is often under-estimated and the efforts of those who invest in it are not valued enough by academic institutions.

Committed individuals are certainly essential for initiating and developing fruitful collaboration, but, without appropriate structures and institutional support, any impact remains necessarily limited and its sustainability is impossible to ensure. Looking for institutional support and creating adequate structures is thus crucial. Moreover, priority should certainly be given to actions where collaboration can make a visible difference while being accessible at a reasonable cost. I would say that teacher education and professional development, teaching and learning at the university, popularization and enriching activities are in some sense natural candidates as many mathematicians already are engaged in them. Also, collaboration is possible at a policy level in curricular commissions and in joint reactions to policy decisions that we think damaging for mathematics education or teacher education.

In looking at my personal experience, many positive examples come to mind that reflect different forms of collaboration. I will briefly evoke six of these and give an extended seventh example:

1. Collaboration at a personal level: collaboration with the mathematician Adrien Douady, and then with Marc Rogalski for instance was essential for my research and engineering work on the teaching and learning of differential and integral processes, and of differential equations (Alibert et al. 1988; Artigue 1992). Conversely they benefited from our collaboration for their teaching at university. There is no doubt that my research interests in the area of Calculus and Analysis favored such a form of collaboration.
2. Collaboration in innovative university programs, teacher education and professional development. As I explained above, this form of collaboration was something normal for me as I have spent more than three decades working both in a mathematics department and in an IREM. For instance, in the early eighties, together with mathematicians and physicists, I was involved in the development of a very innovative and successful experimental mathematics and physics program for first year university students. I also worked with Jean-Luc Verley, an historian of mathematics leading the corresponding group at the IREM, in the creation of an experimental course combining history of mathematics and didactics in the master's program of mathematics. More recently, I worked with François Sauvageot, a mathematician colleague, in the creation of a course on modeling in a master's program devoted to the education of mathematics teacher educators. There is no doubt that the existence of the IREM structure was essential for initiating these innovations and making them successful.
3. Collaboration in national groups and commissions, such as the CREM (Commission of Reflection on Mathematics Teaching), asked in 1999 by the Ministry of Education to reflect on what should be taught in mathematics, why and how,

and whose main reports published in 2001 still constitute a reference in France (Kahane 2001).
4. Collaboration in the supervision of doctorate students. René Cori, who leads the IREM group on logic, and I are jointly supervising a thesis on the teaching of logic at senior high school (some elements of logic were recently reintroduced from grade 10). Even if logic was my initial research area, this was a long time ago and I feel my collaboration with René is necessary.
5. Collaboration in popularization and dissemination activities. This form of collaboration that I personally experienced in the conception of the UNESCO travelling exhibition "Experiencing Mathematics!" is increasingly developing.
6. Collaboration in the organization and management of actions, but also at a reflective level that is required for more systematic study of the functioning and effects of such activities. Doctoral theses such as the recent thesis by Nicolas Pelay in France, which was co-supervised by the mathematical historian Jean-Pierre Crépel and the didactician Viviane Durand-Guerrier (Pelay 2011) are very promising from this perspective.

My seventh example involves collaboration at the ICMI level. One of the main ambitions of ICMI is to organize the collaboration of all those who can contribute to the improvement of mathematics education worldwide, to guide and support their efforts, and to disseminate their outcomes. As I explained in the Rome lecture, my election as ICMI vice-president in 1998 occurred at a moment of tension between ICMI and its mother institution, the International Mathematical Union. We had to reflect on what we wanted to achieve and how it could be possible. At that time, voices were being raised in the community of mathematics education asking ICMI to allow it to become independent. The ICMI Executive Committee resisted these voices while acknowledging that the *status quo* was not acceptable. Thanks to mutual efforts among mathematicians and mathematics educators, the situation progressively improved to a quality of relationships and collaboration today that were difficult to imagine in 1998. Along the years, I experienced the decisive role that can be played by influential and respected individuals (members of the two executive committees, especially their presidents and secretaries, such as Hyman Bass, ICMI President from 1998 to 2006, and Bernard Hodgson, ICMI General-Secretary from 1998 to 2009, Jacob Palis, John Ball and László Lovász, the successive IMU Presidents during that period).

I also witnessed how collaborative work on common projects is essential for overcoming mistrust and bad experiences. I could mention many different experiences. I will limit myself to two of them. The first one is the Felix Klein project, whose aim is to make the mathematics developed since the Klein era accessible and source of inspiration for teachers (focusing on senior high school teacher in the first phase of the project). This project was inspired by the work of Felix Klein himself, who a century ago gave a series of lectures for German teachers that led to the famous series *Elementary mathematics from an advanced standpoint*. The Klein project is now a joint project of ICMI and IMU, led by a team made of mathematicians and mathematics educators (http://blog.kleinproject.org).

The realization of the Klein project cannot be envisaged without a collaboration between mathematics and mathematics education, and we observe that the interest such collaboration raises can reinforce further collaboration between communities and even provoke it. A paradigmatic case is that of Brazil where a Klein project in the Portuguese language has been launched at the initiative of the Brazilian Mathematical Society with the participation of all societies in charge of mathematics and mathematics education, and a strong support from the government (http://klein.sbm.org.br).

I must also mention the recent CANP project (CApacity and Networking Programme)[3] jointly launched by ICMI and IMU with the support of UNESCO and ICIAM (International Council for Industrial and Applied Mathematics) for the development of teacher educators in developing countries. One realization of this project is a conference held annually in a different part of the world. The first conference took place in September 2011 in Bamako (Mali) for Francophone sub-Saharan Africa, the second held in August 2012 in Costa Rica for Central America. The next conferences are planned in Cambodia (2013) and Tanzania (2014). As explained in the description of CANP, this project is jointly led by mathematicians and mathematics educators, and addresses all those involved in teacher education: expert teachers, didacticians, mathematicians, inspectors and advisors. The conferences aim to foster the collaboration among all communities engaged in teacher education in a given country as well as regional collaboration. Conferences take the form of a two-week workshop in which didactic and mathematics themes are combined together with study of questions of specific interest for the region. Once again, the first conferences look very promising, showing that collaboration between mathematicians and didacticians, even if it is unusual in many countries, is nevertheless possible and can be rewarding when carefully organized and planned in a spirit of mutual respect. Such projects also show how the ICMI spirit of collaboration among communities can disseminate and impact local situations.

This contribution to the reflection undertaken in our panel will perhaps appear too optimistic to some whose experience is quite different from mine. This vision is nevertheless realistic. Fruitful collaborations indeed exist at different levels and in many different contexts. Even in difficult contexts, actions are possible which in the long term can move positions and visions, by relying on the individual forces that always exist, and by patiently cultivating these. However, these positive descriptions and outcomes should not hide that collaboration is always costly. Whatever are its conditions, collaboration needs personal efforts of decentration, the building of intermediate languages, the building of an appropriate semiosphere where communication is made possible between members of different communities. Mathematicians, historians, teachers and didacticians have all to learn a lot for productively collaborating; they have to accept the limitation of their knowledge and expertise but also to be convinced of its value and importance for others. There is no alternative because, as already stressed above, each community alone will not produce sustainable and large-scale improvement of mathematics education.

[3] http://www.mathunion.org/icmi/other-activities/outreach-to-developing-countries/canp-project/.

Establishing a collaborative culture requires patience and determination. It is easier to destroy than to build. But when collaboration works, it is so rewarding!

Collaboration Between Mathematics and Mathematics Education: Some Personal Experiences and Remarks

Günter Törner

To paraphrase Euclid, *There is no royal road to collaboration between mathematics and mathematics education.* I am fully aware that each country has its own mathematical tradition and culture, which has been lived for many centuries by mathematicians. The TED-symposium confirmed that situations differ among different countries.

Nevertheless, there might be a chance that each reader reflects on the variables that I identify here for Germany. Eventually in the reader's country there is some latitude, which might be filled and framed. Also, I am aware that my viewpoint is a personal one.

From 1997 I was a member of the Executive Board of the German Mathematical Society (DMV). In the late 1990's the alarming results of TIMSS were published, which showed German students performing much lower than we anticipated, and soon the question arose: Who is "guilty"—the teachers, the society of mathematics educators, or the DMV through teacher education at universities?

Soon we realized that our Society is not very large and thus our influence is limited, since at that time we counted 4,000 members. Thus it was not straightforward for the Society to be invited to public hearings to present our views. Our Society therefore joined efforts with two other learned societies covering the field of mathematics education. After that we started to produce jointly-authored declarations on various occasions when mathematics at school was discussed in the press. Through this effort, we spoke in the name of more than 10,000 members. Next we argued against the claim that schoolteachers were responsible for the poor outcomes of German mathematics teaching and for the claim that the problems were systemic.

We also realized that there is no such thing as *the* mathematician, *the* mathematics educator, or *the* mathematics teacher. Rather, there is great variability in values and perspectives within each group. Further, we had to confess that few people in the mathematics community were intimate with the processes in the ordinary mathematics classroom. Upon getting in contact with mathematics educators and coming to truly appreciate their research, we understood that we do not have a deficit in research on teaching and learning, but rather an implementation problem. However, contrary to the expectations of administrators and politicians, carrying out sustainable and effective research in mathematics education about implementation is much more difficult than just publishing some interesting new results in mathematics.

But mathematics education is not the only area facing an implementation problem. Mathematics itself is facing many implementation problems—cooperation of pure mathematics with applied is just one—and these are often ignored in the daily practice of mathematics departments! To summarize: To change mathematics education, so as to develop collaboration between mathematics and mathematics education, is not solely a personal problem or endeavor, but should be classified as a communal challenge. Collaboration between mathematics and mathematics education should be regarded as a task of cooperation between societies.

Lest I be misunderstood, I must say that in there are fruitful local collaborations between mathematicians and schools or teachers in numerous places in Germany, including mathematics departments. These efforts are important, and should be emphasized publicly. However, though such projects are necessary, they are not sufficient. *Bottom-up approaches* must be complemented by *top-down initiatives*.

Looking back to earlier times at the beginning of the last century, in the glorious time of Felix Klein, mathematics education at school was in accordance with the insights of mathematics education at university. It might sound incredible today, but Felix Klein worked to strengthen school education in the lower secondary grades and in kindergarten. In a seminar (circa 1910) he invited famous educational researchers to share their thoughts about the implications of Pestalozzi's research for mathematics education at school. Pestalozzi was a famous (primarily non-mathematics orientated) educationalist. This was Klein's idea—to learn from an educationalist. I am sorry to say that, today, I do not know any mathematician, internationally recognized as a leading researcher in his or her field, who is so convincingly and simultaneously engaged in mathematics and school mathematics education.

The New Math movement of the 1960's and 1970's provoked a separation of teachers and mathematics educators in the DMV. Mathematicians, sometimes with hubris, ignored teachers' needs, their proposals, and their contributions. Thus there was a "divorce" within the DMV of mathematics and mathematics education. So far mathematics education was a session of some few mathematicians at their annual meetings, now 'Gesellschaft für Didaktik der Mathematik' (GDM) was established as an independent learned society, also attracting primary school teachers. Some members of DMV left the mathematical society and joined GDM. GDM organized its own annual meetings and there was no correspondence between these societies, no discussion on common topics, nearly hostile neighborhood.

Forty years later, both sides recognized that this splitting was a mistake. Unfortunately, there is now so much accumulated divergence that it makes little sense to propose a reunification. We have to accept that, for example, primary school teachers are very far from DMV intellectually and probably would not join DMV, but there are at least more than one hundred persons which are members in both societies. Also, some of the DMV members, in parallel with mathematics education, are aware that foundations for mathematics are laid in the primary grades and thus feel also responsible for that group of teachers. Meanwhile, about 2000, in Germany we started some gentle cooperation between mathematicians and mathematics educators in various projects financed by foundations (see Hoechsmann and Törner 2004). It is a partnership on a level playing field.

Nevertheless, we struggled (and continue to struggle) with closely held beliefs among mathematicians and among mathematics educators. One problematic belief held by many mathematicians is that the problem of learning and teaching is trivial or unimportant. Mathematicians do not entertain elaborate models for learning. Rather, they like to generalize their private opinions to the general case—even though they would never extend knowledge from one mathematical example to all related examples.

Finally, there is too much hubris among mathematicians. Mathematicians often speak as if mathematical objects are real, available for inspection. On the other hand, it is too often the case that mathematics educators do not have a command of the larger body of mathematics that mathematicians see as arising from the ideas under discussion. Mathematicians then form the impression that educators have underdeveloped worldviews of mathematics—on the role of systems, the role of axiomatics, and on formal aspects of mathematics. It is also true that mathematicians overemphasize these same aspects, without appreciating the nature and role of meaning and understanding in students' mathematical learning.

Over the years I have become aware of multiple Do's and Don'ts. Here I offer an incomplete list:

- STEP 1: Avoid philosophical discussions. Although some of our problems are rooting in the philosophy of mathematics, be careful to start discussing them. Philosophies are often deeply anchored and hard to change, hence such discussions might not affect our daily practice.
- STEP 2: Cooperation must first be grounded in communication.
- STEP 3: You may solve the "problem" under discussion in your department, but this will—if successful at all—lead only to a local solution.
- STEP 4: Be sensitive in communication processes. Don't act as a missionary. Don't try to convince. Your collaborators can contribute many experiences and insights.
- STEP 5: Be modest in your expectations. Expect to invest years of effort.

While it is easy to call for win-win collaborations, they are not easy to accomplish in our context. Thus, the DMV tried to define win-win situations on a large scale where groups and societies are involved. Certainly it is a win-win situation for a country when mathematicians and mathematics educators are willing to cooperate. Math wars create losers: teachers, students and finally mathematics itself.

Successes in Germany can also be attributed to additional factors.

- The International Congress of Mathematicians (ICM 1998) at Berlin provided an opportunity for an inventory in the field, and the DMV made use of it. Magnus-Enzensberger, an internationally respected essayist, gave a famous talk (Enzensberger 1999): *Draw-Bridge Up, the Ivory Tower*, portraying misconceptions of mathematics.
- The presidency of Martin Grötschel, who is now serving as the Secretary of IMU, changed the self-view of the DMV by bringing many applied mathematicians into the society and into offices of the DMV. Groetschel also be-

gan an initiative in 1993 that granted a seat for mathematics education on the Executive Committee of the DMV. Today it is no longer disputable that there should be a mathematics education representative in the DMV's internal discussions.

- The 2007 Joint annual conference of mathematicians and mathematics educators in Berlin was a success. However, the 2010 Joint annual conference in Munich was not as successful. We learnt from this conference that success is highly dependent upon the local organizers. As a consequence, at this moment there is no plan for a further joint conference.

We aimed to establish a culture of a reciprocal appreciation among mathematicians and mathematics educators, and we are practicing it. We came to understand that blaming the other side does not improve the situation. We accepted that poor textbooks do exist (in school and in university mathematics) and that poor teaching exists (at school as well as at university).

We are also convinced that transparency and openness generate confidence. We try to abolish envy and jealousy. We are practicing graciousness: Invite math education representatives to all Executive Committee meetings of our mathematical society. It is also important to note that we invite our mathematics education colleagues into our "private homes" and "temples" like The Mathematical Research Institute Oberwolfach, the Fields Institute, and the Banff International Research Station. We know this is not easy, since we have to reject a mathematically oriented conference topic to host mathematics educators for a week; but it is paying off.

Meanwhile there are well-established projects:

- A joint commission on issues of teacher education with delegates from the German Mathematical Society (DMV) and two more societies representing the mathematics educators and teachers of mathematics.
- A joint commission dealing with the transition problems of students starting to study mathematics after leaving school, and who have a high drop rate—which must be lowered.

We are widening our views and are eager to gain more friends in mathematics education. We are convinced they do exist. Together with Celia Hoyles (NCETM, London) and the Deutsche Telekom Foundation[4] (DTS) we are inviting charity foundations, NGOs and institutions to the FOME-conference (Friends of Mathematics Education) in Berlin (March 2013)—organizations that are parallel to mathematics and which are sponsoring projects for mathematics classrooms. The International Mathematical Union (IMU), the International Commission on Mathematical Instructions (ICMI) and the European Mathematical Society (EMS) will support us. All this will serve to improve mathematics education, not least by the help of mathematicians. Better school education improves the success of students at university.

[4]http://www.telekom-stiftung.de/dtag/cms/content/Telekom-Stiftung/en/396336.

Collaboration Between Mathematics and Mathematics Education: Two Personal Examples

Ehud de Shalit

The session on collaboration between mathematics and mathematics education at the TED conference has given me the opportunity to reflect upon two very rewarding experiences I have had in recent years, and to share my thoughts about them with the other members of the panel. I would like to precede my description of these two enterprises, though, with a confession. Coming from the side of mathematics, I often feel unsure in the company of math educators. Math education is by now a mature field that has its own paradigms and methodology, and its own language that I do not speak. In a paradoxical way, serving mathematics for over 30 years blinded me to some very basic truths about math education. I often find the observations made by math educators eye-opening and awe inspiring. I can only regret the fact that despite a somewhat growing trend toward collaboration in recent years, in Israel at least, the two communities still remain largely disjoint.

The Meet Math Exhibition

Some 8 or 9 years ago I was recruited by Prof. Hanoch Gutfreund to participate in a full-scale, 400 square-meters math exhibition. The *Meet Math* exhibition, an Italian-Israeli-Palestinian co-production, opened two years later for 3 months in the Città della Scienza in Napoli, before moving for 8 more months to the Bloomfield Science Museum in Jerusalem, finally settling in its permanent residence at Al-Quds university in Abu-Dis.

I will not say anything here about the very interesting experience of working with people from other nationalities to enhance peace in the region. I will only focus on the scientific experience per se. Our team was incredibly large. It included mathematicians, curators, designers, educators, carpenters, as well as financiers. Everything had to be done from scratch—defining the goals, the target audience and the concepts, and of course, building the exhibits and writing the texts. Focus shifted rather early from History of Mathematics (with emphasis on Arab contributions in the Middle Ages) to the subject matter itself, with hands-on exhibits. To the surprise of the Italians, it was the Palestinian members of the team who preferred an exhibition that would benefit their school children directly, over a learned historical exhibition that would pay tribute to their heritage but attract fewer viewers, mostly adult. Perhaps at my insistence, the target audience was set at junior-high and high school children. I felt that too often science museums catered either to the very young, or to adults and professionals, leaving out the formative years in which the child chooses his or her future direction.

Some of the messages that we wanted to communicate were obvious—the usefulness of mathematics, its role as a language for other sciences, that doing math can be fun, etc. Other messages were subtler and not always easy to explain. Does the mathematician discover or invent the mathematical world? What is the difference between an illustration of a mathematical fact and its proof? We used a rather standard exhibit of Pythagoras' theorem, in which liquid flows from one square to fill up the other two, to address this point. Next to it, we also had a tangram-based proof of the theorem, and the activity around the exhibit focused on which of the two was more convincing and why.

Is there room for ugly mathematics (to paraphrase G.H. Hardy)? What is an algorithm and what is algorithmic complexity? We used the Tower of Hanoi to illustrate this last point. What is an open problem? We presented a computer game in which the visitor chose a number x, and then successively applied to it the transformation $3x + 1$ if what they had at hand was odd, or $x/2$ if it were even. It is an open problem (called the Collatz conjecture) to show that this game always ends with 1. The statistics can be quite amazing—some very high numbers are reached, and the game lasts for quite a long time, before it finally ends with 1.

Some exhibits dealt with fundamental notions encountered in school. Against a background of Leonardo's Vitruvian man, children measured their heights and arm-spans, and a computer recorded their measurements and calculated their ratio, showing that it was almost constant. A toy car moved on a rail by one's hand, produced on a screen a graph of distance versus time, allowing the visitor to "feel" what constant-speed motion or acceleration meant, and relate it to the graph. Other exhibits dealt with more advanced subjects—tiling the floor with "darts" and "kites" to produce a non-periodic Penrose tiling (fun and aesthetic), classifying knots (learning about chirality), or following Euler's path across the Koenigsberg bridges with a rope.

The organization of the exhibition was basically thematic. Its core was arranged in four halls, called Number, Shape, Pattern and Computing respectively. Nevertheless, the unity of mathematics and relations between the various areas were constantly emphasized. Balance between computer-based exhibits and mechanical ones was another issue. Whenever possible, we had a preference for the latter.

As a mathematician, I had to set aside my preconceptions and listen to the experience of curators and educators. Nevertheless, I believe that some of the messages, and the ways in which they were presented, would not have come across, if not for the involvement of the mathematicians. As much as it is important to present science in a friendly, appealing and accessible way, it is also important to adhere to its true nature and meaning, as perceived by the scientist. Resolving the potential conflict between these two goals is possible when Scientist and Educator work together in harmony.

Fundamental Issues of Math Education—Building a New Teacher Education Course at the Hebrew University

High school math teachers in Israel are required to hold both a B.Sc. in math, or in a related area, and a teaching certificate. Unfortunately, the two programs at the

Hebrew University (and to the best of my knowledge at most other universities in Israel) are not coordinated. The prospective teacher takes the same math classes, from logic to topology and differential equations, which any other undergraduate in mathematics would take. In fact, nobody at the department of mathematics takes notice of which of the 100 students in each cohort intends to become a teacher.

They then start, in their third year, certificate studies at the School of Education, where they focus mostly on pedagogy and general education courses. Very little is done to address didactical issues pertaining to mathematics. The practicum is conducted in the fourth year in selected participating schools, but it is often left to the older teachers in those schools to guide the would-be teachers in their first field experience. The old practices of teaching-to-the-test and emphasizing technique at the expense of understanding are then instilled from day one, and whatever spirit of reform the new teacher brings is washed away. Mathematicians, or researchers from the science teaching unit at the university have not been involved with the School of Education's teacher education program.

To add insult to injury, the mathematics department and the school of education at the Hebrew University are located in different campuses, separated by a 30-minute bus ride.

Changing this unfortunate scenario was the ultimate goal of Prof. Baruch Schwarz from the school of education, Prof. Abraham Arcavi from the Weizmann Institute Department of Science Teaching, and myself, when we met 2 years ago with the idea of upgrading the teacher education program, and in particular, forming collaboration between educators, science teaching experts, and scientists.

As a pilot for such a program, we devised and ran a year-long seminar on *Fundamental Issues of Math Education*, which met every Sunday for 2 hours at the school of education. All three of us were present in every class, as well as some 14 students of variable background and age. This was a unique experience. I am not aware of a similar joint effort in Israel, although courses dealing with didactical issues are probably well established worldwide. Every week we met for several hours to discuss between ourselves the coming weeks and the division of labor. The issues were discussed in depth, each of the three organizers contributing his particular angle. I have been exposed to articles and examples that enriched my understanding of mathematical teaching, and I hope my colleagues have profited here and there from my perspective as a mathematician.

The course was structured in such a way as to facilitate the collaboration. It was divided into 7 sections, and 4 weeks were devoted to each of them. Five of the sections were thematic: they dealt with the teaching of (1) arithmetic, (2) algebra and functions, (3) geometry and trigonometry (4) probability and data analysis and (5) calculus. Two sections were "horizontal"—dealing with (6) mathematical modeling and (7) problem solving.

Within each section, each of the organizers gave one 2-hour lecture. Naturally, I would speak about the mathematics of the concept, often in historical perspective, and my colleagues would discuss studies related to didactical questions, or the cognitive and psychological development of mathematical thinking. The last week within each section was a workshop conducted by one or two of the students, who were assigned tasks related to the material discussed in class. Typically, they had to prepare an activity or a school lesson, followed by a discussion among all of us.

The material did not necessarily overlap school curriculum, and no attempt was made to cover every aspect. Some of the students, having years of teaching experience behind them, contributed important insights. At other times we were surprised to see them miss what seemed to us obvious didactical points.

Examples of issues that were discussed included:

- Components of good teaching: understanding math, skill-building, developing mathematical sense and intuition,
- How to avoid compartmentalization: the unity of mathematics,
- Revisiting ideas and making connections among ideas in teaching,
- Procedural vs. conceptual learning,
- Order, pace and age adaptation,
- Application of advanced technologies in teaching.

All were discussed in-context within the sections, and not abstractly. As an example, the section on Functions and Algebra Teaching included:

Lecture 1: Ehud de Shalit: *Evolution of the function concept* (following Kleiner 1989)
Lecture 2: Baruch Schwarz: *Different presentation of functions—a didactic analysis*
Lecture 3: Abraham Arcavi: *Dynamical software and its use in teaching functions*
Lecture 4: Student workshop: *Four schemes for grade adjustment* (a class presentation).

We believe that courses of this sort can serve as a model for collaboration between mathematicians and mathematics educators in teacher education programs in the future.

What Can Mathematics Education Bring to Mathematics?

Patrick W. Thompson

The editors asked our group to address the matter of collaboration between mathematics and mathematics education. Collaboration between the two often is viewed from the perspective that mathematical content is within the purview of mathematicians and pedagogy is within the purview of mathematics education. A consequence of this view of collaboration is that discussions of collaboration assume that each field brings to a collaboration what is in their purview. I would like to pursue a different perspective—one in which mathematics education actually can contribute to mathematics in regard to the mathematical preparation of mathematics majors as well as to the mathematical preparation of future mathematics teachers.

I will develop this thesis through two examples at Arizona State University. The first is the development of a calculus course; the second is the development of a B.Sc. Mathematics degree program with a concentration in mathematics education.

Calculus

Calculus courses, in the US at least, are plagued by an orientation that calculus is nothing but procedures and facts. Students' understandings of calculus are often incoherent when viewed as a body of ideas; any coherence in their understandings is too often about just connections among procedures.

At ASU we have designed an experimental introductory calculus course (differential and integral calculus of one variable) that aims from the beginning to have mathematics and non-engineering science majors learn the calculus as a coherent body of ideas that are necessitated intellectually (Harel 1998, 2008a, 2008b). The course is necessitated by two fundamental problems: (1) You know how fast a quantity is changing and you want to know how much of it there is. This problem leads to the idea of accumulation functions and of an indefinite integral as an accumulation function. (2) You know how much of a quantity there is and you want to know how fast it is changing. This problem leads to the idea of instantaneous rate of change and properties of a function's behavior that can be discerned from its rate of change. A detailed description of the course appears in Thompson et al. (2013).

The course's curriculum emerged from a combination of mathematics education research on students' understandings of accumulation and rate of change (Carlson et al. 2003b; Schnepp and Nemirovsky 2001; Thompson 1994; Thompson and Silverman 2008; Yerushalmy and Swidan 2012), insights into the ways that curricula and instruction can be designed to motivate students' mathematical interest (Harel 2008a; Harel and Sowder 2005), and research on students' quantitative reasoning and uses of notation to represent it (Carlson et al. 2003a; Ellis et al. 2012; Gravemeijer and Doorman 1999; Johnson 2012; Kaput et al. 2007; Schoenfeld 2007; Selter et al. 2000; Smith and Thompson 2007; Thompson 1993, 1995). This body of research did not dictate to us what should constitute a calculus curriculum. Rather, it provided a way to think about meanings that belong evidently to the calculus as emerging from a mosaic of understandings that students typically build in school.

On one hand the design and experimentation of this course could be seen solely as a mathematics education effort. On the other hand, however, it is a true example of collaboration between mathematics and mathematics education. Our effort could not have happened without the support and trust of the Department's director and of the mathematics faculty. Several members of the first-year mathematics faculty are trying this new approach. More are participating with us in planning a grant proposal to investigate what students learn from in-principle different curricular and instructional approaches to major ideas in the calculus.

B.Sc. Mathematics with Mathematics Education Concentration

The Bachelor of Science degree in mathematics is the primary undergraduate degree in ASU's School of Mathematical and Statistical Sciences ("the School"). The School had already created concentrations within the B.Sc. in computational mathematics and statistics. Most recently it created a concentration in mathematics education. Moreover, with the support of ASU's Mary Lou Fulton Teachers College (MLFTC), Arizona's Department of Education (AZDoE) granted graduates of the Math Education concentration what is called *institutional recommendation*, meaning that graduates of the B.Sc. Mathematics/Mathematics Education will automatically receive a license to teach secondary mathematics. The School's B.Sc. Math/Math Education is Arizona's first program not housed in a college of education whose graduates receive an institutional recommendation for licensure.

ASU's Bachelor of Science in Math/Math Education resulted from a long collaboration among the School's mathematicians and mathematics educators (Luis Saldanha, Pat Thompson). In particular Fabio Milner (applied mathematics) was instrumental in obtaining university approval for the program. Bruno Welfert (mathematics) and Matthias Kawski (applied mathematics) supported our effort in their successive terms as Director of Undergraduate Studies. In addition, ASU's MLFTC was instrumental in assisting us to prepare proper documentation to support an application to the AZDoE for institutional recommendation of our program's graduates, and for including our application as part of theirs. MLFTC's support was essential, as it is the only body within ASU from which AZDoE will accept such proposals.

We designed the B.Sc. Math/Math Education degree so that it focuses deeply on its graduates' *Mathematical Knowledge for Teaching secondary mathematics* (MKTsm). Specifically:

(1) Students in our Math/Math Ed program take the School's standard program in mathematics for its B.Sc. Mathematics degree.
(2) They take a subset of the MLFTC program for secondary education majors. Students are not required to take MLFTC's general education courses that overlap with the specialized math education courses described below, in (3).
(3) Students take five courses that the School designed specifically to draw connections between mathematics and mathematics education. The five courses are:

    (a) *Algebra and Geometry in the High School* (Year 1). This is a conceptual overview of the secondary mathematics curriculum. At the same time that students take this course, they enroll in a field experience course called *Mentored Tutoring*. Mentored Tutoring has students in the review course work with students in remedial mathematics courses under the guidance of the review course's instructor.
    (b) *Technology and Mathematical Visualization* (Year 2). The TMV course is designed to have students re-conceive the mathematics they know so that symbolic representations have imagistic content. This is not a programming class. We use software, primarily Geometer's Sketchpad (GSP) for geometry and Graphing Calculator (GC; Avitzur 2011) for everything else. The idea of the course is that students need to engage in mathematical thinking to create visualizations that might help them convey a particular mathematical idea to students. As a simple example, we set the problem of how to define a function that takes two points (of dimension 2 or 3) as input and

produces a graph of the segment that connects them.[5] They must not only define the function, they must explain why the function produces what was requested. A second example is that students must define a function g whose graph will be a plane that is tangent to the graph of an arbitrary function of two variables f at an arbitrary point on f's surface. A user of a student's project must have control over the definition of f and the location of the arbitrary point. The plane must adjust dynamically as the user moves the point of tangency.

(c) *Curriculum and Assessment in Grades 7–12.* In this course we introduce students to curricula from various countries and to principles of assessing school students' understandings of the mathematics in them. We have not yet offered this course, but we anticipate that, for US students, it will be an eye-opening experience for them to see the mathematics that other countries expect their students to learn in high school.

(d) *The Development of Mathematical Thinking.* In essence, this course will introduce students to research on the development of additive and multiplicative reasoning. This is another course we have yet to offer, but our intent is for students to become consciously aware of different ways that school students' might understand mathematical ideas that teachers often take as unproblematic. We also see this course as helping our students conceptualize the school mathematics curriculum as entailing their students' development of systems of ideas over time.

(e) *Research Project in Mathematics Education.* This is a seminar in which students will design, conduct, and interpret a teaching experiment with one or two high school students. We see the Project course as a culminating experience through which our students will draw from what they learned in the courses described above.

We counsel students enrolling in the B.Sc. Math/Math Ed to enroll from the start in the experimental, conceptually oriented calculus that we designed. We feel that moving future teachers from a procedure-oriented mathematics to an idea-oriented mathematics is a long process, and that it is unlikely to happen if their university mathematics continues their school practice of mathematics as memorization.

Mathematics educators in the School are also engaged in the design of curriculum for students who are not in education. Kyeong Hah Roh, who has a Ph.D. in mathematics education and a Ph.D. in mathematics (differential geometry), worked with mathematicians on our faculty to redesign *Mathematical Structures*, a course required of all students in any mathematics concentration. The course gives an introduction to proof and higher mathematics. Dr. Roh also redesigned our undergraduate advanced calculus and real analysis courses based on research on students'

---

[5]One solution: $f(X, Y) = (1 - t)X + tY$, $0 \leq t \leq 1$. GC produces a graph that is a segment in 2- or 3-space, depending upon the dimension of vectors X and Y. An explanation of why this works necessarily involves two things: imagining the value of t varying in small increments and describing the role of proportionality in traversing the hypotenuse of a right triangle.

learning of proof, functions, and limits. Marilyn Carlson led a 10-year research and development project to transform the School's precalculus course, which is a remedial course for students who are unprepared to take calculus. The redesign is rooted firmly in developmental research on students' difficulties in learning mathematical ideas that are essential for students to succeed when they reach calculus, such as deep understandings of linearity, rate of change, and the concept of function.

Comments on Collaboration and Its Outcomes

It might be useful to discuss the nature of the collaborations I've described and about places of friction where things did not go smoothly. The calculus redesign was an outgrowth of my and Marilyn Carlson's research. The School's contribution was to allow the redesign on an experimental level. Actual collaboration began with the attempt to increase the number of course sections using the redesigned curriculum. Jay Abramson and Mark Ashbrook have been instrumental in that effort. However, other instructors have been reluctant to adopt this new curriculum and approach.

The redesign of *Mathematical Structures*, advanced calculus, and real analysis were an outgrowth of Kyeong Hah Roh's research on teaching and learning mathematics. Her redesign was successful in terms of outcome measures, but other instructors of these courses have been slow to pick up Roh's changes.

Marilyn Carlson's redesign of precalculus, also an outgrowth of her research, has been the most successful of the innovations. After several years of resistance among perennial instructors of precalculus, all sections at ASU are now using her curriculum. Fabio Milner and several first-year mathematics instructors were integrally involved in the redesign, giving substantive input regarding the mathematical treatment of ideas, and were important supporters in the politics of curriculum change.

Regarding the B.Sc. Mathematics/Mathematics Education, the School faculty voted to approve this concentration. So the general acceptance among mathematicians that mathematics has an important stake in mathematics education is evident.

The friction in all these moving parts comes from the fact that few mathematicians understand aims, methods, and results of mathematics education as a discipline. The comment, "So, you train teachers how to teach math, right?" is not uncommon. It is a revelation to many who spend time working with us that we take mathematics seriously—in some ways more seriously than they do. Conceptual coherence in the mathematics that is actually conveyed through discourse is of central importance in mathematics education, and we find that it is less important in mathematics. By "less important in mathematics" I mean that language and actions in a mathematician's classroom often have little chance of being interpreted by students as anything remotely resembling what the instructor intended. When we address this problem (intended meaning is the meaning actually conveyed) in curriculum, mathematicians are often puzzled by what we are trying to teach. They are accustomed to discourse in which their personal mathematical language is the language in which ideas are offered to students. They fail to realize that the courses we designed often are more conceptually rigorous than the versions they teach, because our courses are designed with the goal (and expectation) that students actually understand the mathematical ideas taught. As I say to my mathematics colleagues, "Mathematics education is easy—until you take student learning seriously."

Though mathematics education as a discipline is sometimes understood poorly, good things happened nevertheless. There is enough trust, little enough mistrust, and enough shared commitment to address problems in our students' learning to let innovation blossom.

Conclusion

The four discussions of collaboration between mathematics education and mathematics highlight many ways that collaboration can happen and many levels of social organization at which it can happen. Sometimes collaboration is between professional societies; sometimes collaboration is between individuals engaged in a shared task.

Running through the authors' examples is the theme laid by Artigue when she said, "... no substantial and sustainable improvement of mathematics education can be obtained without building on the complementarity of [math and math ed] expertise, without their common engagement and coordinated efforts." Törner illustrated this in his discussion of the separation of mathematics and mathematics education in Germany decades ago and the subsequent realization that, to have an influence at a national level, the two disciplines needed each other to address the problem of systemic sources of unmet expectations about students' mathematical learning.

This chapter's examples also illustrate that, at all levels of collaboration, individuals matter and institutions matter—simultaneously. At a level of collaboration between societies, it is important that individual players have vision and commitment to address problems of mathematics education—and a standing within their respective fields that allows them to exert influence with others in their societies. At a level of personal collaboration, collaborators' efforts happen within institutions whose structures either enhance or obstruct their efforts. The physical separation of education and mathematics at Hebrew University constrained collaborative efforts to improve the University's teacher education program. The inclusion of mathematics education within ASU's School of Mathematical and Statistical Sciences afforded collaboration in the design of a program in which mathematics and mathematics education are often addressed simultaneously within individual courses. The location of mathematics education within the School, and the School's support of it, was also a major factor in the University's approval of the program.

The chapters' examples also point to a shared a trait noted by Törner: successful collaboration requires mutual trust and respect among collaborators in the context of a shared commitment to solving a problem. This is not to say that there cannot be misunderstanding of each other's values, commitments, or competence regarding the nuances of the problem. Rather, the nature of trust is that collaborators carry a commitment to listen respectfully to each other and be open to modifying their positions. The examples by de Shalit of the Meet Math Exhibit and the design of a teacher education program at Hebrew University illustrate this point well. In Thompson's example of calculus redesign there were deep and prolonged discussions of the meanings of rate of change and of differential that would prove foundational for students' future learning, which led to sustained conversations of how the course might be shaped to support students development of those meanings and how it might be shaped to build upon those meanings.

We end by emphasizing a comment by Törner and illustrated by the other three authors. It is that successful collaboration between mathematics and mathematics education is most probable when collaborators have a shared commitment to a problem and believe that others in the effort have something to contribute to its solution.

References

Alibert, D., Artigue, M., Courdille, J. M., Grenier, D., Hallez, M., Legrand, M., Menigaux, J., Richard, F., & Viennot, L. (1988). Le thème "différentielles"—un exemple de coopération maths-physique dans la recherche. In G. Vergnaud, G. Brousseau, & M. Hulin (Eds.), *Didactique des acquisition des connaissances scientifiques. Actes du colloque de sèvres*, May 1987 (pp. 7–45). Grenoble: La Pensée Sauvage.

Artigue, M. (1992). Functions from an algebraic and graphic point of view: cognitive difficulties and teaching practices. In E. Dubinski & G. Harel (Eds.), *MAA notes: Vol. 25. The concept of function—aspects of epistemology and pedagogy* (pp. 109–132). Washington: Math. Assoc. of America.

Artigue, M. (2009). ICMI: a century at the interface between mathematics and mathematics education. In M. Menghini, F. Furinghetti, L. Giacardi, & F. Arzarello (Eds.), *The first century of the International Commission on Mathematical Instruction (1908–2008). Reflecting and shaping the world of mathematics education* (pp. 185–198). Roma: Istitturo della Enciclopedia Italiana.

Avitzur, R. (2011). *Graphing calculator (version 4.0)*. Pacific Tech., Berkeley, CA.

Brousseau, G. (1997). *Theory of didactical situations in mathematics* (trans: Balacheff, N., Cooper, M., Sutherland, R., & Warfield, V.). Dordrecht: Kluwer Academic.

Carlson, M. P., Larsen, S., & Lesh, R. A. (2003a). Modeling dynamic events: a study in applying covariational reasoning among high performing university students. In R. A. Lesh & H. Doerr (Eds.), *Beyond constructivism in mathematics teaching and learning: a models and modeling perspective*. Hillsdale: Erlbaum.

Carlson, M. P., Persson, J., & Smith, N. (2003b). Developing and connecting calculus students' notions of rate-of-change and accumulation: the fundamental theorem of calculus. In N. Patemen (Ed.), *Proceedings of the 2003 meeting of the international group for the psychology of mathematics education—North America* (Vol. 2, pp. 165–172). Honolulu: University of Hawaii Press.

Ellis, A. B., Ozgur, Z., Kulow, T., Williams, C., & Amidon, J. (2012). Quantifying exponential growth: the case of the Jactus. In R. Mayes, R. Bonillia, L. L. Hatfield, S. Belbase (Eds.), *WISDOMe monographs: Vol. 2. Quantitative reasoning: current state of understanding* (pp. 93–112). Laramie: University of Wyoming.

Enzensberger, H. M. (1999). *Drawbridge up—mathematics—a cultural anathema*. Nattick: AK Peters.

Gravemeijer, K., & Doorman, M. (1999). In *Educational studies in mathematics: Vol. 39. Context problems in realistic mathematics education: a calculus course as an example* (pp. 111–129).

Harel, G. (1998). Two dual assertions: the first on learning and the second on teaching (or vice versa). *The American Mathematical Monthly, 105*, 497–507.

Harel, G. (2008a). DNR perspective on mathematics curriculum and instruction, part I: focus on proving. *ZDM. The International Journal on Mathematics Education*, *40*, 487–500.
Harel, G. (2008b). What is mathematics? A pedagogical answer to a philosophical question. In R. B. Gold & R. Simons (Eds.), *Current issues in the philosophy of mathematics from the perspective of mathematicians* (pp. 265–290). Washington: Math. Assoc. of America.
Harel, G., & Sowder, L. (2005). Advanced mathematical-thinking at any age: its nature and development. *Mathematical Thinking and Learning*, *7*, 27–50.
Hoechsmann, K., & Törner, G. (2004). Schisms, breaks, and islands—seeking bridges over troubled waters: a subjective view from Germany. In D. McDougall (Ed.), *Proceedings of the 26th conference for the international group for the psychology of mathematics education (PME)—North America chapter (PME-NA)* (Vol. 3, pp. 993–1001). Toronto: PME-NA.
Johnson, H. L. (2012). Reasoning about variation in the intensity of change in covarying quantities involved in rate of change. *The Journal of Mathematical Behavior*, *31*, 313–330.
Kahane, J. P. (2001). *L'enseignement des sciences mathématiques*. Paris: Odile Jacob.
Kaput, J. J., Blanton, M. L., & Moreno, L. (2007). Algebra from a symbolization point of view. In J. J. Kaput, D. W. Carraher, & M. L. Blanton (Eds.), *Algebra in the early grades* (pp. 19–56). New York: Erlbaum.
Kleiner, I. (1989). Evolution of the function concept: a brief survey. *The College Mathematics Journal*, *20*, 282–300.
Pelay, N. (2011). *Jeux et apprentissage mathématique. Elaboration du contrat didactique et ludique en contexte d'animation scientifique*. Doctoral thesis, Université de Lyon 1. http://bit.ly/XaywDi. Accessed 9 June 2013.
Schnepp, M., & Nemirovsky, R. (2001). Constructing a foundation for the fundamental theorem of calculus. In A. Cuoco & F. Curcio (Eds.), *The roles of representations in school mathematics*, Reston: National Council of Teachers of Mathematics.
Schoenfeld, A. H. (2007). Early algebra as mathematical sense making. In J. J. Kaput, D. W. Carraher, & M. L. Blanton (Eds.), *Algebra in the early grades* (pp. 479–510). New York: Erlbaum.
Selter, C., Verschaffel, L., Greer, B., & De Corte, E. (2000). Making sense of word problems. *Educational Studies in Mathematics*, *42*, 211–213.
Smith, J., & Thompson, P. W. (2007). Quantitative reasoning and the development of algebraic reasoning. In J. J. Kaput, D. W. Carraher, & M. L. Blanton (Eds.), *Algebra in the early grades* (pp. 95–132). New York: Erlbaum.
Thompson, P. W. (1993). Quantitative reasoning, complexity, and additive structures. *Educational Studies in Mathematics*, *25*, 165–208.
Thompson, P. W. (1994). Images of rate and operational understanding of the fundamental theorem of calculus. *Educational Studies in Mathematics*, *26*, 229–274.
Thompson, P. W. (1995). Notation, convention, and quantity in elementary mathematics. In J. Sowder & B. Schapelle (Eds.), *Providing a foundation for teaching middle school mathematics* (pp. 199–221). Albany: SUNY Press.
Thompson, P. W., & Silverman, J. (2008). The concept of accumulation in calculus. In M. P. Carlson & C. Rasmussen (Eds.), *MAA notes: Vol. 73. Making the connection: research and teaching in undergraduate mathematics* (pp. 43–52). Washington: Math. Assoc. of America.
Thompson, P. W., Byerley, C., & Hatfield, N. (2013). A conceptual approach to calculus made possible by technology. *Computers in the Schools*, *30*, 124–147.
Yerushalmy, M., & Swidan, O. (2012). Signifying the accumulation graph in a dynamic and multi-representational environment. *Educational Studies in Mathematics*, *80*, 287–306.

Postscript

Chapter 19
We Must Cultivate Our Common Ground

Jeremy Kilpatrick

Abstract The present volume addresses some valuable themes relevant to the scholarly contributions of Ted Eisenberg to the teaching, learning, and doing of mathematics. In this reflective postscript, I raise some issues associated with the identity of, and the tension between, the academic fields of mathematics and mathematics education. I argue that, far from having drifted apart, those fields continue to make productive contact with and complement one another. Their common preoccupation with mathematics as it is created through teaching keeps them together. The challenge to mathematicians and mathematics educators is to make fertile the common ground they share.

Keywords Mathematics · Mathematics education · Academic fields · Community · Teaching · Ted Eisenberg

Communities of Mathematics and Mathematics Education

To judge by the chapters in the present volume, the spring 2012 symposium honoring Ted Eisenberg must have been a fascinating, frustrating occasion: fascinating because of the wealth of challenging ideas put forward by these distinguished mathematicians and mathematics educators; frustrating because so many of the ideas were developed only sketchily and, despite the best efforts of the synthesizers for each panel, not always well integrated. The themes of the plenaries and panels—mutual expectations, history, problem solving, mathematical literacy, visualization, justification and proof, policy, and collaboration—appear to have been chosen not merely, as Michael Fried says in his introduction, because they reflect "commonality and difference joining and dividing the communities of mathematics and mathematics education" but also because they all touch on issues that Ted has addressed at one point or another throughout his long, productive career. An elaboration of any one of those themes and the issues it raises concerning the common ground between mathematics education and mathematics could easily have filled a book this

J. Kilpatrick (✉)
University of Georgia, Athens, GA, USA
e-mail: jkilpat@uga.edu

M.N. Fried, T. Dreyfus (eds.), *Mathematics & Mathematics Education: Searching for Common Ground*, Advances in Mathematics Education,
DOI 10.1007/978-94-007-7473-5_19,

size. With eight themes plus the "dialogue on the dialogue" regarding the identity of mathematics education as a field as it relates to mathematics as a field, the book cannot possibly do justice to the many points the authors raise. Nor can I in this reflection do anything like adequate justice to any of those points. Instead, I address questions of identity and challenge the contention, made in the introduction, that there is a "growing divide between the mathematics community and the mathematics education community." I see those communities as closely intertwined as they have ever been—even though from some angles of vision they might appear to be moving apart. We have common ground whether we know it or not.

Identity Issues

In her contribution to the volume, Norma Presmeg cites the 1998 book edited by Anna Sierpinska and me, *Mathematics Education as a Research Domain: A Search for Identity*. That book was the report of a multi-year study undertaken for the International Commission on Mathematical Instruction (ICMI) in an effort to answer the question: What is research in mathematics education, and what are its results? The question had been raised by mathematicians concerned with issues of mathematics education and uncertain as to the contributions that research was making to the enterprise. The study began in August 1992 with the production of a discussion document describing the reasons for the study and raising a number of questions the study would attempt to address. That document, which called for papers addressing those questions to be submitted by 1 September 1993, was published in early 1993 in several bulletins and journals concerned with mathematics education. The papers and other expressions of interest were used to develop a program for an invitational study conference of more than 80 people held in and near Washington, DC, in May 1994. Preliminary results of the study were reported at the International Congress of Mathematicians in Zürich in 1994 and at the Eighth International Congress on Mathematics Education in Seville in 1996.

Anna and I used the subtitle "A Search for Identity" because it had become clear at the outset of the study that it would not resolve the question of what research in mathematics education is or should be and that, instead, the study would lead to further questions such as what it means to be a researcher in mathematics education and whether we have a common identity. Our strategy for choosing the subtitle, however, proved a bit foolhardy: Reviewers of the book made clever remarks about an adolescent field seeking selfhood and pointed out that research in mathematics was nowhere near as uncertain about its sense of self as research in mathematics education appeared to be.

In a paper prepared for the centennial of the journal *L'Enseignement Mathématique*, I asked whether some of the self-questioning about the field's identity might be understood in light of history:

> Perhaps the questioning comes from the history of the field, and in particular, from the way mathematics education has developed internationally, as illustrated in the ICMI. One can

> also ask, what other field would have the officers of its premier international organization appointed by a group outside the field? The ICMI is a commission of the International Mathematical Union (IMU) and is therefore subject to IMU's oversight. Could it be that the insecurity and apparent disarray of the field, despite its growth and accomplishments through the twentieth century, might stem in part from the way it has been treated by mathematicians? (Kilpatrick 2003, p. 328)

I went on to make the following argument, which conflicts with some—although certainly not all (e.g., those of Norma Presmeg and Steve Lerman)—of the arguments in the present book:

> Mathematics education is not a branch of mathematics, nor does it belong among the arts and sciences. It is a separate field with very different traditions, foundations, problems, methods, and results. It is much more contingent on history and culture than mathematics could ever be, and that is part of the reason for what outsiders perceive as a field in disarray. (p. 329)

I found it remarkable to observe how the protagonists in the present collection of papers are so often taken to be a triad: mathematicians, mathematics educators, and researchers in mathematics education. Examples include Norma Presmeg's model of "complex human worlds," Jonas Emanuelsson's characterization of the groups whose gap in perception of the nature of mathematics and mathematics education needs to be bridged, and Michael Fried's characterization of different communities, to which he added the community of historians of mathematics.

One issue that selection of protagonists raises is the question of what a mathematics educator is. Is a teacher of mathematics in primary or secondary school a mathematics educator? Some would say yes, but others reserve the term *mathematics educator* for people in tertiary education or serving in other educational enterprises that are not schools. In Europe, the term *didactician* is often used in place of *educator* (see, e.g., the paper by Michèle Artigue), and there the distinction between didactician and teacher is clear.

A second issue is the asymmetry of the two classifications: two kinds of workers in mathematics education, and one kind of worker in mathematics. Is it only in mathematics education that there is a distinction between researchers and non-researchers? Is it possible to be a mathematician and not be a researcher in mathematics? In some articles in the present collection, *mathematician* is preceded by *research*, but there seems to be throughout an unspoken assumption that all mathematicians are (or could be) doing research.

Another issue is what it takes to be an expert in mathematics. In testimony before the Texas legislature regarding the Common Core State Standards in Mathematics (National Governors Association Center for Best Practices, Council of Chief State School Officers 2010), Jim Milgram (2011), professor emeritus of mathematics at Stanford, made the following observation: "I was … one of the 25 members of the CCSSO/NGA Validation Committee, and the only content expert in mathematics." Three of the 25 members of that committee were eminent teachers of school mathematics, and five were university mathematics educators. Although several of the latter worked in departments of mathematics, and although many of the eight possessed graduate degrees in mathematics, they did not, by Jim's standards, qualify as experts on the content of school mathematics—presumably because they lacked a doctorate in mathematics.

That assumption raises the further issue of who a mathematician is. Is having earned a doctorate in mathematics the only criterion? Or is having done research in mathematics beyond the doctorate also a criterion? One can become a mathematics educator simply by declaring one's interest in the field; might that also be true of becoming a mathematician? I simply note here that nothing in the book at hand addresses, let alone resolves, the issue.

Expanding; Not Moving Apart

Many of the chapters in the book suggest that there has been some movement of mathematics education away from mathematics, a movement propelled by the ways in which research in mathematics education has become less concerned with mathematics and more concerned with such matters as, according to Michael Fried's introduction, "psychology of learning, cultural differences, and social justice." My view is that any such movement is something of an illusion caused by the enormous growth of the field of mathematics education.

In 2008, in connection with the ICMI centennial, I searched the Web for "mathematics education." Using Google, I got 1,280,000 hits; using Google Scholar, I got 129,000 (Kilpatrick 2008). Five years later, in February 2013, I conducted a similar search. Using Google, I got 3,100,000 hits; using Google Scholar, 287,000. In each case, the number had more than doubled in just 5 years. Although the literature base of mathematics education has certainly expanded, and perhaps especially in recent years, I think it is the wrong metaphor to say that the fields of mathematics and mathematics have moved apart. The centers of gravity may be further apart than they used to be, but I do not think it is reasonable to conclude that the fields are now more separated than in the past.

Some years ago, a conference was held to examine US doctoral programs in mathematics education (Reys and Kilpatrick 2001). A survey had determined that there was a huge array of such programs, large and small, with some located in departments of mathematics and others in colleges or schools of education. Some institutions were granting from 4 to 7 mathematics education doctorates a year; others were averaging fewer than 1 in 4 years. A surprising characteristic of a number of the programs was their minimal requirements or expectations concerning the advanced study of mathematics. That characteristic, however, was not true of the two largest programs—those at Teachers College, Columbia, and at the University of Georgia. At both institutions, by the time doctoral students finished their program, they were expected to have either a master's degree in mathematics or its equivalent.

It would be easy to conclude from the survey that many of the graduates of US doctoral programs are being minimally prepared in mathematics. That may be true for some programs. But as several authors in the present book observe, and others have observed as well (Bass and Hodgson 2004), mathematics education is a multidisciplinary field. Doctoral students in mathematics education need to acquire knowledge, skills, and abilities in a variety of fields beyond mathematics, among them: educational research methods, philosophy, psychology, sociology, anthropology, history, linguistics, semiotics, and educational policy.

At the University of Georgia, we insist that whatever topic students choose for their dissertation research, it needs to involve mathematics in a serious and intensive way. Others may do research in which mathematics is taken as a placeholder and might be easily replaced by another school subject such as biology or history. For example, an education researcher might examine the effects of homework assignments on mathematics learning. If that study does not, however, take seriously the questions of what sorts of assignments are being made and what mathematics is being learned—in other words, if it takes mathematics as a black box—then we do not consider it an appropriate study for a doctorate in mathematics education. The study might be useful to mathematics educators, but if it does not treat mathematics as problematic and open to analysis, then it does not belong to the field. The field of mathematics education depends on the maintenance of strong connections with mathematics, its many applications, and its history and cultural contexts.

I would argue that mathematics and mathematics education are bound together like yin and yang:

> Mathematics and mathematics education have a synergistic relation, and neither can exist without the other. In my view, mathematics education has not attained the status of a discipline, and it is not completely a profession. But as an academic field, it is linked to mathematics through a mutual concern with teaching. (Kilpatrick 2008, p. 36)

That concern connects the two fields better than one might at first think. As Michael Fried says in his introduction, when Andrew Wiles was working on his proof of Fermat's last theorem, he ended up running a one-student seminar with Nick Katz. Michael notes that "the difference between doing and teaching mathematics is actually never very great in that mathematicians must always communicate their thinking." Jens Høyrup (1994) made essentially the same point when he observed that "one aspect of mathematics as an activity... is to be a reasoned discourse; ...as an organized body of knowledge [it is] the *product of communication by argument*" (p. 3). Consequently, "teaching is not only the *vehicle* by which mathematical knowledge and skill is transmitted from one generation to the next; it belongs to the *essential characteristics* of mathematics to be *constituted through teaching*" (p. 3).

Cultivating; Not Searching

In 2004 and 2005, Richard Schaar, a mathematician with Texas Instruments, brought together three mathematicians (Jim Milgram, Wilfried Schmid, and Richard) and three mathematics educators (Deborah Lowenberg Ball, Joan Ferrini-Mundy, and me) to engage in "constructive discourse between mathematicians and mathematics educators in order to seek common ground in their mutual efforts to improve K–12 mathematics teaching and learning" (Mathematical Association of America 2013).

Given the task of sitting down together to craft a consensus statement, we soon recognized that, once we had put aside stereotypes and become better acquainted, there were a surprising number of issues in school mathematics teaching and learning on which we could agree. Here is our account of how the process worked:

> We tried to bring clarity to key perspectives on K–12 mathematics education. We began by exploring typical "flashpoint" topics and probed our own positions on each of these to determine whether and where we agreed or disagreed. For the first meeting, held in December 2004, we began with summary statements drawn from prior exchanges among the members of our group. We affirmed some agreements in this meeting and "discovered" others. We listened closely to one another, frequently asking for clarification or for examples. We tested our understanding of others' points of view by proposing statements that we then examined collectively. We drafted this document as a group, composing actual text as we worked. One of us typed, and our emerging draft was projected onto a screen in the meeting room. The process enabled us to take issue with particular words and terms and then reshape them until all of us were satisfied. We were forced to look closely at our own language and to seek common ground, not only in the terms we used but even in their nuanced meaning. (Ball et al. 2005, p. 1055)

The resulting document contains some basic premises that underlay our claims together with brief paragraphs on seven areas on which we found common ground—ranging from the automatic recall of basic facts to teacher knowledge.

My experience with the Common Ground Committee, whose work was funded by the National Science Foundation and Texas Instruments, Inc., was similar to an experience I had some three decades earlier with Morris Kline, one of the most outspoken critics of the new math in US schools. When you have a conversation with people who have expressed strong views about a topic like school mathematics, and you get them away from microphones and reporters—or in today's world, away from tweets and blogs—they moderate their language substantially. You discover that there are many areas of mutual agreement that might not have surfaced previously. Consequently, I strongly believe that mathematicians and mathematics educators already occupy considerable common ground.

The problem seems to be largely one of communication and mutual respect, both of which are enhanced by collaboration. The account by Deborah Ball and Hy Bass in the present volume provides dramatic affirmation of the value of sustained collaboration, affirmation that is well supported by the comments of Pat Thompson, Michèle Artigue, Günter Törner, and Ehud de Shalit. For many years, the US National Science Foundation, in awarding grants for work in mathematics education, has encouraged grantees to include mathematicians among their advisors. That practice has helped many projects avoid serious mathematical errors as well as developing a better appreciation by mathematicians of the work of their mathematics education colleagues. It is not always the case, however, that a mathematician's critique is valid. For example, Hung-Hsi Wu (1997) once complained about a school mathematics textbook that failed to include the formula relating radians to degrees. His complaint ignored the instructional issue thereby posed: The textbook authors had formulated an exercise in which students were to find the formula, and putting it into the textbook would have made that exercise pointless (Kilpatrick 1997). With Wu as an example, Michael Fried, in his introduction, points up the need for communication, mutual respect, and collaboration in both directions—on the part of both mathematicians and mathematics educators. I could not agree more. That collaboration should address the common ground we share and work to make it bloom. As Candide so wisely said, observing that this is, in fact, not the best of all possible worlds: "We must cultivate our garden."

References

Ball, D. L., Ferrini-Mundy, J., Kilpatrick, J., Milgram, R. J., Schmid, W., & Schaar, R. (2005). Reaching for common ground in K-12 mathematics education. *Notices of the American Mathematical Society*, *52*, 1055–1058. Reprinted in Winter 2007 in *Leadership Information*, *6*(1), 13–15.

Bass, H., & Hodgson, B. R. (2004). The International Commission on Mathematical Instruction. What? Why? For whom? *Notices of the American Mathematical Society*, *51*, 639–644.

Høyrup, J. (1994). In *Measure, number, and weight: studies in mathematics and culture*, Albany: State University of New York Press.

Kilpatrick, J. (1997). Confronting reform. *American Mathematical Monthly*, *104*, 955–962.

Kilpatrick, J. (2003). Scientific solidarity today and tomorrow. In D. Coray, F. Furinghetti, H. Gispert, B. Hodgson, & G. Schubring (Eds.), *Monographie de L'Enseignement Mathématique: Vol. 39. One hundred years of L'Enseignement Mathématique: moments of mathematics education in the twentieth century* (pp. 317–330).

Kilpatrick, J. (2008). The development of mathematics education as an academic field. In M. Menghini, F. Furinghetti, L. Giacardi, & F. Arzarello (Eds.), *The first century of the international commission on mathematical instruction (1908–2008): reflecting and shaping the world of mathematics education* (pp. 25–39). Rome: Istituto della Enciclopedia Italiana.

Mathematical Association of America (2013). Finding common ground in K–12 mathematics education. http://www.maa.org/common-ground. Accessed 30 May 2013.

Milgram, J. (2011). James Milgram on the new core curriculum standards in math. http://parentsacrossamerica.org/james-milgram-on-the-new-core-curriculum-standards-in-math. Accessed 30 May 2013.

National Governors Association Center for Best Practices, Council of Chief State School Officers (2010). Common Core State Standards in Mathematics. National Governors Association Center for Best Practices, Council of Chief State School Officers, Washington, DC. http://www.corestandards.org/Math. Accessed 30 May 2013.

Reys, R. E., & Kilpatrick, J. (Eds.) (2001). *One field, many paths: US doctoral programs in mathematics education*. Washington: Conference Board of the Mathematical Sciences.

Sierpinska, A., & Kilpatrick, J. (Eds.) (1998). *Mathematics education as a research domain: a search for identity*. Dordrecht: Kluwer Academic.

Wu, H. (1997). The mathematics education reform: why you should be concerned and what you can do. *American Mathematical Monthly*, *104*, 946–954.

Appendix A: Homage to Ted Eisenberg

Introduction

As Francis Lowenthal notes in his address below, most speakers at the symposium began or ended their contributions by relating to Ted Eisenberg's work. Where Ted comes into people's lives, one way or another, it always leaves a mark. Rather than reprinting all of these comments here, let these remarks by David Tall suffice to exemplify their spirit:

> ... Finally, and most importantly on this celebratory occasion, I pay tribute to Ted Eisenberg, who has played a subtle role in this long-term development with his desire to encourage students to grasp the aesthetic values of mathematical proof and his criticism of a behaviorist approach to learning. Out of respect for the thinking of others, even when they are different, indeed *because* they are different, we may come to a greater insight into how we can make sense of mathematics in general and, in particular, how we make sense of mathematical reasoning and proof. (David Tall, symposium lecture, May 2, 2012)

With the purpose of giving the reader an overview of the gist of Ted's scientific work over the more than 40 years since he obtained his PhD, we also include, at the end of this appendix, an annotated bibliography of the most important papers he authored.

However, the main part of the appendix consists of two addresses given on the occasion of the festive dinner in honor of Ted that took place during the symposium on May 2, 2012.

M.N. Fried, T. Dreyfus (eds.), *Mathematics & Mathematics Education: Searching for Common Ground*, Advances in Mathematics Education,
DOI 10.1007/978-94-007-7473-5,

Ted as Advisor and Colleague

Tommy Dreyfus

This may surprise you, but I met Ted in something like a prison cell. It was in a nice park, but then you had to go into one of the older buildings, to the basement, through long corridors until you arrived at a door that led into something like a small shelter or darkroom. There was a light bulb and two desks, one for Ted and one for me. Luckily for us, the whole thing was located in the Weizmann Institute and belonged to a wonderful mathematics education team. And luckily for me, Ted was there.

That was in 1977. We both spent a lot of time there. We both had small children at home (I think his were 1, 3 and 5, and mine were 2 and 4) and it was not easy to work at home. Home was, respectively, the responsibility of Polly and Marianne. This was my first and best opportunity to get to know Polly, her devotion to the family, to education, and to teaching, whether it was mathematics or English. All those of us who have known her are very sad that she is not with us this week.

So at that time, Ted and I spent a lot of time in the prison cell, and I can tell you, you get to know each other quite well in this way. It was an incubator—no, not for hatching chicks but for ideas and papers.

The ideas ranged from empirical research on intuitions about functions via issues of visualization, to issues concerning the acceptance of mathematical theories, and aesthetics in mathematics. Writing with Ted was a lot of fun and very satisfying. He did well where I was weak: After we collected data, analyzed, and discussed things thoroughly, sometimes for months, he could sit down for just a few hours and come up with a complete draft for a paper that was quite unintelligible to anybody except me. Everything was there but it was often cryptic, and the arguments had to be pieced together from shreds. I hope I then did reasonably well doing what Ted was less good at.

T. Dreyfus (✉)
Department of Education in Mathematics, Science and Technology, Tel Aviv University, Tel Aviv, Israel
e-mail: tommyd@post.tau.ac.il

M.N. Fried, T. Dreyfus (eds.), *Mathematics & Mathematics Education: Searching for Common Ground*, Advances in Mathematics Education,
DOI 10.1007/978-94-007-7473-5,

There was commitment from both of us: When there was a deadline, we were there in the prison cell at all hours of the day and night; I remember one instance when we stayed there until about four hours before Ted's 7 am flight to the US, where he hand delivered the paper the next day (no, there was no email at the time).

We also had fights. The fights between us arose after the paper was written: Who would be the lead author? If you know Ted, you also know what form the fights took. You may know the story of the woman who conceived but didn't give birth for many years; when the doctors finally decided to operate, they found two bearded men, each bowing to the other one and saying "After you, please". So Ted always fought for being *second* author. And guess who won: We have over 30 common publications, and on two thirds of them I am first author. We have 14 common peer reviewed journal articles, and on only 3 of them is Ted the first author.

Well, not only did this help my promotions, but having come to the Weizmann Institute in 1977 as a theoretical physicist, it was also the collaboration with Ted that introduced me to mathematics education as a scientific domain in the first place, to designing and carrying out research studies, writing papers based on them, presenting the work at conferences, editing special issues of journals, and so on. In other words, in the prison cell and afterwards, in a selfless and deliberate way, Ted put me on the path of my career. Thank you, Ted.

Thank You, Ted!

Francis Lowenthal

If I were a really nice guy, I would first address myself to the authorities: the Dean of the Faculty of Human and Social Sciences and the Dean of the Faculty of Natural Sciences, the chairperson of the Department of Mathematics, the organizers of this excellent symposium and Ina who helped so much. But I am not a nice guy, as everybody knows; I will thus start with those words: "Thank you Ted".

Ted, I am not one of your former students, nor did we spend long hours in the same tiny office. I am only a friend. I will thus start with the words others used to end their speeches.

Ted, we met a very long time ago, during your first PME conference. It was in Grenoble, in a bus. We immediately sympathized. You are only a few years older than I am, but you started speaking using what sounded like fatherly words. It was in 1981 and I was in love, but I had not told anybody, except the lady concerned of course. I don't know why, but I immediately told you. I do not remember exactly how you answered, but as a consequence of that as soon as I was back in Belgium, I proposed. And the lady accepted. And Christiane is here with me.

Thank you Ted!

We live far away from one another, but using e-mails (the very beginning of e-mails!) we kept in touch. We always remained in contact. There were very good and less good moments, depending on the events. There were also very sad ones. And there were meetings everywhere, including in Israel and stays here in Beer Sheva. This is how I met Polly and the three Eisenberg girls: Rivka, Naomi and Davida.

But one of these stays remains radiant in my memory. In 1998, I came to Tel Aviv University to work, with a colleague, on language and aphasia. The research ended on a Friday morning but I had to wait till Sunday to have an El Al plane. My colleague had booked a room for me in the closest possible place to the hospital where we worked. Unluckily, that place happened to be a home for elderly people

F. Lowenthal (✉)
Cognitive Sciences, University of Mons, Mons, Belgium
e-mail: francis.lowenthal@umons.ac.be

M.N. Fried, T. Dreyfus (eds.), *Mathematics & Mathematics Education: Searching for Common Ground*, Advances in Mathematics Education,
DOI 10.1007/978-94-007-7473-5,

Fig. A.1 The square grid

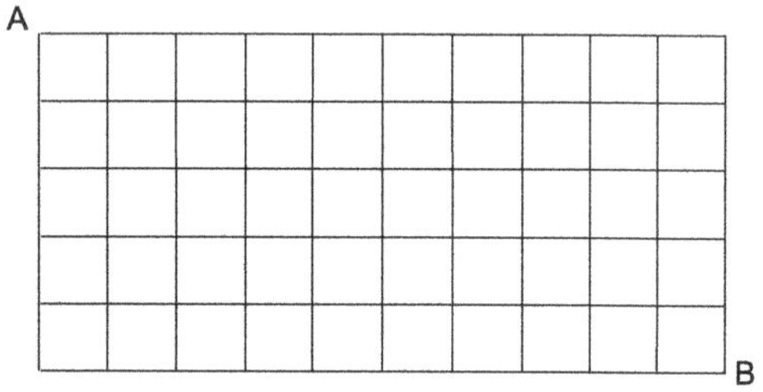

Fig. A.2 How many ways for the taxi driver from *A* to *B*?

How would you proceed to find the number of subsets with an even number of elements of a given set A?

Fig. A.3 How many even subsets?

and I did not feel *that* old. I was alone till Sunday. So I phoned you at home. Your wonderful Polly answered. I told her: "I am alone till Sunday, may I come one of these days and visit you?" And Polly told me: "I come and fetch you. You will spend the full Shabbat with us". It was, at least for me, the non religious Jew, a wonderful Shabbat.

Thank you Ted!

We also worked together, most of the time by e-mails. Once you had the idea to study the "Modeling up, modeling down principle". You had the feeling that many mathematical problems are badly presented in schools, and thus badly solved by the students.

You created a set of exercises such as the one in Fig. A.1, showing a 5 by 5 grid and asking the children "How would you proceed to find how many squares there are, knowing that the answer is more than 26".

The answer is 55! The idea was to ask then the same question but with a 20 by 20 grid. Christiane tried this with very weak high school students in Belgium. She had surprises such as "since it is 55 (five five) for a 5 by 5 grid, it must be 2020 (twenty twenty) for a 20 by 20 grid".

But, your genial idea, Ted, was to ask students "How are you going to solve that?" rather than "Please, find the solution!"

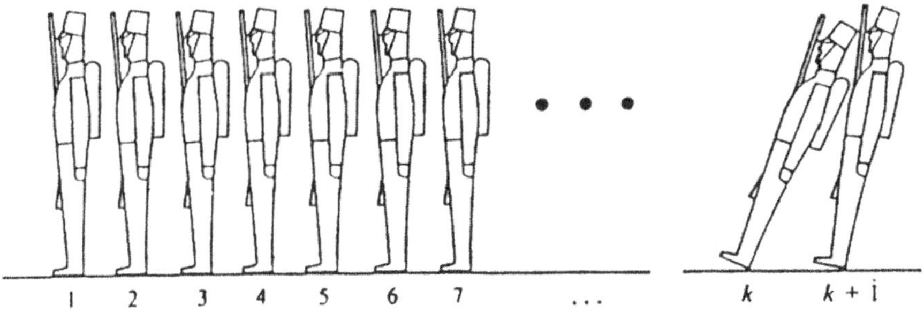

Fig. A.4 Introduction to proofs by induction

Fig. A.5 Shall we prove this by induction?

$$\sum_{i=1}^{n} i + \sum_{i=1}^{n-1} i = n^2$$

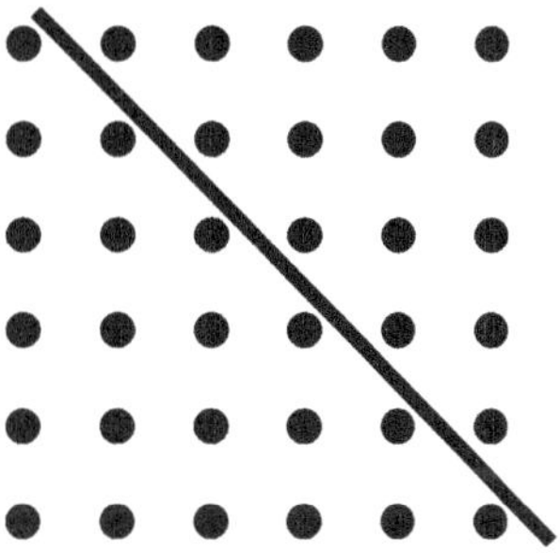

Fig. A.6 Example of a graphic proof, easier to understand

We went on with the number of paths of minimal length for a taxi driver in New York (Fig. A.2): this is a much more difficult problem. And then we went on: we devised together other problems, visual or not.

I must confess, Ted, that I still do not know how to solve the "even number subsets" problem (Fig. A.3).

A few years later, we worked together on proofs by induction.

Ted, you chose the picture in Fig. A.4 for one of our papers! We tried to show that a proof by induction was only acceptable because it was based on a hidden axiom: the acceptance of the principle of induction, which is not granted! This automatically led us to reexamine certain proofs and show that in many cases another type of proof would be better suited.

How can we prove the equality in Fig. A.5? Using a proof by induction? No: in this case, it is better to give a graphic proof, using the partition shown in Fig. A.6, for $n = 6$.

In other cases, the simplest and most efficient proof requires the use of the principle of recursion (Fig. A.7).

Fig. A.7 Tower of Hanoï problem

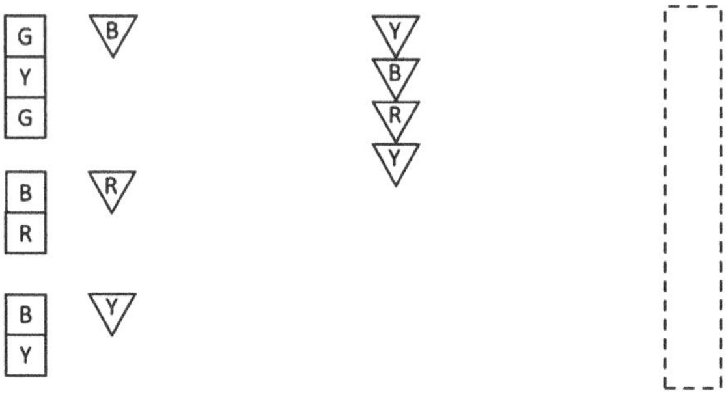

Fig. A.8 Execute the program in the open space

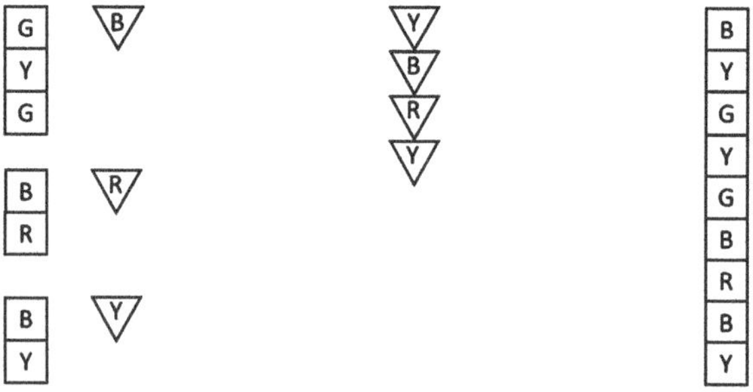

Fig. A.9 The result of executing the program

And this is where our research ways parted. We remained in contact, but you had given me a new idea: "go back to your PhD field, Recursion". And this is what I did.

Thank you Ted!

I had used logico-mathematically inspired tools to observe and favor cognitive development in children.

A simple procedural computer language: on the left the procedures, in the center the program (*in fact the pegs used are colored pegs, but they are represented in this paper by squares and triangles with the initial of the color used*). The children had to manipulate the pegs in order to execute this program by placing little squares on the right: the child had to "do the same (*as in the center*) but only with squares (*as defined on the left*)". To my great surprise, these mathematical tools and situations had a bigger influence on language acquisition than on mathematical knowledge (Figs. A.8, A.9).

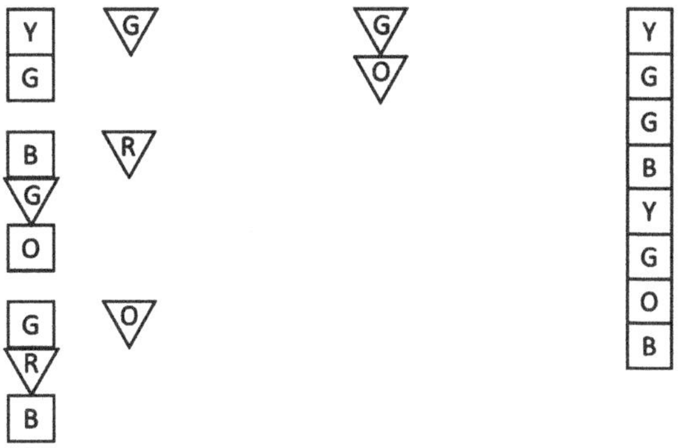

Fig. A.10 Two "embedded" procedures

After our paper on induction and recursion I discovered Hauser, Chomsky and Fitch's 2002 paper. These authors looked at different sentences which, they claim are grammatically correct. One can find some of their examples here. They consider that all of these sentences, even the last one, are understandable by all humans, even when presented in a single font!

1. This is the house **that Jack built**
2. This is the cat **that lives in the house** *that Jack built*
3. The malt **that the rat** that the cat killed **ate** lay in the house *that Jack built*
4. *The boy* **the girl** Peter likes **likes** *likes spaghetti*

They claimed that the main (and only) difference between human and animal communication is precisely the fact that we, human beings, are the only ones able to use full recursion.

Some researchers agreed but others were against this idea: To clarify the situation, I organized last year, in Mons University, an international conference. The main result of our discussions is that full mathematical recursion is probably inaccessible to normal human beings (we mathematicians are not normal beings!). What actually matters for language acquisition is the fact that we can use embedded structures, up to a certain level of complexity. Last year's conference was a great success. This led me to a new research approach:

What would be the influence of manipulations of embedded structures on children's language acquisition and on reading acquisition? In order to observe this I chose to examine the influence of pegboard problems containing embedded logico-mathematical structures, like those shown in Fig. A.10.

And all this started with our discussions about Induction and Recursion, here in Beer Sheva, and by mail.

Thank you Ted!

Recently, Christiane tried to convince me to plan, for the first time in my life, a Seder, the evening meal opening Pessah. It should be a rather special Seder since most of the participants would be non Jewish people: there would be Jews but also Christians, Moslems and nonbelievers. But all these people will have something in common: all of them are sincerely interested in everything that touches the notion of sacred. I felt embarrassed and I turned to you, Ted, the fatherly figure. I asked you if such a Seder would not be sacrilegious, would it be "proper behavior"? And you told me "go ahead with it". So next year, in Belgium, there will be such a Seder, and it is obvious that we hope that you will be one of us.

תודה רבה, טד

Thank you Ted!
Todah rabah Ted!

Annotated Bibliography of Ted Eisenberg's Major Publications

Tommy Dreyfus

Eisenberg, T.A., & McGinty, R.L. (1974). On comparing error patterns and the effect of maturation in a unit on sentential logic. *Journal for Research in Mathematics Education*, *5*, 225–237.
Comment: The authors show that elementary school teachers are no better in logical reasoning that 3rd grade children. That message was picked up by a few national magazines and used as filler by them.

Eisenberg, T.A. (1975). Behaviorism: The bane of school mathematics. *International Journal of Mathematical Education in Science and Technology*, *6*, 163–171.
Comment: In this paper, Ted makes the point that behaviorism equates training with education, and that adoption of such a philosophy for all of school mathematics misses the heart and essence of the discipline of mathematics.

Eisenberg, T. (1977). Begle revisited: Teacher knowledge and student achievement in algebra. *Journal for Research in Mathematics Education*, *8*, 216–222.
Comment: In this paper, Ted showed that there is close to a zero correlation between teacher knowledge and student achievement, and that other factors appear to be responsible for student achievement. Begle had earlier done a similar study with analogous results.

Dreyfus, T., & Eisenberg, T. (1986). On the aesthetics of mathematical thought. *For the Learning of Mathematics*, *6*, 2–10.
This paper has been further discussed in *For the Learning of Mathematics*, *6*, 39–41; reviewed in the Media Highlights Section of the *College Mathematics Journal*, *17*, 368–369; translated into Hebrew in *Alon l'More Hamathematika*, *6*, 5–12 & *7*, 5–13 (1990); and republished in A. J. Bishop (Ed.) (2010). *Mathematics Education* (Vol. 1, pp. 62–77). London: Routledge.

T. Dreyfus (✉)
Department of Education in Mathematics, Science and Technology, Tel Aviv University, Tel Aviv, Israel
e-mail: tommyd@post.tau.ac.il

M.N. Fried, T. Dreyfus (eds.), *Mathematics & Mathematics Education: Searching for Common Ground*, Advances in Mathematics Education,
DOI 10.1007/978-94-007-7473-5,

Comment: The authors attempt to grasp what makes a piece of mathematics beautiful for experts, and argue that an aesthetic sense can and should be nurtured in school mathematics. The paper generated some interest when it appeared and it is still being considered one of the central references on the role of aesthetics in mathematics learning.

Eisenberg, T., & Dreyfus, T. (1991). On the reluctance to visualize in mathematics. In W. Zimmermann and S. Cunningham (Eds.), *Visualization in Teaching and Learning Mathematics* (pp. 25–37). MAA Notes Series, Vol. 19. Washington, DC: Mathematical Association of America.
Comment: The authors examine students' preference for algorithmic-algebraic over visual styles of reasoning, and give some reasons for this preference. This is by far the most cited among Ted's papers, and has created considerable discussion among experts.

Eisenberg, T., & Dreyfus, T. (1994). On understanding how students learn to visualize function transformations. In E. Dubinsky, J. Kaput & A. Schoenfeld (Eds.), *Research on Collegiate Mathematics Education I* (pp. 45–68). Providence, RI: American Mathematical Society—CBMS Issues in Mathematics Education, Vol. 4.
Comment: This is a spin off paper from the previous one, showing what are some of the benefits and difficulties that student may encounter when learning to reason visually.

Eisenberg, T. (2008). Flaws and idiosyncrasies in mathematicians: Food for the classroom? *The Montana Mathematics Enthusiast*, *5*, 3–14.
Comment: This paper raises the question whether aspects of a mathematician's personality, political beliefs, physical handicaps, and the ironies surrounding their life should be mentioned parenthetically or otherwise in our lessons? And what about the political and social norm of the times in the countries in which they lived? The paper seems to have touched a nerve in many of our colleagues and some have written to Ted saying that the paper was being read in their graduate seminars.

Eisenberg, T., & Fried, M. N. (2009). Dialogue on mathematics education: Two points of view on the state of the art. *ZDM—The International Journal on Mathematics Education*, *41*, 143–150.
Comment: This paper, more than any other single paper, gave the impulse for the theme of the symposium and it will provide the topic for the opening session of the symposium. It deals with the fact that the field of mathematics education does not speak with a single voice. There appears to be no firm consensus regarding the scientific character of mathematics education, the research methodologies it deems legitimate, the kinds of questions it addresses, the appropriate preparation for its practitioners, and its relationship with other disciplines, including, ironically, mathematics itself. The paper is reprinted in Appendix B of this volume.

Appendix B: Reprints of the Dialogues Between Presmeg, Eisenberg, and Fried from *ZDM 41*(1–2)

Mathematics Education Research Embracing Arts and Sciences

Norma Presmeg

Abstract As a young field in its own right (unlike the ancient discipline of mathematics), mathematics education research has been eclectic in drawing upon the established knowledge bases and methodologies of other fields. Psychology served as an early model for a paradigm that valorized psychometric research, largely based in the theoretical frameworks of cognitive science. More recently, with the recognition of the need for sociocultural theories, because mathematics is generally learned in social groups, sociology and anthropology have contributed to methodologies that gradually moved away from psychometrics towards qualitative methods that sought a deeper understanding of issues involved. The emergent perspective struck a balance between research on individual learning (including learners' beliefs and affect) and the dynamics of classroom mathematical practices. Now, as the field matures, the value of both quantitative and qualitative methods is acknowledged, and these are frequently combined in research that uses mixed methods, sometimes taking the form of design experiments or multi-tiered teaching experiments. Creativity and rigor are required in all mathematics education research, thus it is argued in this paper, using examples, that characteristics of both the arts and the sciences are implicated in this work.

1 Introduction

'Beauty is truth, truth beauty,'—that is all
Ye know on earth, and all ye need to know (Keats, 1820/1953, p. 234).

As reflected in his famous closing lines to "Ode on a Grecian urn," John Keats had a deep sense of the extent to which the arts and the sciences are intertwined in the human psyche. As mathematics education researchers with interest in improving

This chapter is a reprint of an article published in ZDM—The International Journal on Mathematics Education (2009) 41, 131–141. DOI 10.1007/s11858-008-0136-6.

N. Presmeg (✉)
Mathematics Department, Illinois State University, Normal, IL, USA
e-mail: npresmeg@msn.com

M.N. Fried, T. Dreyfus (eds.), *Mathematics & Mathematics Education: Searching for Common Ground*, Advances in Mathematics Education,
DOI 10.1007/978-94-007-7473-5,

the experiences of those learning and teaching mathematics, we are involved with human beings in all their complexity. The *beauty* of aesthetic experience and the affective issues that accompany this experience or its absence, are counterbalanced and intertwined with the need for mathematical *truth*. "Don't force it! Maths just won't be forced. That's the beauty of it, that's its beauty: where it stands strong against this forcing things into it that don't have any place for it at all," emphasized Mr Blue, in pointing out an error of reasoning to the boys in his grade 12 mathematics class (Presmeg, 2006b, p. 20). Thus I shall argue in this position paper that despite the inevitable fashions that influence modes of research, the *humanism* of our endeavor necessitates the implication of aspects of both the arts and the sciences in investigating issues of mathematics education.

As a means of summarizing where we were a decade ago, I shall revisit a vignette that I described for the International Commission on Mathematical Instruction (ICMI) study, "What is research in mathematics education and what are its results?" in 1994 (Sierpinska & Kilpatrick, 1998). I shall view this vignette in the light of some of the directions taken in our field since this ICMI study, with the lens of aspects of the arts and the sciences as a focus for attention. Throughout this paper, the *arts* and the *sciences* are taken broadly as ways of viewing the world. I wish to highlight both the creative features of the arts, as epitomized in poetry, painting, and creative writing, and the humanistic features of fields that relate to human beings in all their complexity. At the same time, I acknowledge and celebrate the rigor and certainty (albeit contingent) of the methods of the sciences. The main thrust of the argument is that these contrasting aspects are not necessarily mutually exclusive, and that an integrated, unified whole is possible in mathematics education research—as it is in individual human beings—with an appreciation of the strengths, limitations, and purposes of each facet on its own terms, but in relation to the whole.

2 A Vignette

In the 1990s, I taught a course on informal geometry to students at The Florida State University who are prospective middle grades and high school mathematics teachers. One goal of the course was to introduce the students to ways that manipulatives and real world experiences might undergird the learning of geometry in grades 5–8. In the first week I asked them to bring or wear to the next class, something that had geometry in it, and to come to class prepared to tell why they had chosen that particular item and to talk about its geometry. In an interview, one of the students, Dena (who wanted to teach algebra rather than geometry), told me about her reactions to this task, as follows (Presmeg, 1998a, pp. 57–58).

> *Dena.* I noticed when you said, for us to bring something to class or wear something that had geometry in it, for a little while I was having a difficult time, because, everything I picked up had geometry in it. And, I said, maybe there's something I misunderstood about the directions. Y'know.
> *Interviewer.* In fact, even just the shape of a piece of clothing, any clothing.

Dena. Yeah. Anything, has geometry in it. So, for a little while I was confused. I didn't know what to bring to class, until, until I realized that, everything is going to have. I said to myself, everything, of course everything is going to have geometry to it because, y'know, anytime... You're going to make a desk. I mean, you draw, y'know. Your plans, for making the desk, involves geometry. And everything, that is, just everywhere. I think that geometry is taught as something abstract, sketching things with proofs and rules and, not as very, everyday.

Dena's recollections of her high school geometry experiences were negative ones. "I didn't like it at all!" she concluded.

It is important to note that in this vignette Dena is using the word "abstract" as a placeholder for the rote, and for her meaningless, way that she learned school geometry. This usage in no way implies that abstraction is unimportant in mathematics education. On the contrary, I believe that, along with generalization, it is essential in meaningful teaching and learning of mathematical content, a difficult and deep topic that I have addressed in more detail elsewhere (Presmeg, 1997b, 2008).

Implicit in this episode are several points that are relevant to the emergence of mathematics education as a field in its own right, separate from but not unrelated to other disciplines such as mathematics, psychology, sociology, philosophy, linguistics, history, and anthropology. It is significant that in coming of age, mathematics education research broke away from its primary reliance on psychometric research and emulation of the hard sciences. After all, in the complex worlds of human beings learning mathematics in group settings, all aspects of the arts and the sciences that might have bearing on the improvement of this learning are relevant.

2.1 The Many Fields Implicated in Learning and Teaching Mathematics

Firstly, the disciplines of mathematics in its research aspect and mathematics education research are related by their common interest in mathematics. However, these fields differ substantially because their subject matters and goals are different. The subject matter of research mathematicians is the content of mathematics, and without this content there would be no mathematics education. On one level, because mathematicians teach, they are also engaged in mathematics education. However, in mathematics education it is the complex "inner" and "outer" worlds of *human beings* (Bruner, 1986), as they engage in activities associated with learning of mathematics, that form a primary focus of the enterprise, and therefore also of its research. Dena's agonizing over the nature and boundaries of geometry is fruitful and provocative subject matter to a mathematics education researcher interested in the teaching and learning of geometry. The avenues along which this research may lead depend not only on the data, but also on the interests and interpretations of the researcher. The tendency of such hermeneutic research to use progressive focusing rather than pre-ordinate design (Hartnett, 1982) makes this kind of research as interesting as a mystery story, even if the mystery is to some extent self-created. In this respect, mathematics education research may have elements in common with mathematics research. Certainly, the humanism of the arts and the rigor of the sciences are implicated in both, despite their different goals.

A second point is that the inner and outer worlds of a student relate to concerns of the disciplines of psychology and sociology, respectively, and to the interactions between their elements. A balance between elements of these two disciplines is required in mathematics education, as witnessed in 1990s debates on the necessity of steering a course between Piaget and Vygotsky, representing individual and social aspects of learning, respectively, in constructing theory for mathematics education research (Confrey, 1991; Ontiveros, 1991). It is significant that Confrey believed that neither Piaget's nor Vygotsky's theory alone was adequate to model the complex processes of human learning. She suggested that proposing an interaction between the two lenses would necessitate significant changes in both theories.

Confrey's analysis prefigures the point—well expressed by Cobb (2007)—that in the face of incommensurable theories one way of proceeding is to find out how practitioners in the discipline of the parent theory view the canons of their research. This perspective enables the mathematics education researcher to bring a broader vision to the construction of home-grown theories that will be useful in addressing problems of mathematics education. Cobb (2007) explained the benefit of this attitude as follows.

> The openness inherent in this stance to incommensurability has the benefit that in coming to understand what adherents of an alternative perspective think they are doing, we develop a more sensitive and critical understanding of some of the taken-for-granted aspects of our own perspective (p. 32).

In the creativity literature it has long been a well-accepted principle that new views may be garnered by *making the familiar strange*, and by *making the strange familiar* (e.g., De Bono, 1970). However, Cobb (2007) went much further than that. He compared four theoretical perspectives that have been influential in mathematics education research. The first of these is experimental psychology, whose methodologies have been advocated again—as in the 1950s and 1960s—by funding agencies in the USA recently as the only form of *scientific* research in mathematics education (US Congress, 2001). Next is cognitive psychology, viewed from the *actor's* perspective rather than the *subject's*. The final two are Vygotskian sociocultural theory, and distributed cognition. In comparing these four perspectives with regard to their characterization of the individual learner, and in their usefulness for design research in mathematics classrooms, Cobb came to the balanced conclusion that each perspective has merits *for certain purposes*, but not necessarily for designing effective mathematics teaching. In his view, scientific randomized experiments are useful to and serve the administrative and political purposes of policy makers. He makes a strong case that insistence on the hegemony of scientific research in the form of randomized statistical experiments would be short-changing the community of classroom teachers of mathematics. As he shows clearly, all theories are based on philosophical premises, although those advocating a particular stance may not acknowledge the limiting effect of these choices. This analysis suggests that although the scientific and the humanistic aspects of mathematics education research are both legitimate and integral to the enterprise, they address different questions, have different purposes, and are useful to different stakeholders.

A third point implicit in Dena's pondering in the initial vignette is that philosophy is ubiquitous in all questions which are of concern to mathematics education researchers. The nature of geometry is an ontological issue, while how it was taught in Dena's school experience relates to issues of epistemology. Both components are essential in mathematics education theory building, since one's beliefs about the nature of mathematics and mathematical knowledge are the 'spectacles' through which one looks at its teaching and learning. These ontological and epistemological issues are still being debated in research that concerns the beliefs of teachers and students regarding the nature of mathematics and its teaching and learning (Leder, Pehkonen, & Törner, 2002).

Tension between the view that "Everything is mathematics" (as Dena expressed it, "Everything is going to have geometry to it"), and the rigorous mathematical position that "Only formal mathematics is valid", was well expressed by Millroy (1992) in her monograph on the mathematical ideas of a group of carpenters, who did not consider their practice to involve *mathematics*. This tension still plays out in mathematics classrooms. On the basis of her research results, Millroy argued strongly for the broadening of traditional ideas of what constitutes mathematics. She wrote, "We need to bring nonconventional mathematics into classrooms, to value and to build on the mathematical ideas that students already have through their experiences in their homes and in their communities" (p. 192). Steen's (1990) view of mathematics as the science of pattern and order opens the door to this lifting of the limiting boundaries of mathematics. Millroy's recommendation is consonant with those in the National Council of Teachers of Mathematics (NCTM)'s (2000) recent calls for connected knowledge in mathematics education. A related point is that a "mathematical cast of mind" may be a characteristic of students who are gifted in mathematics (Krutetskii, 1976). This mathematical cast of mind enables these students to identify and reason about mathematical elements in all their experiences; they construct their worlds with mathematical eyes, as it were. But unless teachers are aware of the necessity of encouraging students to recognize mathematics in diverse areas of their experience, only a few students will develop this mathematical cast of mind on their own. Many more will continue to regard mathematics as "a bunch of formulas" to be committed to short term memory for a specific purpose such as an examination, and thereafter forgotten (Presmeg, 1993).

The foregoing sets the scene for a fourth point which emerges from these considerations, namely, the links which mathematics education research has been building with various branches of anthropology, particularly with regard to methodology and construction of theory. Millroy's (1992) study was ethnographic. Entering to some extent into the worlds of Cape Town carpenters in order to experience their "mathematizing" required that Millroy become an apprentice carpenter for what she called an extended period, although the four-and-a-quarter months of this experience might still seem scant to an anthropologist (Eisenhart, 1988). But the point is that the ethnographic methodology of anthropological research is peculiarly facilitative of the kinds of interpreted knowledge that are valuable to mathematics education researchers and practitioners. After all, each mathematics classroom may be considered to have its own culture (Nickson, 1992). In order to understand the learning, or, sadly, the prevention of learning which may take place there, the ethnographic mathematics education researcher needs to be part of this world, interpreting its events for an extended period, and then documenting the culture of this world, making the familiar strange and the strange familiar while walking the tightrope of being in but not totally of the world that is observed. In this kind of research the humanities are implicit.

3 Recent Trends in Research Foci and Methodologies

While the history of mathematics goes back several millennia, mathematics education as a field of study in its own right is barely half a century old (Sierpinska & Kilpatrick, 1998). The oldest fully international journal in this field, *Educational Studies in Mathematic*, a few years ago celebrated its 50th volume. (The journal was founded by Hans Freudenthal in 1968. *Journal for Research in Mathematics Education* was started shortly thereafter.) All of the emphases identified in the foregoing section are still relevant to mathematics education research (Lester, 2007). However, in the last decade there have been some developments that emphasize the integrated nature of all the aspects of being human that play out in the learning of mathematics. I shall mention just a few of these trends here. The recent work of Luis Radford and his collaborators epitomizes two such strands, namely, an expanding emphasis on semiotics as a theory for mathematics education research, and the place of gestures, not as an adjunct but as part of an integrated semiotic system for learners to make sense of mathematical concepts (Radford, Bardini, & Sabena, 2007). Radford et al. use a "semiotic-cultural" theoretical framework as a lens for interpreting the learning taking place in a micro-analysis of a video segment in which a group of three grade nine students are trying to generalize the pattern in a sequence of geometrical figures. The video technology is indispensable in this fine-grained work, because the researchers aim to document the role of their gestures as semiotic means for students to grasp the ways that they are seeing the patterns, not merely for the purpose of communication, but in order to reify these patterns and give them meaning. This research emphasizes the integrated nature of human learning. The rigor of the careful documentation certainly has scientific qualities, while the humanities are implicit in the goals and methods of the investigation. Another recent indicator of the significance of attention to the whole learner rather than an emphasis on cognition, is evident in research that addresses the mathematical *identities* of learners, and the way in which culture and experience shape these identities (Sfard & Prusak, 2005).

In recent years it has become acceptable in mathematics education research to use a methodology of mixed methods (Johnson & Onwuegbuzie, 2004), in which the scientific rigor of statistical research is perceived as complementary to the intuitive insights that are possible in fine-tuned qualitative research. Each addresses different questions, and serves different functions. In mixed-methods research, going beyond the significance for different stakeholders that Cobb (2007) identified, an investigation may address the details of some educational phenomenon and attempt to generalize by identifying, for instance, how widespread the phenomenon is. Johnson and Onwuegbuzie present an eight-step process for conducting such research, which they consider superior to mono-method research. As more mixed-method investigations appear in mathematics education research it will be interesting to see whether they have significance for both the groups identified by Cobb (2007)—policymakers and administrators, as well as classroom teachers of mathematics. What counts as "good" educational research? Hostetler (2005) encourages researchers to move beyond questions of qualitative and quantitative paradigms, and to consider the ethical and moral values entailed in research methodologies.

The foregoing account downplays the contestations that accompany changes in any field, and these have certainly been present in the changing paradigms of mathematics education research too (Sriraman, 2007; US Congress, 2001). In what follows, I shall use a first-person-singular account of my own experiences (characterized as a war between the arts and the sciences in my own nature) in parallel with a narrative description of some elements of the changing field of mathematics education research during the last four decades. The different and sometimes conflicting voices in these accounts find a rationale in some elements of hermeneutics and phenomenology, which are addressed briefly in the next section.

4 A Conceptual Framework for a Narrative Account

To some extent a first-person narrative account finds conceptual underpinnings in a hermeneutic-phenomenological theoretical framework such as that used by Roth (2008) to justify his personal voice in analyzing editorial power and its role in authorial suffering in science education research journals, exacerbated by the demands of promotion and tenure processes in academia. As he points out (citing Ricoeur and Latour), this framework acknowledges and celebrates the importance of both scientific explanation and personal understanding in interpretation. Thus it is also an appropriate framework for an account that compares personal history and the history of a field, and that posits complementary roles for humanistic and scientific elements in both. The phenomenological dimension draws on lived experience, whereas the hermeneutic aspect relates to the interpretation of parallels between this personal experience and the changing modes of research in mathematics education. These interpretations can never be considered as complete. As Brown (1997) pointed out,

> In emphasizing that mathematics only ever comes to life in human exchanges we highlight [the] selfreflexive dimension. For Derrida, meaning is always in the future, always 'deferred', there is never a closure to a story because this story can always be extended (for example, 1992)... We can always explore further and revise the meanings we have created. The meaning we derive is always contingent.... I cannot disentangle things independently of my history (pp. 70–71).

Thus I undertake to analyze movements in the field of mathematics education research in conjunction with my own history as a mathematics education researcher.

This hermeneutic-phenomenological position resonates with that of Peirce (1992, p. 313) in his construct of *synechism*, "the tendency to regard continuity... as an idea of prime importance in philosophy," the startling notion that knowledge in its real essence depends on future thought and how it will evolve in the community of thinkers.

In the following contingent account, I describe how the "war" between the arts and the sciences in my nature during my teenage years was reconciled to an integrated whole in the conduct of contemporary mathematics education research. I suggest that it is possible for the corresponding "war" between scientific and humanistic elements in the field of mathematics education research to find integration in recent unifying trends that see both quantitative and qualitative methodologies as valuable, although serving different purposes and having different goals.

5 The Arts and the Sciences—At War?

When I was a teenager, a senior in high school, I read Sir James Jeans' books about the universe, and I was also particularly inspired by the life and work of Marie Curie, who was a dedicated woman in the man's world of the hard sciences at the end of the nineteenth century. I was also intrigued by the incomparable life and work of Albert Einstein (1970, 1973, 1976, 1979). At that point it seemed that the arts and the sciences were at war in me, because I was attracted to both and choosing a career was difficult. At last, decades later, I "came home" to mathematics education research, which included elements of both of these two sides of my nature.

Albert Einstein was a visualizer, and his mental imagery was the rich source of his creative insights (Holton, 1973; Schilpp, 1959). In my first career as a high school mathematics teacher, I noticed that there were students in high school mathematics classes who were visualizers, as I knew from the exceptionally high spatial scores they were achieving on the battery of tests they were doing for vocational guidance—and they were achieving poorly in mathematics, as had Einstein in the restrictive environment of the *gymnasium* he attended in Munich before moving to Switzerland. The question of *why* demanded further investigation. Thus the following central research goal, as it concerned mathematics education, became the topic of my doctoral research (Presmeg, 1985):

> *To understand more about the circumstances which affect the visual pupil's operating in his or her preferred mode, and how the mathematics teacher facilitates this or otherwise.*

The research was exciting, absorbing, and full of surprises. In keeping with the phenomenological stance I am adopting, I see parallels between my experience in this investigation and the field of mathematics education research itself, which was starting to emerge as a field of study in its own right.

As suggested in the opening section, initially the study of problems in the learning of mathematics was a small subset of the wider realm of the concerns of psychology. With respect and admiration for the relative certainty of results obtained by researchers in the hard sciences, in which empirical investigation was used to confirm or disconfirm theory, early researchers in mathematics education (especially in the 1960s and 1970s) tried to emulate this research. Psychometric research was the only genre of research in mathematics education that was considered worthy of the name. Of this period, the Soviet psychologist Krutetskii (1976) wrote as follows:

> It is hard to understand how theory or practice can be enriched by, for instance, the research of Kennedy [in 1963], who compared, for 130 mathematically gifted adolescents, their scores on different kinds of tests and studied the correlation between them, finding that in some cases it was significant and in others not. The process of solution did not interest the investigator. But what rich material could be provided by a study of the process of mathematical thinking in 130 mathematically able adolescents! (p. 14).

Indeed, it was lamented that mathematics education research was having little impact, in fact appeared to be irrelevant, in mathematics teachers' classroom practices. Research as epitomized in "Aptitude-Treatment Interaction" studies (ATIs) seemed to have little impact or relevance in mathematics classrooms. The question of relevance is still an issue in mathematics education research, but more recent developments in this growing field as it embraces mixed methods and welcomes teachers as researchers (Kemmis, 1999) may have the capacity to address this issue.

In the early 1980s, when I was engaged in my doctoral research, qualitative, hermeneutic research under banners such as "illuminative evaluation" (McCormick, Bynner, Clift, James, & Brown, 1977) was starting to be viewed as legitimate in mathematics education because it could address questions about details of teaching and learning that were inaccessible to purely statistical research. My study involved both quantitative and qualitative methods, but it was the fine grain of transcribed interview data that enabled the insights into *why* some students who liked to visualize were not achieving their potential in mathematics. At about the same time, research carried out by teachers in their own classrooms (later widely accepted as "action research", e.g., Ball, 2000) was gaining currency. It was recognized that methods from other disciplines might need adaptation to the particular requirements of mathematics education research, but that there was a rich variety of methodologies that could be valuable. In the last three decades, mathematics education journals and conferences have proliferated, and universities internationally have established programs in mathematics education, housed either in schools of education or more rarely in mathematics departments. These changes accelerated in the 1990s. In a search for identity in its own right (Sierpinska & Kilpatrick, 1998), mathematics education and its research became recognized as a legitimate field, distinct from, yet informed by, the disciplines of mathematics, psychology, sociology, anthropology, philosophy, and even linguistics (Sfard, 2000; Dorfler, 2000). Mathematics education, as a human science, embraces human concerns as well as the need for abstraction and rigor. Various qualitative research methodologies adapted from the humanities became recognized as legitimate in addition to the previously dominant psychometric paradigms. In particular, following Bishop's (1988, 2004) seminal work, there was increasing recognition of cultural and social aspects of the classroom learning of mathematics, complementing the psychological emphasis of cognitive theories of learning. Despite some movements that resisted the changes (cf. the "math wars" in the USA), in this field there is no need for war between the arts and the sciences—both are important. I have come home!

6 Creativity in the Arts and in the Sciences: Mathematics Education Creativity Spanning Both

As mentioned, the heart of Albert Einstein's immensely creative thought was his capacity to visualize (Schilpp, 1959). Mathematics has an obvious visual component, not only overtly, as in geometry or trigonometry, but also in the mental imagery that by self-report enhances the thinking of many creative mathematicians (Sfard, 1994). Why, then, were there visualizers in high school mathematics classes who were finding this subject so difficult that they were obtaining failing grades in examinations (Presmeg, 1985)?

The purpose of my doctoral research was to investigate the strengths and limitations of visual processing in mathematics in a classroom context at senior high school level, and to investigate the effect on learners who are visualizers of the preferred cognitive modes, attitudes, and actions of their mathematics teachers. (For a fuller account, see Presmeg, 2006a, b.) Selection of students and teachers required the development of a new mathematical processing instrument to measure preference for visual thinking in mathematics. I still use this instrument to understand more about the visualization styles of students in my classes. On the basis of the preference for mathematical visualization (MV) scores obtained using this instrument, 13 mathematics teachers were chosen to represent the full range of scores available. In the senior classes of these teachers, 54 visualizers (23 boys and 31 girls) were chosen from 277 high school students. Visualizers were taken to be those who scored above the median score for this population, on the preference test.

The research methodology included participant observation in the classes of the teachers over an eight-month period, and tape-recorded interviews with teachers and students, as well as sparing use of non-parametric statistics to identify trends in the data from the visualization instrument. As a framework for observation in lessons, 17 classroom aspects (CAs) were identified that the literature suggested were facilitative of formation and use of visual imagery in mathematics. The teaching visuality scores obtained by triangulation of viewpoints (teacher's, students', and researcher's) on the basis of the CAs were only weakly correlated with the teachers' MV scores from the preference instrument. It made sense that a good teacher who feels little need of visual supports might recognize the need of mathematics learners for more of these supports. After item analysis and refinement of the CAs, teaching visuality scores divided the teachers neatly into three groups, namely, a nonvisual, a middle, and a visual group according to their styles of teaching. Analysis of 108 transcripts of lessons revealed 45 further classroom aspects that differentiated the three groups of teachers, and that suggested that the visual teachers manifested traits associated with creativity, such as use of humor in their teaching. (Einstein had a marvelous sense of humor—see Dukas & Hoffmann, 1979.)

One of the biggest surprises in this research was that it was the teaching of the middle group of teachers, not the visual group, which was optimal for the visualizers in the study. All the difficulties experienced by the visualizers in their learning of mathematics related in one way or another to the generality of mathematical principles. An image or a diagram, by its nature, is one concrete case, and students need to learn how to distinguish the general elements from the specific ones in learning mathematics. Visual teachers, who had mastered these distinctions, were not cognizant of the difficulties experienced by their students. In my data, there were two ways in which a mental image or related diagram could represent generalized mathematical information. Firstly, the image itself could be of a more general form, which I designated *pattern imagery*. Secondly, a concrete picture (mental or represented on paper or a computer screen) could be used *metaphorically* to stand for a general principle. This latter result of this research led me to the fascinating study of the use of metaphor and metonymy in mathematics education, during the decade of the 1990s (Presmeg, 1992, 1997a, b, 1998b). However, I also became involved in another compelling research agenda, which I shall describe in the next section.

It is noteworthy that the need for rigor, including equivalents of validity and reliability, respectively, was never absent in the qualitative research paradigms that were gaining ground in the 1980s. But the pendulum swung too far away from the previous quantitative paradigm in the 1990s, occasioning a necessary backlash in the 2000s from proponents of statistical methodologies—suggesting that in this field the war was not yet over.

7 Different Bridges: Semiotic Chaining Linking Mathematics in and out of School

In the last two decades, two strands of significance have been developing in the mathematics education research community. On the one hand, there have been increasing calls that teachers facilitate the construction of *connected* knowledge in mathematics classrooms (National Council of Teachers of Mathematics, 1989, 2000). These connections entail not only the linking of various branches of mathematics that have been taught as separate courses at high school level, but also the linking of classroom mathematics with other subjects in the curriculum. And particularly, the importance is stressed of linking school mathematics with the experiential realities of learners. On the other hand, the importance of symbolizing and discourse in the teaching and learning of mathematics has come to the fore (Cobb, Yackel, & McClain, 2000), along with recognition of the significance of sociocultural aspects of the learning of mathematics (Bishop, 1988).

I set out to link these two significant strands by exploring answers to the following question: *How can teachers use semiotic theories to help them facilitate the construction of connections in the classroom learning of mathematics?* In particular, semiotic chaining presented a fruitful method of bridging the formal mathematics of the classroom and the informal out-of-school mathematical experiences of learners.

The significance for mathematics education of theories originating in linguistics was becoming apparent to me. At first in this research I used chaining of signifiers based on Lacan's inversion of Saussure's dyadic model of semiosis (Saussure, 1959). I investigated how teachers and graduate students could use these chains to link the cultural activities of learners with mathematical principles. Working with two research assistants and a doctoral student, Matthew Hall, we interviewed students and taught teachers to build such chains and use them in the mathematics classroom (Hall, 2000). There was the potential for the celebration of diversity and equity. We had some success, but the research suggested the need for a more complex model, because not just signifiers and signifieds, but *interpretation*, were endemic in the activities. Thus I was led to development a nested model of chaining based on the triadic theory of Peirce (1992, 1998). Some of his many constructs illuminated the research, like searchlights, and I am still excited and involved in the exploration of the repercussions of this work. Many instances of the potential of semiotic chaining to foster connected knowledge of mathematics illustrated its significance (e.g., Presmeg, 2006c), and the research is ongoing. Recently, I have been using a triadic Peircean lens to investigate ways that students connect, or fail to connect, the various registers (Duval, 1999) of school trigonometry (Presmeg, 2006b).

There are clearly intertwined elements of the arts and the sciences in this mathematics education research. In the wider field of research methodologies accepted as useful in the twenty-first century, a renewed interest in statistical research to counter the pendulum swing of the 1990s is evident. It will be a pity if another (counter) pendulum swing prolongs the war, because both qualitative and quantitative methodologies have a role to play in the complex field of research on the teaching and learning of mathematics.

In the next section I invoke Habermas's (1978) *knowledge-constitutive interests* to argue this case further.

8 Knowledge-Constitutive Interests Invoking Arts and Sciences

Using Ewert (1991) and Grundy (1990) as sources, in Fig. 1 I have summarized the three types of knowledge and their philosophical bases posited by Habermas (1978). This triad comprises not merely three different ways of looking at knowledge, but three different ways of characterizing what *counts* as knowledge. It is beyond the scope of this paper to discuss Habermas's theory in depth. (Interested readers should consult the original sources.) In this paper I shall use this summary to argue that there is room in mathematics education research for all three kinds of knowledge.

Of Habermas's three types of interests that constitute knowledge, it is the technical one that epitomizes knowledge in the hard sciences. Literary creativity and research are examples of the seeking for knowledge of the second type, in which interpretation of the human condition is paramount. The enterprise seeks to understand that condition, but not necessarily to change it. The critical reflection called for in the third category, by way of contrast, has the goal of changing the human

Technical	Practical	Emancipatory
Social media:		
labour	*interaction*	*power*
Conditions for the three sciences:		
empirical-analytic	*hermeneutic*	*critical*
→ procedures for basic activities:		
control of external conditions	*communication*	*reflection*
Trichotomous division between sciences:		
natural science	*cultural science*	*critical science*
Forms of knowledge:		
instrumental rationality	*subjective meaning*	*critical theory*
Philosophical basis:		
positivism	*phenomenology*	*critical theory*
**		
Eidos and disposition:		
specific, definable ideas - techne (skill)	*the Good - phronesis (judgement)*	*liberation - critique (critical community)*
Action and outcome:		
poietike → product	*practical action → interaction*	*emancipatory action → praxis*
**		

Fig. 1 Three Knowledge-constitutive Interests

condition in some way—hence its designation as emancipatory. In contemporary mathematics education research, examples are found of all three types of interests. In broad categories, the *technical* interest is ongoing in large-scale statistical studies, the *practical* interest is evident in hermeneutic studies that aim for understanding of the mathematical thinking of individual students or small groups of students, and the *emancipatory* interest is apparent in studies that address issues of social justice and critical issues such as access to the study of mathematics. It is beyond the scope of this paper to characterize the landscape of mathematics education research in detail, but the following are examples of research in each of these three categories.

As an example of research in the first category, the investigations of Gagatsis and his co-researchers at the University of Nicosia seek new knowledge of issues in the teaching and learning of mathematics through the statistical investigation, using large samples, of such topics as "Students' improper proportional reasoning" (Modestou and Gagatsis, 2007), or "Exploring young children's geometrical strategies" (Gagatsis, Sriraman, Elia, & Modestou, 2006). Because it is not feasible to assign children randomly to the classes in these studies, the studies may be characterized as of pseudo-experimental design. The methodology enables group trends and relationships to be uncovered, without seeking to ascertain the reasons *why* these trends and relationships are significant. In-depth investigation of the question of "Why?" would entail research in the second category. In my own research on visualization, the construction and validation of an instrument for preference for visualization involved interests in the technical category: validity and reliability were established using non-parametric statistics (Presmeg, 1985). Large samples showed that there was no statistically significant difference between the boys and the girls with regard to their preference for visual thinking in mathematics; however, there was a significant difference between the preference for visualization of the teachers in this part of the study, and their students, who needed far more visual supports than they did.

Again, the question of *why* was deferred to Habermas's second category. Insights into the difficulties and strengths of visualization in teaching and learning mathematics came from interpretive research involving a whole school year of classroom observation and interviews with 54 high school "visualizers" and their 13 mathematics teachers. All of the problems experienced by these learners related in one way or another to the need for mathematical abstraction and generalization, as indicated in an earlier section of this paper. Whereas this kind of research provided insights, it did not have the overt goal of changing classroom practice, although teacher awareness of the results might in fact result in "practical action"—*praxis*—in the classroom (Grundy, 1990). Emancipatory interests, in contrast, have the goal of praxis.

Examples of research involving emancipatory interests can be found in the chapters of the monograph on *International perspectives on social justice in mathematics education* (Sriraman, 2007). After a useful historical introduction to issues of social justice by the editor, Sriraman, several of the chapters describe projects that in one way or another attempt to address the issues of equity that are implicit in social justice applied to mathematics education. For instance, Merrilyn Goos, Tom Lowrie, and Lesley Jolly describe a framework for analyzing key features of partnerships amongst families, schools, and communities in Australian numeracy education. Iben Maj Christiansen contributes a thoughtful and exploratory chapter based on her experiences introducing mathematical ideas to university students in South Africa and Denmark, through social data that highlight inequity. Her analysis leads her to the startling question, "Does our insistence on these 'critical examples' end up being 'imposition of emancipation'?" Tod Shockey contributes the positive influence of a culturally appropriate curriculum for Native Peoples in Maine, USA. Libby Knott explores issues of status and values in the professional development of mathematics teachers in Montana, USA. Eric Gutstein provides a companion piece to his recent influential book on social justice in a Chicago school classroom (Gutstein, 2006). These chapters and others have the more or less explicit goal of changing praxis in mathematics education. Although the monograph also contributes useful empirical and theoretical ideas to the ongoing conversation about social justice in mathematics education (practical interest), its emancipatory interest places it squarely in Habermas's third category. My own research on ways that teachers may incorporate the cultural practices of students in their classes into the praxis of school teaching and learning of mathematics also embraces this category to some extent (Presmeg, 2006a).

9 Final Thoughts

Although I am positing a balance among Habermas's categories, and the necessity of embracing all three interests in various aspects of the complexities of mathematics education and its research, Habermas in his formulation suggested a movement in the direction of the critical theory component (Brown, 1997). Brown described succinctly the educational implications of movement towards the emancipatory interest, as follows.

> If we were to follow Habermas in defining more 'emancipatory' forms of educational practice we would need to differentiate more clearly between *teacher's intention* and *significance for the student* and stress the developing critical powers of the individual student. Such moves towards emphasizing interpretive aspects of mathematical activity, however, inevitably result in placing less stress on the conventional categories of mathematics, as may be represented in the teacher's input or school curriculum.... In doing this we may hope to achieve a style of teaching which enables students to critically examine the purpose and scope of the mathematics they meet, while at the same time recognizing its grounding in their personal experience (pp. 97–98, his emphasis).

It is my contention in this paper that it is not necessary to abandon the "conventional categories" of mathematics in striving for students' individual critical thinking and personal interpretation. Of the three categories of Habermas's (1978) knowledge-constitutive interests, the technical one pertains to the sciences, whereas the practical and emancipatory belong to the concerns and complexities of human life and its interpretation, to the integrated thoughts and feelings of human beings. The discipline of mathematics itself, with its inexorable logic and *instrumental rationality*, resides as a content domain in the technical category, although the creative domain of mathematicians doing research in mathematics might arguably relate better to the *subjective meaning* of the practical category. In contrast, because the teaching and learning of mathematics are practices engaged in by human beings, subjective meaning is all-important if mathematics is to be learned meaningfully, and *critical theory* relates to the improvement of this teaching and learning in mathematics classrooms. However, the content of mathematics with its historically constituted canons is the subject of this teaching and learning.

Thus I argue that both the sciences and the arts are inevitably implicated in mathematics education, whose research also requires the full gamut of methodologies available in the arts and the sciences.

References

Ball, D. L. (2000). Working on the inside: using one's own practice as a site for studying teaching and learning. In A. E. Kelly & R. A. Lesh (Eds.), *Handbook of research design in mathematics and science education* (pp. 365–402). Mahwah: Lawrence Erlbaum Associates.

Bishop, A. J. (1988). *Mathematical enculturation: A cultural perspective on mathematics education*. Dordrecht: Kluwer Academic Publishers.

Bishop, A. J. (2004). Mathematics education in its cultural context. In T. P. Carpenter, J. A. Dossey, & J. L. Koehler (Eds.), *Classics in mathematics education research* (pp. 201–207). Reston, VA:

National Council of Teachers of Mathematics. Reprinted from *Educational Studies in Mathematics, 19* (1988), 179–191.
Brown, T. (1997). *Mathematics education and language: Interpreting hermeneutics and post-structuralism.* Dordrecht: Kluwer.
Bruner, J. (1986). *Actual minds, possible worlds.* Cambridge: Harvard University Press.
Cobb, P. (2007). Putting philosophy to work. In F. K. Lester Jr. (Ed.), *Second handbook of research in mathematics teaching and learning* (pp. 3–38). Charlotte: Information Age Publishing.
Cobb, P. Yackel, E., & McClain, K. (Eds.) (2000). *Symbolizing and communicating in mathematics classrooms: Perspectives on discourse, tools, and instructional design.* Mahwah: Lawrence Erlbaum Associates.
Confrey, J. (1991). Steering a course between Piaget and Vygotsky. *Educational Researcher, 20*(8), 28–32.
De Bono, E. (1970). *Lateral thinking: A textbook of creativity.* London: Pelican.
Dorfler, W. (2000). Means for meaning. In P. Cobb, E. Yackel & K. McClain (Eds.), *Symbolizing and communicating in mathematics classrooms: Perspectives on discourse, tools, and instructional design* (pp. 99–131). Mahwah: Lawrence Erlbaum Associates.
Dukas, H. & Hoffmann, B. (Eds.) (1979). *Albert Einstein: The human side.* Princeton, NJ: Princeton University Press.
Duval, R. (1999). Representations, vision and visualization: Cognitive functions in mathematical thinking. Basic issues for learning. In F. Hitt & M. Santos (Eds.), *Proceedings of the 21st Annual Meeting of the North American Chapter of the International Group for the Psychology of Mathematics Education* (Vol. 1, pp. 3–26). Columbus: ERIC Clearinghouse for Science, Mathematics, & Environmental Education.
Einstein, A. (1970). *Out of my later years.* New York: Greenwood Press.
Einstein, A. (1973). *Ideas and opinions.* London: Souvenir Press.
Einstein, A. (1976). *Relativity: The special and the general theory.* London: Methuen.
Einstein, A. (1979). *The world as I see it.* New York: Citadel Press.
Eisenhart, M. A. (1988). The ethnographic research tradition and mathematics education research. *Journal for Research in Mathematics Education, 19*(2), 99–114.
Ewert, G. D. (1991). Habermas and education: A comprehensive overview. *Review of Educational Research, 61*(3), 345–378.
Gagatsis, A., Sriraman, B., Elia, I., & Modestou, M. (2006). Exploring young children's geometrical strategies. *Nordic Studies in Mathematics Education, 11*(2), 23–50.
Grundy, S. (1990). *Curriculum: Product of praxis.* New York: The Falmer Press.
Gutstein, E. (2006). *Reading and writing the world with mathematics: Toward a pedagogy for social justice.* New York: Routledge.
Habermas, J. (1978). *Knowledge and human interests.* London: Heinemann.
Hall, M. (2000). *Bridging the gap between everyday and classroom mathematics: An investigation of two teachers' intentional use of semiotic chains.* Unpublished doctoral dissertation, The Florida State University.
Hartnett, A. E. (1982). *The social sciences in educational studies: A selective guide to the literature.* London: Heinemann.
Holton, G. (1973). *Thematic origins of scientific thought: Kepler to Einstein.* Cambridge: Harvard University Press.
Hostetler, K. (2005). What is "good" educational research? *Educational Researcher, 34*(6), 16–21.
Johnson, R. K., & Onwuegbuzie, A. J. (2004). Mixed methods research: A research paradigm whose time has come. *Educational Researcher, 33*(7), 14–26.
Keats, J. (1880/1953). Ode on a Grecian urn. In H. B. Forman (Ed.), *The poetical works of John Keats* (pp. 233–234). London: Oxford University Press.
Kemmis, S. (1999). Action research. In J. P. Keeves & G. Lakomski (Eds.), *Issues in educational research* (pp. 150–160). New York: Pergamon.
Krutetskii, V. A. (1976). *The psychology of mathematical abilities in schoolchildren.* Chicago: University of Chicago Press.

Leder, G. C., Pehkonen, E., & Torner, G., (Eds.) (2002). *Beliefs: A hidden variable in mathematics education?* Dordrecht, The Netherlands: Kluwer Academic Publishers.
Lester, F. K. Jr. (Ed.) (2007). *Second handbook of research on mathematics teaching and learning*. Charlotte: Information Age Publishing.
McCormick, R., Bynner, J., Clift, P., James, M., & Brown, C. M. (Eds.) (1977). *Calling education to account*. London: Heinemann.
Modestou, M., & Gagatsis, A. (2007). Students' improper proportional reasoning: A result of the epistemological obstacle of "linearity". *Educational Psychology, 27*(1), 75–92.
Millroy, W. L. (1992). *An ethnographic study of the mathematics of a group of carpenters*. Reston: National Council of Teachers of Mathematics, Monograph 5.
National Council of Teachers of Mathematics. (1989). *Curriculum and evaluation standards for school mathematics*. Reston: The Council.
National Council of Teachers of Mathematics. (2000). *Principles and standards for school mathematics*. Reston: The Council.
Nickson, M. (1992). The culture of the mathematics classroom: An unknown quantity? In D. A. Grouws (Ed.), *Handbook of research on mathematics teaching and learning* (pp. 101–114). New York: Macmillan.
Ontiveros, J. Q. (1991). Piaget and Vygotsky: Two interactionist perspectives in the construction of knowledge. In *Paper presented at the fifth international conference on theory of mathematics education*, Paderno Del Grappa, 20–27 June 1991.
Peirce, C. S. (1992). *The essential Peirce* (Vol. 1). In N. Houser & C. Kloesel (Eds.) Bloomington: Indiana University Press.
Peirce, C. S. (1998). *The essential Peirce* (Vol. 2). In The Peirce Edition Project (Ed.). Bloomington: Indiana University Press.
Presmeg, N. C. (1985). *The role of visually mediated processes in high school mathematics: A classroom investigation*. Unpublished Ph.D. dissertation, University of Cambridge.
Presmeg, N. C. (1992). Prototypes, metaphors, metonymies, and imaginative rationality in high school mathematics. *Educational Studies in Mathematics, 23*(6), 595–610.
Presmeg, N. C. (1993). Mathematics—'A bunch of formulas'? Interplay of beliefs and problem solving styles. In I. Hirabayashi, N. Nohda, K. Shigematsu, & F.-L. Lin (Eds.), *Proceedings of the 17th annual meeting of the international group for the psychology of mathematics education, Tsukuba, Japan*, July 18–23, Vol. 3 (pp. 57–64).
Presmeg, N. C. (1997a). Reasoning with metaphors and metonymies in mathematics learning. In L. D. English (Ed.), *Mathematical reasoning: Analogies, metaphors, and images* (pp. 267–279). Hillsdale: Lawrence Erlbaum Associates.
Presmeg, N. C. (1997b). Generalization using imagery in mathematics. In L. D. English (Ed.), *Mathematical reasoning: Analogies, metaphors, and images* (pp. 299–312). Hillsdale: Lawrence Erlbaum Associates.
Presmeg, N. C. (1998a). Balancing complex human worlds: Mathematics education as an emergent discipline in its own right. In A. Sierpinska & J. Kilpatrick (Eds.), *Mathematics education as a research domain: A search for identity* (Vol. 1, pp. 57–70). International Commission on Mathematical Instruction Study Publication. Dordrecht: Kluwer.
Presmeg, N. C. (1998b). Metaphoric and metonymic signification in mathematics. *The Journal of Mathematical Behavior*, **17**(1), 25–32.
Presmeg, N. C. (2006a). Research on visualization in learning and teaching mathematics. In A. Gutiérrez & P. Boero (Eds.), *Handbook of research on the psychology of mathematics education: Past, present and future* (pp. 205–235). Rotterdam: Sense Publishers.
Presmeg, N. C. (2006b). A semiotic view of the role of imagery and inscriptions in mathematics teaching and learning. Plenary Paper. In J. Novotna, H. Moraova, M. Kratka, & N. Stehlikova (Eds.), *Proceedings of the 30th annual meeting of the international group for the psychology of mathematics education*, Vol. 1 (pp. 19–34). Prague, 16–21 July 2006.
Presmeg, N. C. (2006c). Semiotics and the "connections" standard: Significance of semiotics for teachers of mathematics. *Educational Studies in Mathematics, 61*(1–2), 163–182.

Presmeg, N. C. (2008). Trigonometric connections through a semiotic lens. In L. Radford, G. Schubring & F. Seeger (Eds.), *Semiotics in mathematics education: Epistemology, history, classroom, and culture*. Rotterdam: Sense Publishers.

Radford, L., Bardini, C., & Sabena, C. (2007). Perceiving the general: The multisemiotic dimension of students' algebraic activity. *Journal for Research in Mathematics Education*, 38(5), 507–530.

Roth, W.-M. (2008). Editorial power/authorial suffering. *RISE* special issue on peer review (to appear).

de Saussure, F. (1959). *Course in general linguistics*. New York: McGraw-Hill.

Schilpp, P. A. (Ed.) (1959). *Albert Einstein: Philosopher-scientist*. Vol. I & II, Library of Living Philosophers. London: Harper & Rowe.

Sfard, A. (1994). Reification as the birth of metaphor. *For the Learning of Mathematics*, *14*(1), 44–55.

Sfard, A. (2000). Symbolizing mathematical reality into being—or how mathematical discourse and mathematical objects create each other. In P. Cobb, E. Yackel & K. McClain (Eds.), *Symbolizing and communicating in mathematics classrooms: Perspectives on discourse, tools, and instructional design* (pp. 37–98). Mahwah: Lawrence Erlbaum Associates.

Sfard, A., & Prusak, A. (2005). Telling identities: In search of an analytic tool for investigating learning as a culturally shaped activity. *Educational Researcher, 34*(4), 14–22.

Sierpinska, A., & Kilpatrick, J. (Eds.) (1998). *Mathematics education as a research domain: A search for identity*. Dordrecht: Kluwer.

Sriraman, B. (Ed.) (2007). *International perspectives on social justice in mathematics education*. The University of Montana Press: Monograph 1, The Montana Mathematics Enthusiast (http://www.math.umt.edu/TMME/Monograph1/).

Steen, L. A. (1990). *On the shoulders of Giants: New approaches to numeracy*. Washington, DC: National Academy Press.

US Congress. (2001). *No child left behind Act of 2001*. Washington, DC: Author.

Dialogue on Mathematics Education: Two Points of View on the State of the Art

Theodore Eisenberg and Michael N. Fried

Abstract On many fronts, the field of mathematics education does not speak with a single voice. There appears to be no firm consensus regarding the scientific character of mathematics education, the research methodologies it deems legitimate, the kinds of questions it addresses, the appropriate preparation for its practitioners, and its relationship with other disciplines, including, ironically, mathematics itself. Our field seems to be going through a new phase of self-definition, a crisis from which we shall have to decide who we are and what direction we are going. The authors of the present paper themselves tend towards different positions on these questions. The paper, then, takes the form of a letter in which one of us raises issues about the current state of mathematics education and the other responds. We see this as an attempt to initiate a dialogue on our field, which we consider urgently needed.

Keywords Nature of mathematics education research · Legitimate methodologies · Interdisciplinary influences · Mathematical content

1 Introduction

Recently, I (T.E.) had the opportunity to read a version of the plenary paper delivered by Norma Presmeg at the Second International Symposium on Mathematics and its Connections to the Arts and Sciences (MACAS2). My view of the paper was quite critical and left me feeling that all is not well in mathematics education

This chapter is a reprint of an article published in ZDM—The International Journal on Mathematics Education (2009) 41, 143–149. DOI 10.1007/s11858-008-0112-1.

T. Eisenberg (✉)
Department of Mathematics, Ben-Gurion University of the Negev, Beersheba, Israel
e-mail: eisen@cs.bgu.ac.il

M.N. Fried
Program for Science and Technology Education, Ben-Gurion University of the Negev, Beersheba, Israel
e-mail: mfried@bgu.ac.il

M.N. Fried, T. Dreyfus (eds.), *Mathematics & Mathematics Education: Searching for Common Ground*, Advances in Mathematics Education,
DOI 10.1007/978-94-007-7473-5,

research. Rather than responding directly and with a single voice, the idea arose, partly through the urging of Bharath Sriraman, that I and a colleague, Michael Fried, write a dialogue taking off from various points in Norma Presmeg's paper. It made sense to engage Michael in this, since while our positions are not entirely opposed they are opposed enough to highlight the concerns of others in the field, and in actual fact we do often discuss these issues via e-mail. An exchange of letters as the chosen form for our dialogue, therefore, seemed natural. Our hope, needless to say, is not to settle all the questions involved, but to initiate a genuine and broad dialogue on the nature of mathematics education and mathematics education research.

2 Dear Michael

Although we teach at the same university our paths seem not to cross as often as I would like. Anyway, I wanted to raise several issues Norma Presmeg discusses in her paper "Mathematics Education Research: Embracing Arts and Sciences." Norma Presmeg's long history of deep and serious commitment towards understanding how we learn and teach mathematics is everywhere evident in her paper, though we did not need her paper to tell us that! But it is just this that makes her views on the current situation in the mathematics education community and her interpretations and conclusions so troubling for me. She states several times in the paper she has found her academic home, and perhaps she is right; but I feel that because of the path mathematics education research (hereafter mathematics education research = MER) has taken over recent years, I have been banished from my academic home; these days, I feel like a fish out of water. Let me raise a few of the items, prompted by the paper that have made me reflect on the nature of mathematics education and get your take on them.

2.1 MER: What It Should Be, What It Was, and What It Is

Research in mathematics education should be about the teaching and learning of mathematics; it's definition is that simple to state. And at one time it seemed to be just that; individuals worrying about better ways to teach and learn mathematics. Forgetting the roots of our discipline, which one can easily trace back to ancient Greece, as I see it, modern mathematics education research was born in 1957 with the launching of Sputnik. The world was going crazy in those years with fear that another world war was around the corner and that the democratic world was going to lose the race to claim outer space for its own. I was just a schoolboy then, but I remember well the movements around the world to close the educational gap between the good guys, us, and the bad guys, them (the Sputnik launchers). New curricula were popping up everywhere, and their construction and implementation were heavily financed with governmental money. To an outsider, the world of mathematics education in those days must have looked like a bowl of acronyms; the alphabet soup of projects and programs from those years and before included SMSG, UICMS, Minimast, the Madison Project, the PRIMES project, SMP, AAAS, and the list goes on and on. But the common denominator of all these projects was that mathematics educators were working with mathematicians and classroom teachers to make mathematics accessible to larger segments of the school population and to do this in an intellectually honest way. Indeed, the ideology "That we can teach any topic to any individual in an intellectually honest way" became one of the mantras of the time and the core of the belief system of young teachers and students like myself. But make no mistake about it, the driving forces behind the reforms were fear of the USA and its allies not being number one in the world and competition. Mathematics and its teaching and learning are driven by fear and competition. The fact that we made a mess of things in implementing new curricula at the school level is another matter.

Although Norma Presmeg elaborates in MER that esthetics is the main force that drives mathematics forward—and one can easily find many statements in the literature supporting her on this point—when we get right down to it, it is nonsense. Competition with our self and amongst ourselves is the driving force behind mathematics. It is not money per se that breeds ingenuity, but the challenge of succeeding where others have not, with professional recognition, kudos, and fame being some of the wonderful by-products of success. Beauty follows proof and invention, it does not drive it. Recognition as being the first to have accomplished something others could not is the real motivator. I have yet to find a single mathematician willing to delay publication of a proof for, say, the Riemann Hypothesis because the proof, though correct, was not esthetic. Not one mathematician has ever admitted that he or she would delay publication of a proof in hope of finding a route that is more esthetic; not one. Esthetics? Nice to muse about, but not in today's academic world.

(Recent TIMSS evaluations have placed our country into non-enviable rankings and our Ministry of Education has been jolted, sending delegations to Singapore and Finland to see if we can adapt their curricula and teaching methods here. Do you think that the esthetics of mathematical thought has played any role whatsoever in our Ministry pushing teachers to improve the performance of their students? We only participate in TIMSS and other such comparisons with the hope that we will be rated number one. I once wrote a letter to someone high in the Ministry's hierarchy suggesting that we should stop participating in such evaluations because we are never going to do well in them; my letter has yet to be answered, or its receipt even acknowledged.)

2.2 MER, "Our" Background in Statistics

Norma Presmeg's paper is right on target when she states that the beginning years of mathematics education research studies were statistical in nature; I recall in those days texts similar to Campbell and Stanley's (1963) classic *Experimental and Quasi-Experimental Designs* standing at the heart of many mathematics education programs. Mathematics educators of my generation were skilled in applied statistics; we could talk about ANOVA, ANCOVA, regression and correlation, factor analysis, biased estimators, etc. Today's math educators seem to know none of these notions; but what is worse is that these notions seem to be pooh-poohed by the MER community. Michael, at our own university, even basic statistics is no longer a required course at the MA level in most of the mathematics education programs offered on campus. Worse, I have seen some instructors on staff instructing students to skip over statistical tables in articles, telling them that if anything important is buried in those tables, the author will state it in a more readable format elsewhere (and as you might guess, the instructors who practice this are amongst the most popular on campus). Norma mentions that we have "moved on" from statistical studies, but to what have we moved? As I see it, we have moved to an amorphous sea of gobbledygook, which has in it islands of sanity.

2.3 MER, "Our" Language

The professional vocabulary of those who claim to be doing mathematics education research today is foreign to me. Words and phrases like hermeneutic research, semiotics, ethnographics, metaphor and metonymy, are sprinkled like spice throughout Norma's paper; but I venture to say that most math educators in the world would be hard pressed to define these notions in an inclusive and exclusive way. And as you well know things get much worse if we look at the papers of others: therein we will see words like reification, sensorial stimuli, an epistemic stance, knowledge objectification, apodeictic calculations, and ontogenetically speaking. I agree that these are bona fide words and notions. I have looked-up their definitions in dictionaries many times; why many times? Because the definitions do not stick. But my real concern with this sort of language is: who are we trying to impress? A discipline should be built around simple words, not ones that most in the field can barely pronounce, let alone define. It is as though authors who use such words believe that this type of language elevates their paper into being "real science." But to those in the field like me, using this sort of language has the opposite effect; using such words gives the impression that mathematics education is a pseudo-discipline; that is putting on the airs of trying to be more than what it is. Using such a vocabulary and coining new words is detrimental to the discipline. Einstein used to say, things should be made as simple as possible, but not simpler; I think as a profession we should follow his advice.

Norma Presmeg is also correct in saying that MER has switched its emphasis from mathematics to the individual. But whereas she embraces it, I abhor it. I am well aware that discussions in the classroom, non-verbal reinforcement, alternate methods of evaluation, making peace with one's deficiencies, feeling good about yourself, etc. are important in life, and that they have become the focal points of many studies in our discipline. But I believe that they are missing the point. The first and foremost goal of a mathematics teacher is to help students learn mathematics, not to make them feel good about not knowing mathematics. And in order to help students learn mathematics, the teacher must know mathematics; it is as simple as that.

2.4 MER and Formal Mathematics

In general, the mathematics requirement in most math education programs today is appalling. We have talked about this with one another before. I do not have any minimal levels of competence in mathematics that I can recommend, to set the math level for everyone in our profession at level X, or $X+20$, or $X+50$, or even higher. But as a profession we seem to be lost. We are wandering looking for identity; and I don't like it. As I see it, mathematics education is a sub-domain of mathematics, and as such, mathematics educators belong in departments of mathematics, not in schools of education or in units of science teaching. And if mathematics educators do sit in mathematics departments, then they must know some mathematics. I am not saying that all mathematics educators should be mini-mathematicians nor that knowing mathematics is all they need to be good teachers, but they should have taken a good chunk of formal mathematics; they should like mathematics and they should maintain an interest in mathematics. This last sentence was Polya's mantra, and I agree with it totally. But I am embarrassed to say that I know mathematics educators who denigrate mathematics (and in so doing, they also unwittingly denigrate themselves.) Martin Gardner, of *Scientific American* fame, would often be asked how he could write and explain things so clearly. And his answer was, because he had to work so hard to understand them. We should follow in his footsteps.

The world wide web has made it is easy to find the academic background of many professors at major universities who have identified themselves as being mathematics educators. Mathematics education groups at some of the most prestigious universities have many members without even a first degree in mathematics. At one university that I checked their mathematics educators were trained in elementary education, and psychology, and there was even one on that staff whose academic degrees were in social work! I am not saying that these individuals are stupid, far from it. But what I am saying is that these individuals have not studied higher-level mathematics; they have not experienced the abstraction that boggles so many students in our classes, they have not "been there." Yet these individuals seem more than willing to tell the MER community how mathematics is learned and how it should be taught! I find it incredible that they have the nerve to speak and sad that we listen to them and often hang on their every word. Yes, MER has moved away from the teaching and learning of mathematics, and I feel as though we are not going to return to it anytime in the near future. This is a tragedy of our own making. Norma relishes the openness of the MER, I do not. (And for those mathematics educators who sit in departments of education or elsewhere, I again repeat my definition of mathematics education; it deals with the teaching and learning of mathematics, not the zillion things around it.)

2.5 Making MER More Humane

Norma Presmeg's paper talks about making mathematics more humane and I think we both know what she means by this. I imagine that most everyone in the profession has been at one time or another humbled by mathematics, how it is taught and how difficult some topics are to grasp, etc. When I hear of such testimonials or even experience difficulties myself, I try to think of the mantra that any topic can be taught to anyone in an intellectually honest way. All that is needed is time, patience, desire, and empathy.

2.6 MER Some Last Words

It seems to me that the job of math educators is to build environments where learning can occur. It is not to continually look for easy solutions, quick fixes and panaceas that are just not there. We should not continually look for ways to get students around the abstraction, that just happens to be the fiber of mathematics today, but rather we should search for ways to bring students to the abstraction. There comes a time when we have to buckle down and master the abstraction, not to continually look for ways to circumvent it. Norma has given us a wonderful snapshot of where MER is today. But I seem to see things very differently. What's your take on it all?

Sincerely,
Ted.

3 Dear Ted

Many thanks for your letter and for the opportunity to put on paper some things we have often spoken about over coffee (and, I agree, those coffee conversations are far too rare and short!). First, I should say that while I do tend towards the view of things conveyed in Norma Presmeg's paper, I think she can defend her own ideas well enough. So, I will only speak for myself. And, as for that, there are many points on which we see eye to eye, as you know. Let me begin with Sect. 1, then, "MER, what it should be, what it was, and what it is."

First I want to say that "Research in mathematics education should be about the teaching and learning of mathematics" is truly, as you say, a simply stated definition; however, it is also one that hides many complexities. For one, there is the question of what mathematics should be taught and learned or, alternatively, why we should learn mathematics in the first place, the "why" having much to do with the "what."

You state that "... modern mathematics education research was born in 1957 with the launching of Sputnik." Although I am not sure modern mathematics education research was actually *born* with Sputnik,[1] you are certainly right to point to that time as a watershed period in our field. The shock of Sputnik gave a new weight and urgency to mathematics and science education and presented a warning that the neglect of mathematics and science education has a price; indeed, that message was still echoing in the 1983 report "A Nation at Risk" (which even mentions Sputnik!). As motivations, Sputnik or the space-race, which followed, or the drive for technological and scientific prowess suggest that the goal of mathematics education should be to produce good mathematicians, scientists, or engineers. The force of that goal is clear in the strength with which Bruner denies it in the book that so reflects the mood after 1957, Bruner's (1960) *The Process of Education* [the same book containing the statement you quote: "any subject can be taught effectively in some intellectually honest form to any child at any stage of development" (p. 31)]: "The intention (of the 1959 Woods Hole conference) was not to institute a crash program, but rather to examine the fundamental processes involved in imparting to young students a sense of the substance and method of science. Nor was the objective to recruit able young Americans to scientific careers, desirable though such an outcome might be" (p. vii). So, even then, in the late 1950s and early 1960s, what was born, therefore, was a question: what is the goal of mathematics education, and therefore of mathematics education research? What does it mean to teach, in our case, the "substance and method" of mathematics?

Questions on the teaching and learning of mathematics surely rest on how we conceive the goals of mathematics education. Put differently, as educators, we ought to think about what it means to be educated, and, as *mathematics* educators, what it means to be *mathematically educated*. The "new math" that emerged from meetings such as that at Woods Hole (though the movement, of course, has a much longer history) was one answer to what mathematically educated means; the social utility movement,[2] reincarnated, in a sense, in the "back to basics" movement of the 1980s and 1990s, was another; sensitivity to social and cultural issues connected to mathematics is yet another. I do not mean to say that one of these is the true and correct view, nor that anything goes (I will say more about that later, when I speak about your Sect. 4). However, what I do want to emphasize is that what it means to be mathematically educated is not an issue that is completely clear, nor is it one that is going to be answered adequately by mathematicians, historians, industrialists, or social reformers alone. In this connection, it is worth pointing out that the Woods Hole meeting engaged not only scientists and mathematicians, but also psychologists, historians, cinematographers, and even a classicist!

Treating the question of what it means to be mathematically educated demands this kind of integrative view, and mathematical education research may, collectively, provide it.

---

[1]Considerably earlier origins of research in mathematics education in European traditions are discussed in (Sriraman, B., & Torner, G. 2008).

[2]See pp. 17–18 of (Kilpatrick, J. 1992).

This brings me to the issue of esthetics. You are probably right in claiming that esthetics is not the "main force that drives mathematics forward" and that "competition with our self and amongst ourselves" must also be taken into account. It is true indeed that the history and, particularly, sociology of science and mathematics has shown how far scientists and mathematicians are influenced by institutional forces, a zeal for recognition, and, yes, competition.[3] But, surely, this cannot be the whole story, and, surely, one cannot discount also an esthetic force, or, more generally, what Polanyi (1964) called "intellectual passions"[4] in scientific and mathematical work. Esthetics may not be the *main* force in mathematical work, but it is *a* force, and certainly a part of what makes a mathematician tick. There are enough accounts of this from mathematicians themselves. Speaking from my own experience, I can say my own love of mathematics has been charged by a few, but powerful, mathematical-esthetic moments; they are rare, but one holds on to them like talismans. Being mathematically educated must somehow take in this aspect of mathematical experience: having felt at least once or twice the beauty of mathematics seems to me essential for knowing what mathematics is about. In this way, I agree with Davis and Hersh (1981) when they write: "Blindness to the esthetic element in mathematics is widespread and can account for a feeling that mathematics is dry as dust, as exciting as a telephone book, as remote as the laws of infangtheif of fifteenth century Scotland. Contrariwise, appreciation of this element makes the subject live in a wonderful manner and burn as no other creation of the human mind seems to do" (p. 169).

On one level, it is hard to disagree with your sentiments regarding issue Sect. 2 "MER, 'our' background in statistics." It is so easy to be fooled by numerical data, to think what is negligible, significant, and what is significant, negligible; it is hard, then, to argue against the necessity of researchers possessing at least some degree of statistical sophistication. But, on another level, I think such sophistication serves researchers best when it helps them develop, rather, a sense of the limits of statistics, a kind of healthy suspicion, in general, of quantitative approaches in our field. Like many other fields connected to the social sciences, we suffer not a little from "physics envy." What we must realize is that physics can be physics only because, at bottom, its objects are simple—particles can move in six directions, can rotate, mutually attract, mutually repel (they cannot chase one another!), and so on. Schrodinger (1967) said something like that in his wonderful essay "What is Life?" in comparing physics to biology. But our objects— "learning," "understanding," "affect," etc.—are infinitely more complex even than biological concepts. In fact, we are still trying to understand what these things are, let alone quantify them. For statistics to be useful, on the contrary, one must satisfy two conditions: (1) one must know what it is one is interested in measuring; (2) one must measure what one is interested in. So, statistics can be very useful in telling whether more cars pass through a certain intersection between 9:00 am and 11:00 am than

[3]See, for example, Hagstrom, (1974); Merton (1957) and Fisher (1973).

[4]Especially, p. 192 of (Polanyi, M. 1964).

between 1:00 pm and 3:00 pm; one is interested in a certain number of cars, and what one measures is just that. All too often, what happens in social science research is that there is some idea, say "learning," which we really do not completely understand, that is, we do not really know what it is, but we are interested in measuring it; we then measure some numerical quantity, taken to be related somehow to "learning," run statistical tests on our measurements of that quantity, and draw our conclusions about "learning." In doing this, we violate both (1) and (2). What is worrisome, to my mind, is that, with our thirst to be "scientific" and the, often concomitant, aversion towards "mere philosophical talk," we may forget to ask basic questions such as what we truly mean by "learning." I do not mean to reject all such statistical studies, but only to say that ours is a field that is at least equally—and, perhaps, even primarily—interpretative in character, that is, beyond the making of measurements, mathematics education rests on explorations of meaning. I think this is what Norma has in mind when she refers to MER embracing arts and sciences.

Much of what you say about "'our' language" in (4) I applaud heartily. Mathematics education research, of course, is not the only academic field guilty of using language which obfuscates more than it illuminates, but that does not free us from the sin! Interestingly enough, the language problem goes hand-in-hand with what I wrote above about statistics; both statistics and jargon have the appearance of being precise without truly being so. And it is not only jargon. It is also simple words, like "powerful," "rich," and "meaningful," when they are over used or used thoughtlessly. Our research will only teach us something if we constantly ask ourselves what our words mean and not just throw them around to fill up the page. You know, the trap of seductive pseudo-technical or otherwise imprecise language is not new; it is what Bacon had in mind when he speaks about the "idols of the market." He says that these particular idols (there are three other "idols of the mind" that get in the way of knowledge) "... are the most troublesome of all... which have entwined themselves round the understanding from the associations of words and names."[5] In this regard, an injection of true philosophy—which, when it is true, has everything to do with being precise and saying what we mean—could do us a great deal of good. On the other hand, we ought not judge a word, necessarily, by its length. There are words imported from other fields—say, from anthropology, linguistics, or philosophy itself—that are as precise in those fields as "manifold," "variety," "diffeomorphism," and "surgery" are in mathematics. In the right context, and used in the right way, such words can be very useful and often necessary. The golden rule, I think, is, as in your quotation from Einstein, to use words "... as simple as possible, but not simpler..." But one must keep in mind that both halves of the rule are equally essential.

In response to your very first point, I argued that grasping what it means to be "mathematically educated" requires our going outside the mathematicians bailiwick. Still, we must not exclude mathematicians nor, as absurd as it sounds, to

[5] Francis Bacon, *Novum Organum*, Book I, Sect. 59.

exclude mathematics itself. So, here I certainly agree with you that something is terribly wrong when mathematics gets pushed out of mathematics education.[6] As I have said on many occasions, it is a sad fact that most of Polya's writings on mathematics education would probably be rejected by many professional mathematics education research journals (probably—going back to my rant about statistics—on the grounds that they were not scientific enough!). Saying how or what mathematics should be taught and, obviously, teaching mathematics should rest, it seems to me, on mathematical experience that is broad and deep. By deep, I mean, first of all, that one has seen enough mathematics to know that there are different levels at which a mathematical concept or procedure can be framed. For example, one's first encounter with the derivative might be as the slope of the tangent to the graph of a function or as the rate of change of a function; as one learns more mathematics one learns that the derivative of a function is a matrix whose dimensions depend on the dimensions of the spaces containing its range and domain; later, when one learns about more general linear spaces, one learns to see a derivative of a function as the linear operator best approximating the function at a point in the space. At each level, one goes, literally, deeper into the meaning of "derivative," sees it in a wider context, and gets closer to its foundations; teachers who have not gone through something like this themselves will not truly know what depth means in mathematics and, therefore, I doubt they will be very successful in persuading their students that mathematics is deep. What I said above about esthetics too, I believe, implies that teachers (if they are to inculcate an esthetic sense in their students) need broad enough and, I dare say, deep enough experience doing mathematics to be able to see the true beauty of subject;[7] it is hard to see how someone who has never worked on a difficult problem and suddenly glimpsed how the parts of the problem harmonized into a neat solution would be able to inspire student's delight in mathematics. Mathematics educators, for their research as well as for their task as teacher educators, need such experience even much more than teachers do. I might add that "broad experience" takes in such things, in my view, as the history of mathematics as well, but I won't ride that particular hobbyhorse of mine right now....

As for humanising mathematics and what you wrote earlier about the shift from mathematics itself to the individual, I will only add this. I think, as people who care about individuals learning mathematics, we have to care about individuals. But—and I am agreeing with you here—this does not mean making mathematics easy and painless. Learning mathematics, learning anything serious, is a little like mountain climbing; it is hard work, but the view at the end is worth it; and the view is all

[6] I do not think this concern is yours only. If you recall, PME30 in 2006 was given the title: "Mathematics in the Center." Such an obvious thing would be unnecessary to say if it were not that mathematics had moved away from the center, or even off to the periphery.

[7] Nathalie Sinclair has argued (e.g. in Sinclair, N. 2004) that we should not assume esthetic experience belongs exclusively to professional mathematicians, but that all students have a kind of esthetic faculty allowing this kind of experience. She might be right about that; however, I would argue that the *depth* of one's mathematical esthetic experience reflects the depth of one's mathematical understanding.

the more beautiful *because* one has worked hard to get there. I am reminded here of what Rilke says, writing to the young poet F. Kappus about the difficulty of sex but more broadly about life in general: "Sex is difficult; yes. But it is difficult things which we have been given to do; nearly everything serious is difficult, and everything is serious."[8] Things which humanize us are not always easy, indeed, are rarely so; there is no cutting corners. In this way, Ted, I think your "MER, some last words" are mine as well, mutatis mutandis. . . .[9]

Best wishes,
Michael.

References

Bruner, J. S. (1960). *The Process of Education*. Cambridge: Harvard University Press.

Campbell, D. T., & Stanley, J. C. (1963). *Experimental and Quasi-Experimental Designs for Research*. Chicago: Rand McNally College Publishing Company.

Davis, P. J., & Hersh, R. (1981). *The Mathematical Experience*. Boston: Houghton Mifflin Company.

Fisher, C. S. (1973). Some Social Characteristics of Mathematicians and Their Work. *American Journal of Sociology*, *78*(5), 1094–1118. doi:10.1086/225424.

Hagstrom, W. O. (1974). Competition in Science. *American Sociological Review*, *39*(1), 1–18. doi:10.2307/2094272.

Kilpatrick, J. (1992). History of Research in Mathematics Education. In D. A. Grouws (Ed.), *Handbook of Research on Mathematics Teaching and Learning* (pp. 3–37). New York: Macmillan Publishing Company.

Merton, R. K. (1957). Priorities in Scientific Discovery: A Chapter in the Sociology of Science. *American Sociological Review*, *22*(6), 635–659. doi:10.2307/2089193.

Polanyi, M. (1964). *Personal Knowledge*. New York: Harper & Row.

Schrodinger, E. (1967). *What is Life? & Mind and Matter*. Cambridge: At the University Press.

Sinclair, N. (2004). The roles of the Aesthetic in Mathematical Inquiry. *Mathematical Thinking and Learning*, *6*(3), 261–284. doi:10.1207/s15327833mtl0603_1.

Sriraman, B., & Torner, G. (2008). Political Union/Mathematics Education Disunion: Building Bridges in European Didactic Traditions. *Handbook of International Research in Mathematics Education* (pp. 656–690). London: Taylor & Francis.

[8]"Das Geschlecht ist schwer; ja. Aber es ist Schweres, was uns aufgetragen wurde, fast alles Ernste ist schwer, und alles ist ernst" (*Briefe an einen jungen Dichter*, #4).

[9]"Having changed what needed to be changed."

The Harmony of Opposites: A Response to a Response

Norma Presmeg

After months of waiting, when authors finally receive the decision letter from an editor for a manuscript that has been submitted to a refereed journal, the decision is often a relatively negative one, rejecting the manuscript but asking for a complete revision and resubmission. In reading the reviews, the author may become discouraged, perceiving that at least some of the reviewers did not interpret the contents of the manuscript in the way that was intended. This phenomenon is consonant with the three kinds of interpretant posited by Peirce (1998), namely, the intensional, the effectual, and the communicational interpretants, which are, respectively, "a determination of the mind of the utterer", "a determination of the mind of the interpreter", and "a determination of that mind into which the minds of utterer and interpreter have to be fused in order that any communication should take place" (p. 478). Peirce called the latter fused mind the commens. As an editor, I attempt to help writers to see that "misinterpretation" by a reviewer is something for which to be grateful, because it shows the author that more clarity is needed in the writing. It is in this spirit that I respond to the letters between Ted Eisenberg and Michael Fried. One could consider my position paper on the topic "Mathematics education research: Embracing arts and sciences" to be the intensional interpretant in this case. Then Eisenberg and Fried's response is the effectual interpretant. My current response to their interpretant is an attempt to establish a commens, a fused mind for the purpose of establishing communication.

This chapter is a reprint of an article published in ZDM—The International Journal on Mathematics Education (2009) 41, 151–153. DOI 10.1007/s11858-008-0133-9.

This paper is a response to Theodore Eisenberg and Michael Fried's critique ("Dialogue on mathematics education") of Norma Presmeg's paper ("Mathematics education research: embracing arts and sciences").

N. Presmeg (✉)
Illinois State University, Normal, IL, USA
e-mail: npresmeg@msn.com

M.N. Fried, T. Dreyfus (eds.), *Mathematics & Mathematics Education: Searching for Common Ground*, Advances in Mathematics Education,
DOI 10.1007/978-94-007-7473-5,

1 Distinctions Among Terms

Before responding directly to the points raised by Eisenberg and Fried,[1] I want to point out that we are talking about three central topics, and in the zeal to establish a point it is easy to slide among these three referents, namely, mathematics, mathematics education, and mathematics education research (MER). These three referents are related, but they are distinct. Mathematics refers to the content itself, the body of established knowledge in this field. Mathematics education refers to the teaching and learning of this content by people at all levels. MER refers to attempts to find out more about what is involved in the latter endeavor, that is, to investigate elements of what Michael rightly points out is a complex phenomenon, human teaching and learning of mathematics.

My paper was not about mathematics, nor about mathematics education as such; it was about MER, as the title indicates. According to the title of Ted and Michael's response paper ("Dialogue on mathematics education"), they are writing about mathematics education rather than MER; however, their text addresses elements of this research. I agree with Ted (in his Sect. 2.1) when he claims, "Research in mathematics education should be about the teaching and learning of mathematics; its definition is that simple to state." However, I would add just three words to clarify what is meant: "Research in mathematics education should be about *the investigation of* the teaching and learning of mathematics." Then it is clear that "individuals worrying about better ways to teach and learn mathematics" are not, simply by that action, doing MER. More is required: if one thinks of research as disciplined inquiry, then some form of systematic investigation is needed—and that is where not only scientific methods, but also all of the methodologies of the human sciences may have a role to play in casting light on the varied and complex phenomena under investigation.

2 Mathematics

Although my paper was not about mathematics itself, let me clarify some aspects of the way that I view the nature of mathematics, because this aspect was a caveat introduced by Ted. It goes without saying that without mathematics there would be no mathematics education, and hence also no MER. However, following research that showed the effects of beliefs about the nature of mathematics on its teaching and learning (summarized well in the book edited by Leder et al. 2002), the way that mathematics itself is viewed becomes a foundational issue, not only affecting mathematics education but therefore also relevant to its research.

Ted claims that "mathematics and its teaching and learning are driven by fear and competition" (Sect. 2.1), rather than by a sense of the beauty of mathematics,

[1] In the friendly spirit of the response paper by Ted Eisenberg and Michael Fried, I shall hereafter refer to them as Ted and Michael.

or an aesthetic sense as attested by the elegance of a proof. It is an unfortunate fact that "fear and competition" have a role to play not only in the press for mathematicians to publish their results, but also in the publish-or-perish syndrome impinging on mathematics education researchers in academia. But the need to attain promotion and tenure should not be put forward as a defining factor of the nature of mathematics. I remember being thrilled when I first read Davis and Hersh's (1981) account (also cited by Michael in Sect. 3) of what is involved in "the mathematical experience", because it resonated with my own perception of how the inexorable logic of mathematics has an austere beauty of its own. Thus, I do believe with Sinclair (2004, 2006) that aesthetics (in both senses—artistic beauty and sensory awareness, as the derivation of the word, aisthesis, implies) have a role to play in experiencing mathematics.

With regard to the nature of mathematics, Ted and I are in agreement that abstraction is important. The form is what matters, without any necessary connections to sensory experience, even if such experience is essential because it is the only way that we can write down or communicate the form to others (in representations—or better, inscriptions, the term that I prefer, e.g., Presmeg 2006). The relationships among what Peirce (1998) would designate the objects of mathematics, the signs that stand for these objects in mathematical symbolism, and the interpretants that we create for them, are a deep subject that is beyond the purpose of this response paper. But the point is that these objects are abstract, no matter what signs are created to stand for them. Perhaps that is why mathematicians resort to metaphors to describe the objects of their cognition (Sfard 1994).

3 Mathematics Education

I also agree heartily with Ted that "watering down" of the content of mathematics in order to remove the challenge of learning mathematics just so that learners will "feel good" is totally unacceptable. The following quotation, by Hiebert and Wearne (2003), captures not only the spirit of the reform initiated by the National Council of Teachers of Mathematics in the USA (2000), but also what I have believed and tried to put into practice in my own teaching, at high school and at college levels, for well over three decades.

> Allowing mathematics to be problematic for students requires a very different mind-set about what mathematics is, how students learn mathematics with understanding, and what role the teacher can play. Allowing mathematics to be problematic for students means posing problems that are just within students' reach, allowing them to struggle to find solutions and then examining the methods they have used. Allowing mathematics to be problematic requires believing that all students need to struggle with challenging problems if they are to learn mathematics deeply. (p. 6)

It is ironic that Ted claims (Sect. 2.5) that "Norma Presmeg's paper talks about making mathematics more humane and I think we both know what she means by this"—implying a watering down of the content. In the first place, my paper was not addressing mathematics itself, nor its teaching directly. In the second place, this claim is in direct opposition to my beliefs, as captured in the foregoing quotation. I do believe that learning and doing mathematics can be enjoyable—another possible concomitant of "humane"—but the joy comes from meeting the challenge in solving a mathematics problem, and recognizing harmony and unity where it was not perceived previously. I can resonate with Martin Gardiner's remark, as quoted by Ted (Sect. 2.4), that he could write and explain things so clearly because he had to work so hard to understand them. I have witnessed many prospective teachers doing well in the classroom because of the effort they themselves had to put in, in order to understand the mathematical content deeply.

4 Mathematics Education Research

MER was the central focus of my paper *Mathematics education research: embracing arts and sciences*. The main thrust of the paper was a parallel between the "war" of the arts and sciences in my own early thinking (as I struggled between aspirations to become a theoretical physicist or to take up music and poetry), and the "wars" among paradigms in the history of MER over the last half century. (See the book by Latterell 2005, for one view of the "math wars" and their impact on mathematics education in the USA.) Although the conclusion of my paper dealt with unity and harmony, the transitions between paradigms were certainly not without contestation, as the term "war" implies. In emulating the hard sciences, statistical research in mathematics education was the only kind of research deemed worthy of the name in the 1960s and 1970s. But this research, at that time, had little or no impact on the teaching and learning of mathematics in actual classrooms (Krutetskii 1976). Thus the paradigm had to change. Ted is right (Sect. 2.2) that a knowledge of statistics became less mandatory in the qualitative research paradigm that gained ascendance in the 1990s as the paradigm "moved on" from statistics: the pendulum swung too far! But the whole point of my paper was that the pendulum swung back again to a far more stable and balanced position in the decade of the 2000s. More and more of the new generation of mathematics education researchers *are* fluent in "ANOVA, ANCOVA, regression and correlation, factor analysis, biased estimators, etc." (Sect. 2.2). The point is that both qualitative and quantitative methodologies are now acknowledged to be important, for different reasons and with different purposes, in investigating the phenomena of mathematics education.

5 Balance and Harmony

"Mixed methods" of research are commonplace in mathematics education investigations now, as researchers try to combine the deep insights of qualitative case study methods and others, with the generalizability and precision afforded by statistical designs and quasi-experimental methods ("quasi" because it is often not possible to include random assignment of students in whole-class sampling). The field can only benefit from the enlarged arsenal of methodological tools now available and accepted in MER. It is not that validity and reliability belong only to statistical research. Variants of these criteria have been hardwon in the decades of the ascendance of qualitative methodologies, and it is now taken for granted that such studies will use elements of quality control such as triangulation, respondent validation ("member checks"), and full reporting ("paper trails"). These terms should not be viewed as mere jargon for the sake of impressing or providing pseudo-scientific quality where it does not exist (Ted's Sect. 2.3). These terms are just as necessary in a qualitative paradigm as are validity, reliability, hypothesis-testing, and significance in a quantitative one. Now the paradigms are no longer at war in the minds of current mathematics education researchers.

There is also no war now, in the arts and sciences in my own nature. I use reason and logic in all aspects of my work: but I also use the more creative artistic human elements of my nature, not only in literary work, editing, and research, but also in my teaching of college level mathematics and mathematics methods courses. It is in that sense that I have "come home". My intent was in no way to drive others, such as Ted, from their homes. I hope that this "response to a response" has helped to clarify what I was trying to express. I am grateful to both Ted and Michael for their deep thought, and for the push to elucidate my meanings.

References

Davis, P. J., & Hersh, R. (1981). The mathematical experience. Boston: Houghton-Mifflin.

Hiebert, J., & Wearne, D. (2003). Developing understanding through problem solving. In H. L. Schoen & R. I. Charles (Eds.), *Teaching mathematics through problem solving: Grades 6–12* (pp. 3–13). Reston: National Council of Teachers of Mathematics.

Krutetskii, V. A. (1976). The psychology of mathematical abilities in schoolchildren. Chicago: University of Chicago Press.

Latterell, C. M. (2005). Math wars: A guide for parents and teachers. Westport: Praeger.

Leder, G. C., Pehkonen, E., & Torner, G. (Eds.). (2002). Beliefs: A hidden variable in mathematics education? Dordrecht: Kluwer Academic Publishers.

National Council of Teachers of Mathematics. (2000). Principles and standards for school mathematics. Reston: National Council of Teachers of Mathematics.

Peirce, C. S. (1998). In Peirce Edition Project (Ed.), *The essential Peirce* (Vol. 2). Bloomington: Indiana University Press.

Presmeg, N. C. (2006). A semiotic view of the role of imagery and inscriptions in mathematics teaching and learning. Plenary paper. In J. Novotna, H. Moraova, M. Kratka, & N. Stehlikova (Eds.), *Proceedings of the 30th annual meeting of the international group for the psychology of mathematics education* (Vol. 1, pp. 19–34). Prague, 16–21 July 2006.

Sfard, A. (1994). Reification as the birth of metaphor. *For the Learning of Mathematics*, *14*(1), 44–55.

Sinclair, N. (2004). The roles of the aesthetic in mathematical inquiry. *Mathematical Thinking and Learning*, *6*(3), 261–284. doi: 10.1207/s15327833mtl0603_1.

Sinclair, N. (2006). Mathematics and beauty: Aesthetic approaches to teaching children. New York: Teachers' College Press.

Index

M.N. Fried, T. Dreyfus (eds.), *Mathematics & Mathematics Education: Searching for Common Ground*, Advances in Mathematics Education,
DOI 10.1007/978-94-007-7473-5,

CPSIA information can be obtained at www.ICGtesting.com
Printed in the USA
LVOW03*0533231213

366423LV00009BA/88/P

9 789400 774728